MySQL数据库
基础与应用

主编◎陈玉勇
副主编◎张 雨 邸柱国

参编◎白 云 张丽梅 吴 进
张亚林 孟祥玉

·上海·

内 容 提 要

MySQL 具有开源、免费、体积小、易于安装、性能高效、功能齐全等特点。本书以 MySQL 数据库管理系统为平台，较全面地介绍了数据库的基础知识及其应用。本书共 7 个项目，包括数据库基础知识、MySQL 的安装与配置、数据库的基本操作、数据类型与约束、数据查询、数据库的高级应用和数据库综合实训。

本书采用案例教学方式，课上项目图书馆管理系统和课后项目图书管理系统贯穿始末。每个项目先通过对一个典型的应用实例进行分析，给出解决问题的完整方案；然后，围绕这个应用案例阐述知识要点，并提供相对应课后项目的任务训练，以便读者在实践中模拟操作；最后，通过实战项目演练来帮助读者巩固所学的内容。本书以面向工作过程的教学方法为导向，合理安排各项目的内容，并设计了大量的课堂实践和课外拓展，符合高职高专教育的特点。

本书配备了丰富的学习资源，包括教学 PPT、微课视频等。本书既可作为职业院校计算机相关专业和非计算机专业数据库基础或数据库开发课程的教材，也可作为计算机软件开发人员、从事数据库管理与维护工作的专业人员、广大计算机爱好者的自学用书。

图书在版编目(CIP)数据

MySQL 数据库基础与应用/ 陈玉勇主编. -- 上海：同济大学出版社，2022.12
ISBN 978-7-5765-0621-1

Ⅰ.①M… Ⅱ.①陈… Ⅲ.①SQL 语言-数据库管理系统 Ⅳ.①TP311.132.3

中国国家版本馆 CIP 数据核字(2023)第 001187 号

MySQL 数据库基础与应用
陈玉勇　主　编

责任编辑	张　莉	
助理编辑	屈斯诗	
责任校对	徐逢乔	
封面设计	渲彩轩	
出版发行	同济大学出版社　　www.tongjipress.com.cn	
	（地址：上海市四平路1239号　邮编：200092　电话：021-65985622）	
经　　销	全国各地新华书店	
排　　版	南京月叶图文制作有限公司	
印　　刷	启东市人民印刷有限公司	
开　　本	787mm×1092mm　1/16	
印　　张	20	
字　　数	499 000	
版　　次	2022 年 12 月第 1 版	
印　　次	2024 年 12 月第 2 次印刷	
书　　号	ISBN 978-7-5765-0621-1	
定　　价	78.00 元	

本书若有印装质量问题，请向本社发行部调换　　　　版权所有　侵权必究

前　　言

习近平总书记在党的二十大报告中强调"推进教育数字化,建设全民终身学习的学习型社会、学习型大国"。教育数字化是教育教学活动与数字技术融合发展的产物,也是进一步深化教育教学改革、推进教育现代化、实现教育高质量发展的驱动力。党的二十大报告为我国新一代信息技术产业发展指明了方向,报告明确指出,需要构建新一代信息技术、人工智能等一批新的增长引擎。随着新一代信息技术高速发展,数据处理已经广泛应用于各行各业,数据库技术是发展最快、应用最广的计算机技术之一,已经成为信息化建设、数据资源共享及各类应用系统的核心技术和重要基础。数据库是任何信息系统都不可缺失的核心系统部件。数据库原理、数据库设计、数据库操作、数据库管理及数据库应用编程等技术知识与方法,是软件从业人员必须掌握的专业技能。

本书是教师团队为深入贯彻落实党的二十大精神,推进教育数字化,实现教育资源开放共享,在长期从事数据库课程教学和科研的基础上,以满足"数据库技术及应用"课程的教学需要为基本要求编写而成,并合力打造优质数字教育资源库。本书同时也是一本校企合作教材。本书采用任务驱动、项目开发与理论教学紧密结合的编写方式,以实践应用为目标,理论阐述主要围绕实际应用技术组织和展开,充分考虑学习者的认知曲线,由浅入深,边讲边练,课上项目学生管理系统和课后项目图书管理系统贯穿始末,项目内容贴近实际生活,实用性强,以教育数字化助力中国式现代化。

本书以 MySQL 数据库技术为背景,介绍 MySQL 数据库系统的技术原理和应用开发方法,突出数据库技术方法的实践应用,给出了大量操作实例,帮助读者掌握 MySQL 数据库应用技术。本书包括 7 个项目,分别从数据库基础知识、数据库基本操作、数据查询等方面进行讲述,其中项目 7 为综合实训,能够很好地帮助读者巩固所学知识,提高操作技能。本书由陈玉勇、冯颖、孟祥玉进行统稿,项目 1～项目 5 由陈玉勇、邸柱国、张雨编写,项目 6～项目 7 由陈玉勇、张亚林、张丽梅、吴进、白云、张雨、孟祥玉编写。

在编著本书的过程中,编者参阅了大量相关文献资料,同时还得到了具有实战开发经验的大连中软卓越信息技术有限公司总经理孟祥玉的大力支持,谨在此对参考借鉴的书刊资料的作者和给予帮助的企业工程师表示真诚的谢意。虽然编者希望能够为读者提供最好的教材和教学资源,但由于水平和经验有限,不足之处在所难免,敬请广大读者批评指正!

编者

2022 年 5 月

教学视频资源

序号	项目	任务	视频名称	二维码	序号	项目	任务	视频名称	二维码
1	项目 1	任务 1	数据库的概念与发展		12	项目 5	任务 2	单表有条件查询	
2		任务 2	关系数据库		13		任务 3	排序与限量	
3		任务 3	E-R 图设计		14		任务 4	分组与聚合	
4		任务 4	数据表设计		15		任务 5	内置函数	
5	项目 2	任务 1	MySQL 的安装与配置		16		任务 6	多表查询	
6		任务 2	MySQL 的使用		17		任务 7	嵌套查询	
7	项目 3	任务 1	数据库操作		18	项目 6	任务 1	索引	
8		任务 2	数据表操作		19		任务 2	外键约束	
9		任务 3	数据操作		20		任务 3	视图	
10	项目 4	任务 1	数据类型		21		任务 4	存储过程	
11		任务 2	表的约束		22		任务 5	触发器	
					23		任务 6	事务	

目 录

前言

项目 1　数据库基础知识 ··· 001
　　任务 1　数据库基本概述 ·· 002
　　任务 2　设计数据库 ·· 012
　　任务 3　设计 E-R 图 ·· 023
　　任务 4　设计数据表 ·· 029

项目 2　MySQL 的安装与配置 ··· 036
　　任务 1　MySQL 的安装 ·· 037
　　任务 2　MySQL 的使用 ·· 055

项目 3　数据库的基本操作 ·· 064
　　任务 1　数据库操作 ·· 065
　　任务 2　数据表操作 ·· 074
　　任务 3　数据操作 ··· 089

项目 4　数据类型与约束 ··· 098
　　任务 1　数据类型 ··· 099
　　任务 2　表的约束 ··· 103

项目 5　数据查询 ·· 111
　　任务 1　单表无条件查询 ·· 112
　　任务 2　单表有条件查询 ·· 120

任务 3　聚合函数的使用 ………………………………………………………… 130
　　任务 4　分组与排序 ……………………………………………………………… 136
　　任务 5　多表连接查询 …………………………………………………………… 140
　　任务 6　子查询 …………………………………………………………………… 149
　　任务 7　集合查询 ………………………………………………………………… 156

项目 6　数据库的高级应用 …………………………………………………………… 158
　　任务 1　索引 ……………………………………………………………………… 159
　　任务 2　外键约束 ………………………………………………………………… 170
　　任务 3　视图 ……………………………………………………………………… 181
　　任务 4　存储过程 ………………………………………………………………… 190
　　任务 5　触发器 …………………………………………………………………… 218
　　任务 6　事务 ……………………………………………………………………… 229

项目 7　数据库综合实训 ……………………………………………………………… 244
　　实训 1　数据库操作 ……………………………………………………………… 244
　　实训 2　数据表操作 ……………………………………………………………… 247
　　实训 3　数据的插入操作 ………………………………………………………… 250
　　实训 4　简单数据查询 …………………………………………………………… 254
　　实训 5　数据的修改删除操作 …………………………………………………… 261
　　实训 6　数据的聚合处理 ………………………………………………………… 263
　　实训 7　多表查询 ………………………………………………………………… 267
　　实训 8　外键约束 ………………………………………………………………… 278
　　实训 9　索引操作 ………………………………………………………………… 286
　　实训 10　视图应用 ……………………………………………………………… 290
　　实训 11　存储过程 ……………………………………………………………… 296
　　实训 12　触发器 ………………………………………………………………… 301
　　实训 13　事务 …………………………………………………………………… 305

参考文献 ………………………………………………………………………………… 309

项目 1　数据库基础知识

　　数据库技术是计算机应用领域中一门非常重要的技术,是计算机科学的重要分支,也是进行信息管理的重要方法,还是现代信息系统的核心和基础,它的出现与应用极大地促进了计算机技术在各领域的渗透。

　　MySQL 是一种开放源代码的关系型数据库管理系统(RDBMS),使用最常用的数据库管理语言——结构化查询语言(SQL)进行数据库管理,它凭借体积小、开放源代码、成本低等优点,被广泛地应用于中小型网站。读者在学习设计和使用数据库之前,需要先理解数据库的基本概念和基本原理。

> **学习目标**
>
> - 了解数据库的发展过程;
> - 理解数据库基本概念;
> - 掌握数据库的三大模型;
> - 理解关系数据库;
> - 掌握设计 E-R 图的方法和技巧;
> - 能够进行 E-R 图与逻辑结构设计的转换;
> - 掌握数据表的设计方法。

任务 1
数据库基本概述

1.1.1 任务描述

数据库技术是现代信息科学与技术的重要组成部分,是计算机数据处理与信息管理系统的核心。数据库技术研究和解决了计算机信息处理过程中大量数据有效地组织和存储的问题,在数据库系统中减少数据存储冗余、实现数据共享、保障数据安全以及高效地检索数据和处理数据。

数据库技术主要研究如何组织和存储数据,如何高效地获取和处理数据。我们需要通过研究数据库的结构、存储、设计、管理以及应用的基本理论和实现方法,并利用这些理论来实现对数据库中数据的处理、分析和理解。

本次任务主要讲解数据库相关基本概念、数据库的发展历程、数据库系统体系结构,以及常见数据库管理系统等内容。

1.1.2 知识准备

1. 数据

1) 数据

数据(Data)是用来记录信息的可识别的符号,是信息的具体表现形式。例如,某人年龄 47 岁,身高 1.74 米,体重 68 千克等,这里的"47""1.74""68"等都是数据,数据可以是文字、数字、声音、图形、图像等形式。

2) 信息

信息是指经过加工以后对客观世界产生影响的数据,具有事实性、时效性、不相关性、等级性等特点。

信息与数据既有联系,又有区别,主要表现在以下四点。

① 信息是一种经过选择、分析、综合、加工后的数据,它使用户可以更清楚地了解正在发生什么事。所以,数据是原材料,信息是产品,信息是数据的含义,数据和信息是相对的。表现在一些数据对某些人来说是信息,而对某些人而言则可能只是数据。

② 信息是观念上的,因为信息是加工了的数据,所以采用什么模型、需要多长的信息间隔来加工数据,是受人对客观事物变化规律的认识制约的。因此,信息是揭示数据内在的含义,是观念上的。

③ 信息是指通信系统传输和处理的对象,泛指人类社会传播的一切内容。在一切通信和控制系统中,信息是事物普遍联系的形式。

④ 数据是事实或观察的结果,是对客观事物的逻辑归纳,是用于表示客观事物的未经加工的原始素材。它可以是连续的值,比如声音、图像等称为模拟数据;也可以是离散的,如符号、文字等称为数字数据。在计算机系统中,数据以二进制信息单元 0、1 的形式表示。

3) 数据处理

数据处理是数据转换成信息的过程,也叫信息处理。数据处理的内容主要包括数据的收集、整理、组织、存储、加工、维护、查询和传播等一系列活动。数据处理的目的是从大量的数据中,根据数据自身的规律和它们之间固有的联系,通过分析、归纳、推理等科学手段,提取出有效的信息资源。

例如:学生的各科成绩是原始数据,可通过计算提取出平均成绩、总成绩等信息,其中的计算过程就是数据处理。数据处理可分为三个方面,即数据管理、数据加工和数据传播。

(1) 数据管理

数据管理的主要任务是收集信息,将信息用数据表示并分类存放。数据管理的目的是快速准确地提供必要的、可能被使用和处理的数据。

(2) 数据加工

数据加工的主要任务是对数据进行变换、抽取和运算。通过数据加工得到更加有用的数据,以指导或控制人的行为或事务的变化趋势。

(3) 数据传播

通过数据传播,信息在空间或时间上以各种形式传递。在数据传播的过程中,数据的结构、性质和内容不发生变化。数据传播会使更多的人得到信息,并且更加理解信息的意义,从而使信息的作用得到更充分的体现。

数据以信息的形式存储在机器中,称为机器世界或数据世界。在这里,每一个实体用记录来表示,相对应的实体属性用数据项(又称字段)来表示,现实世界中的事物及其联系用数据模型来表示。

客观事物及其联系是信息之源,是组织和管理数据的开始。为了能将现实世界中的具体事物抽象、组织为某一数据库管理系统(Database Management System,DBMS)支持的数据模型,人们常常首先将现实世界抽象为信息世界,之后将信息世界转换为机器世界。通俗地说,就是先将现实世界中的客观对象抽象为某一种信息结构,这种信息结构不依赖于具体的计算机系统,不是某一个数据库管理系统支持的数据模型,而是概念级的模型,然后把模仿模型转换为计算机上某一数据库管理系统支持的数据模型,这一过程如图 1-1-1 所示。

2. 数据库的发展

随着计算机的普遍应用,数据处理的量也随之增长,由此产生了数据管理技术。数据管理是对数据进行组织、存储、检索和维护,是数据处理的中心工作。数据管理技术的发展与计算机硬件(主要是外部存储器)、系统软件及计算机应用的范围有着密切的联系。

数据库(Database,DB)是数据管理的工具,但并不是一开始就有数据库技术的,数据管

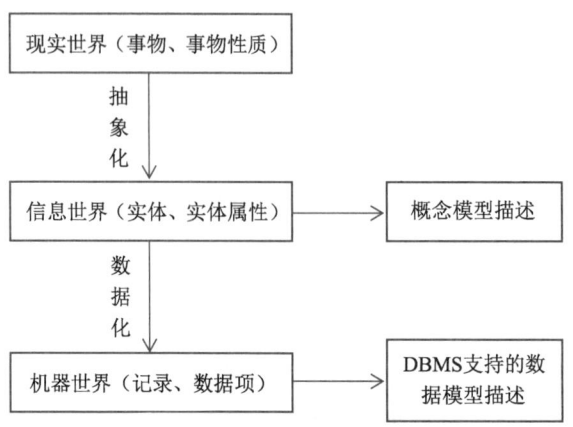

图 1-1-1　现实世界中客观对象的抽象过程

理技术经历了从人工管理阶段、文件系统阶段、数据库系统阶段到高级数据库系统阶段的变迁。

1）人工管理阶段

20 世纪 50 年代中期,计算机的主要用途是科学计算。硬件方面,计算机的外存只有磁带、卡片等,没有磁盘等直接存取的存储设备,存储的数据非常少;软件方面也没有操作系统和高级语言,数据处理的方式也是批处理,也就是一次处理一批数据,直到运算完成为止,然后才能进行下一批数据的处理,中间不能被打断,这是因为这个时候的外存只能顺序输入,工作过程如图 1-1-2 所示。

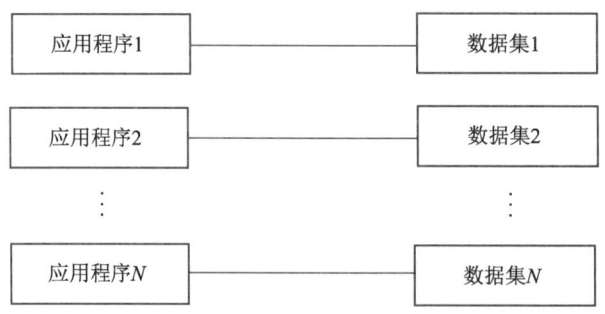

图 1-1-2　人工管理阶段应用程序与数据之间的关系

人工管理阶段的数据具有以下四个特点。

① 数据不保存。这个时期的计算机主要是在科学计算中应用,对数据的保存并没有特别要求,只是在计算某一个课题时才输入数据,数据处理结束后就退出,所以对数据的保存没有要求。

② 数据和程序不具有独立性。数据和程序是一个不可分割的整体,同时提供给计算机运算使用。

③ 数据不共享。这个时期数据是面向应用的,一组数据对应一个程序,不同的应用数

据之间是相互独立、彼此不相联的,即使两个不同应用涉及相同的数据,也必须各自定义,不能相互参照。数据不但高度冗余,而且不能共享。

④ 没有专用的数据管理软件。数据没有专门的应用软件管理,需要应用程序自己进行管理。

2) 文件系统阶段

20世纪50年代后期至60年代中期,也是数据管理发展到文件系统阶段的时期,这时的计算机不仅仅应用于科学计算,还大量用于管理数据。存储设备磁盘的出现可直接用于存取数据。软件方面,操作系统中已有了专门的数据管理软件,称为文件系统。文件系统管理阶段应用程序与数据之间的关系如图1-1-3所示。

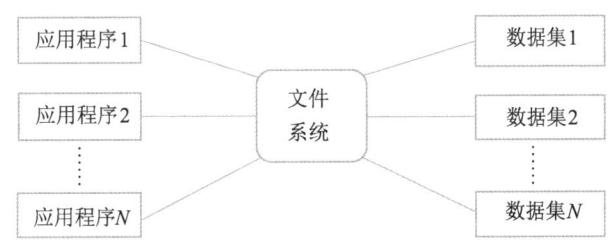

图1-1-3 文件系统管理阶段应用程序与数据之间的关系

文件系统阶段的特点有以下三个方面。

① 数据能够长期保存。计算机可以对数据进行查询、修改和删除等操作,所以数据需要长期保存,以方便用户反复处理。

② 数据和程序具有一定的独立性。操作系统提供了文件管理功能和文件的存取方法,程序和数据之间有了存取的接口,程序可以通过文件名和数据进行存取,而不需要直接找到数据的物理位置。

③ 可以实时处理数据。由于有了直接存取设备,也有了索引文件、链接存取文件、直接存取文件等,可对数据采用顺序批处理,也可以采用实时处理方式。数据的存取以记录为基本单位。

3) 数据库系统阶段

数据库技术的诞生可以以20世纪60年代末和70年代初的三个事件为标志。

① 1968年研制成功、1969年形成产品的美国IBM公司的数据库管理系统IMS (Information Management System)问世,该系统支持的是层次数据模型。

② 美国数据系统语言协会CODASYL(Conference On Data System Language)下属的数据库任务组DBTG(Database Task Group)对数据库方法进行了系统的研究,在20世纪60年代末和70年代初发表了若干个报告(称为DBTG报告)。该报告建立了数据库技术的很多概念、方法和技术,DBTG所提议的方法是基于网状数模型的。

③ 从1970年起,IBM的研究员E. F. Codd发表了一系列论文,提出了数据库的关系模型,开创了数据库关系方法和关系数据库理论的研究,为关系数据库的发展和理论研究奠

定了基础。

从20世纪60年代后期开始,数据管理进入数据库系统阶段,这一时期的数据管理系统规模日益发展壮大,应用也越来越广泛,需要使用的数据量急剧增长,对数据共享的要求也更高。此时的计算机有了大容量的磁盘,计算能力也非常强大。由于硬件价格下降,编制软件和维护软件的费用相对增加,文件系统管理数据的方法已经不能满足需要,数据库技术应运而生。它的出现解决了数据的独立性问题,并且能够实现数据的统一管理与共享。人工管理阶段应用程序与数据间的关系如图1-1-4所示。

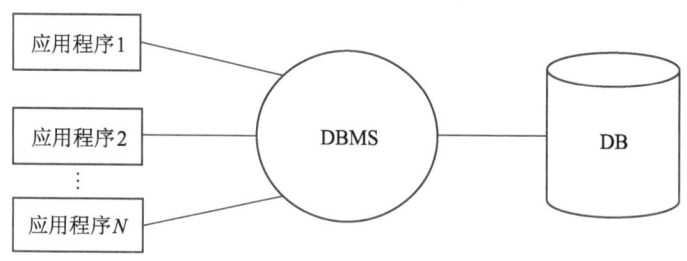

图1-1-4 人工管理阶段应用程序与数据间的关系

数据库系统的主要特点有以下四个方面。

① 数据结构化。它是数据库系统与文件系统的根本区别,也是数据库主要特征之一。

② 数据共享性高,冗余度小,易扩充。数据库是从整体角度建立起来的,数据不再是面向某一应用,而是面向整个系统。这样既减小了数据的冗余,节约了存储空间,缩短了存取时间,又避免了数据之间的不相容和不一致。

③ 数据独立性高。数据库提供数据的存储结构之间的映像或转换功能,使得当数据的物理存储结构改变时,数据的结构可以不变,从而程序也不用改变。

④ 统一的数据管理和控制功能。数据库是面向多个用户和应用的,因此对数据的存取往往是并发的。数据库管理系统提供了数据的并发性控制功能、安全性控制功能、完整性控制功能及数据库恢复功能等。

4) 高级数据库系统阶段

现在的数据库管理已经进入高级数据库系统阶段,具体体现在分布式数据库和面向对象数据库技术的广泛应用。

① 分布式数据库是数据库技术和计算机网络技术相互渗透和有机结合的产物。

② 面向对象数据库系统是面向对象的程序设计与数据库技术相结合的产物。

3. 数据库

计算机的快速发展将人类推进到信息化社会,同时也将人类淹没在信息世界中。简单来说,数据库本身可视为电子化的文件柜——存储电子文件的所在,用户可以对文件中的数据进行新增、截取、更新和删除等操作。

数据库是用来存放数据的仓库,它是以一定的数据结构来组织、存储和管理数据的集合,具有可供多用户共享、尽可能小的冗余度、较高的独立性和易扩展性等特点。

例如,学校的学生管理部门常常要把学生的基本情况(学号、姓名、年龄、性别、籍贯、专业、班级、成绩等)存放在表中,这张表就可以看成是一个"数据仓库"。有了这个"数据仓库",就可以根据需要随时查询某名同学的基本情况,也可以查询在某个范围内满足不同条件的学生人数等。这些工作如果都能在计算机上自动进行,那么学生管理工作就可以达到很高的水平。此外,在财务管理、仓库管理、生产管理中也需要建立众多的这种"数据仓库",使其可以利用计算机实现财务、仓库、生产的自动化管理。

4. 数据库管理系统

数据库管理系统是操纵和管理数据库的软件,它介于应用程序与操作系统之间,为用户或应用程序提供访问数据库的方法,具有数据定义、数据操纵、数据库运行管理及数据库创建与维护等功能。

1) 数据定义功能

DBMS 提供数据定义语言(Data Definition Language,DDL),用户通过它可以很容易地对数据库中的数据对象定义。

2) 数据操纵功能

DBMS 提供数据操纵语言(Data Manipulation Language,DML),用户可以使用 DML 操纵数据以实现对数据库的查询、增加、删除和修改等基本操作。

3) 数据库的运行管理功能

数据库的运行管理功能是指在创建、运用和维护时,DBMS 统一管理、控制,以保证数据的安全性、完整性、多用户对数据的并发使用及发生故障后的系统恢复。

4) 数据库的创建和维护功能

数据库的创建和维护功能包括数据库初始数据的输入、转换功能,数据库的存储、恢复功能,数据库的组织功能和性能监视、分析功能。这些功能通常是由一些程序完成的。

5. 数据库系统

数据库系统由数据库管理员、数据库、数据库管理系统(开发工具)、数据库应用系统和数据库用户组成,是指在计算机系统中引入数据库后的系统。

数据库由数据库管理系统统一管理,数据的增加、修改和检索都要通过数据库管理系统进行,数据库管理系统是数据库系统的核心。数据库系统结构如图 1-1-5 所示。

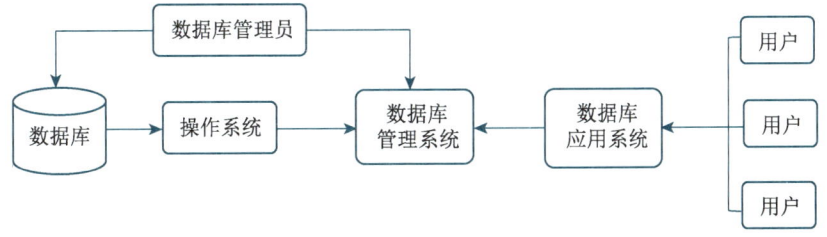

图 1-1-5　数据库系统结构

数据库用户包括数据库管理员、系统分析员、数据库设计人员及应用程序开发人员和终端用户。数据库管理员（Database Administrator，DBA）是高级用户，其任务是对使用中的数据库进行整体维护和改进，负责数据库系统的正常运行；系统分析员负责应用系统的需求分析和规范说明，要和用户及 DBA 结合，确定系统的硬件和软件配置，并参与数据库系统的概要设计；数据库设计人员负责数据库中数据的确定、数据库各级模式的设计；应用程序开发人员负责设计和编写应用程序的程序模块，并进行调试和安装；终端用户是数据库的使用者，主要负责对数据进行增加、删除、修改等基本操作，有两种方式可以实现，一是使用系统提供的操作命令，二是使用程序开发人员提供的应用程序。数据库用户之间的关系如图 1-1-6 所示。

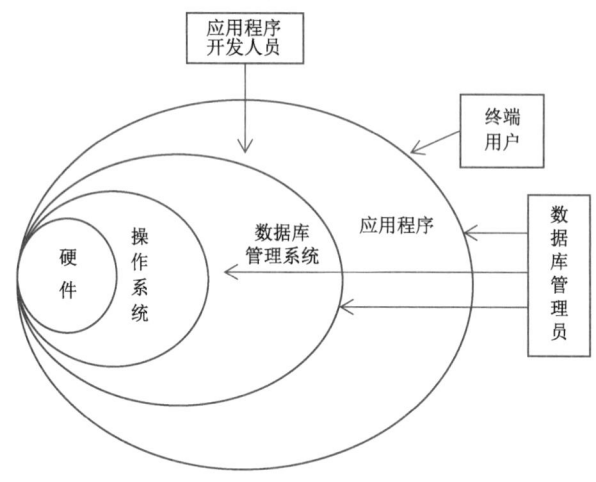

图 1-1-6　数据库系统用户关系图

6. 数据库系统的体系结构

数据库系统的体系结构可以用三级模式和两级映像来表述，如图 1-1-7 所示。

三级模式的组织形式称为数据库的体系结构，包括数据抽象的三个级别：内部层、概念层和外部层。该结构于 1975 年在美国 ANSI/X3/SPARC（美国国家标准协会的计算机与信息处理委员会中的标准计划与需求委员会）数据库小组的报告中提出。它把数据的具体组织留给 DBMS 去处理，用户只要抽象地处理数据，而不必关心数据在计算机中的表示和存储，这样就减轻了用户使用系统的负担。

三级模式之间差别往往很大，为了实现这三个抽象级别的联系和转换，DBMS 在三级模式之间提供了两级映像（Mapping）：外模式/概念模式映像，概念模式/内模式映像。两级映像保证了数据库系统中的数据能够具有较高的逻辑独立性和物理独立性。

1）三级模式

为了在计算机系统中实现数据的三级组织形式，必须用计算机可以识别的语言对其进行描述。DBMS 提供了这种数据描述语言（DDL）。用 DDL 精确定义数据视图的程序为模

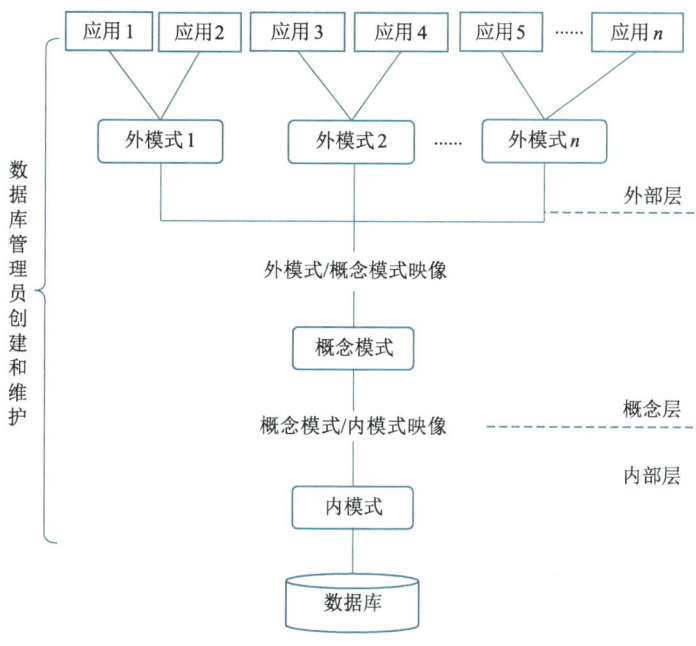

图 1-1-7　数据库体系结构

式(Schema)。三级模式包括概念模式、外模式和内模式。

(1) 概念模式

概念模式是对数据库全部数据的逻辑结构和特征的描述,是所有用户的公共数据视图,也称为逻辑模式,简称模式。它是数据库系统模式结构的中间层,既不涉及数据的物理存储细节和硬件环境,也不涉及具体的应用程序及所使用的应用开发工具和高级程序设计语言。

一个数据库只有一个概念模式,它通常以某种数据模型为基础,统一综合地考虑所有用户的需求,并将这些需求有机地结合成一个逻辑整体。它由对全局视图中全体数据文件的逻辑结构描述以及和存储视图中文件的对应关系的描述组成,由 DBMS 提供的 DDL 定义。定义概念模式时既要定义数据的逻辑结构,如数据记录的型(组成记录的数据项名、类型、取值范围等),又要定义数据项之间的联系、不同记录之间的联系,以及和数据有关的完整性、安全性等要求。

完整性包括数据的正确性、有效性和相容性。数据库系统应提供有效的措施,以保证数据处于约束范围内。

安全性主要是指保密性。不是任何用户都可以存取数据库中的数据,也不是每个合法用户可以存取的数据范围都相同,一般采用口令和密码的方式对用户进行验证。

数据库管理系统提供概念模式描述语言(模式 DDL)来定义模式。

(2) 外模式

外模式(External Schema)也称子模式(Subschema)或用户模式。它是对数据库用户

(包括程序员和终端用户)能够看见和使用的局部数据的逻辑结构和特征的描述,即个别用户涉及的数据的逻辑结构。它由对用户的数据文件逻辑结构的描述以及和全局视图中文件的对应关系的描述组成。

外模式通常是模式的子集,一个数据库可以有多个外模式。外模式是用户根据自己对数据的需要从局部的角度进行设计,因此如果不同的用户在应用需求、看待数据的方式、对数据保密的要求等方面存在差异,则其外模式描述也不同。一个子模式可以由多个用户共享,而一个用户只能使用一个子模式。

外模式是保证数据库安全性的一个有效措施,每个用户只能看见或访问所对应的外模式中的数据,数据库中的其余数据是不可见的。

数据库管理系统提供外模式描述语言(外模式 DDL)来定义外模式。

(3) 内模式

内模式(Internal Schema)也称存储模式(Storage Schema)或物理模式,一个数据库只有一个内模式。三级模式所描述的仅仅是数据的组织框架,而不是数据本身。在模式这个框架中填上具体数据来构成物理数据库,它是外部存储器上真实存在的数据集合。该模式框架下的数据集合是概念数据库,它仅是物理数据库的逻辑映像。子模式框架下的数据集合是用户数据库,它是概念数据库的逻辑子集。例如,记录的存储方式是顺序存储、按照 B 树结构存储还是按 Hash 方法存储;索引按照什么方式组织;数据是否压缩存储,是否加密;数据的存储记录结构有何规定等。

内模式的设计目标是将系统的模式(全局逻辑结构)组织成最优的物理模式,以提高数据的存取效率,改善系统的性能指标。

数据库管理系统提供内模式描述语言(内模式 DDL)来定义内模式。

2) 两级映像

外模式/概念模式映像。概念模式描述的是数据的全局逻辑结构,外模式描述的是数据的局部逻辑结构,同一个概念模式可以对应任意多个外模式。对于每个外模式,数据库系统都有一个外模式/概念模式映像,它定义了该外模式与概念模式之间的对应关系。这些映像定义通常包含在各自外模式的描述中。

概念模式/内模式映像。数据库中只有一个概念模式,也只有一个内模式,所以概念模式/内模式映像是唯一的,它定义了数据库全局逻辑结构与存储结构之间的对应关系。该映像模式定义通常包含在内模式描述中。

3) 两级数据独立性

数据独立性(Data Independence)是指数据库结构之间、应用程序和数据库之间是相互独立的,不受影响。

(1) 逻辑数据独立性

当概念模式增加新的关系、属性,或改变属性的数据类型时,数据库管理员可以对各个外模式/概念模式映像作相应改变,使外模式保持不变。应用程序是依据数据的外模式编写的,因而应用程序不必修改,保证了数据与程序的逻辑独立性,简称逻辑数据独立性。

(2) 物理数据独立性

当数据库的存储结构改变时,数据库管理员可以对概念模式/内模式映像作相应改变,使概念模式保持不变,因而应用程序也不必改变。保证了数据和程序的物理独立性,简称物理数据独立性。

数据库的两级映像保证了数据库外模式的稳定性,从而从底层保证了应用程序的稳定性,除非应用需求本身发生变化,否则应用程序一般不需要修改。特定的应用程序是在外模式描述的数据结构上编制的,它依赖于特定的外模式,与数据库的模式和存储结构相独立,不同的应用程序可以共用同一外模式。

(3) 数据与数据之间的独立性

在应用程序中可使数据的定义和描述分离出去。由于数据的存取由数据库管理系统管理,用户不必考虑存取路径等细节,从而简化了应用程序的编写,对应用程序的维护及修改工作也相应地减少了。

7. 常见数据库管理系统

1) Oracle

Oracle 是美国 ORACLE(甲骨文)公司提供的以分布式数据库为核心的一组软件产品,是一个开放式商品化关系型数据库管理系统,是商用关系型数据库管理系统中的典型代表。Oracle 作为一个通用的数据库管理系统,采用标准的 SQL 结构化查询语言,提供面向对象存储的数据支持,具有完整的数据管理功能。它还是一个分布式数据库系统,支持各种分布式功能,支持 Unix、Windows NT、OS/2、Novell 等多种平台,是目前最流行的客户/服务器体系结构的数据库之一,广泛应用于电信、邮政、金融、电力、医院及工业生产等领域。作为一个开发环境,Oracle 提供了一套界面友好、功能齐全的数据库开发工具。它使用 PL/SQL 语言执行各种操作,具有可开放性、可移植、可伸缩性等特点。

2) SQL Server

SQL 即结构化查询语言(Structured Query Language)。SQL Server 最早出现在 1988 年,当时只能在 OS/2 平台上运行。2000 年 12 月,微软发布了 SQL Server 2000,该软件可以运行于 Windows NT/2000/XP 等多种操作系统,是支持客户机/服务器结构的数据库管理系统。同时,SQL Server 也是一种比较典型的数据库管理系统,广泛应用于电子商务、电力、银行、教育等行业,它使用 Transact-SQL 语言完成数据操作。其版本的不断升级,使得该 DBMS 具有可用性、可管理性、可伸缩性、可靠性等特点。

3) MySQL

MySQL 由瑞典 MySQL AB 公司开发一款开放源代码的数据库管理系统,目前属于 Oracle 旗下产品,因此 MySQL 成为了 Oracle 公司又一重量级数据库产品。MySQL 具有跨平台的特性,可以在 Windows、Linux、Mac OS 和 UNIX 等平台上使用。由于其开源免费,运营成本低,而且具有快速、可靠和易于使用的特点,受到许多公司的青睐,如新浪、网易、百度、Google 等企业都使用 MySQL 作为数据库。MySQL 是现今最流行的开放源代码的数据库管理系统。

4）DB2

DB2 是美国 IBM 公司开发的一套关系数据库管理系统，是基于 SQL 的关系数据库产品。它具有较好的可伸缩性，适用于各种硬件与软件平台，各种平台上的 DB2 有共同的应用程序接口，运行在一种平台上的程序可以很容易地移植到其他平台上。DB2 提供了高层次的数据利用性、完整性、安全性、可恢复性，以及小规模到大规模应用程序的执行能力，具有与平台无关的基本功能和 SQL 命令。它的主要用户分布在金融、商业、铁路、航空等领域，其中在金融系统中应用最为突出。

5）Access

Access 是在 Windows 操作系统下工作的关系型数据库管理系统，采用了 Windows 程序设计理念，以 Windows 特有的技术设计查询、用户界面、报表等数据对象，内嵌了 VBA（Visual Basic Application）程序设计语言，具有集成的开发环境。Access 提供图形化的查询工具和屏幕、报表生成器，用户建立复杂的报表、界面无须编程和了解 SQL 语言，它会自动生成 SQL 代码。它被集成到 Office 中，具有 Office 系列软件的一般特点，如菜单、工具栏等。与其他数据管理系统软件相比，更加简单易学。

本书选用的关系数据库产品是 MySQL。

1.1.3 任务总结

数据库技术是现代信息科学与技术的重要组成部分，是计算机数据处理与信息管理系统的核心。数据库技术研究和解决了计算机信息处理过程中大量数据有效地组织和存储的问题，在数据库系统中减少数据存储冗余、实现数据共享、保障数据安全以及高效地检索数据和处理数据。数据库技术涉及许多基本概念，主要包括信息、数据、数据处理、数据库、数据库管理系统以及数据库系统等。

在本任务中，我们介绍了数据库相关基本概念、数据库的发展历程、数据库系统体系结构，以及常见数据库管理系统等内容。通过本任务的学习，读者需理解数据库系统体系结构的三级模式和两级映像，能够说出常见数据库管理系统的特点和应用。

设 计 数 据 库

1.2.1 任务描述

对学校而言，学生成绩管理是管理工作中重要的一环，学生的成绩管理工作量大、繁杂,人工处理非常困难。因此,借助于计算机强大的处理能力,能够把人从繁重的成绩管理工作中解脱出来。

本任务主要以学生成绩管理系统为例,讲解数据库设计的六个阶段、数据模型的要素和种类等内容。

1.2.2 知识准备

在设计一个简单、规范、高效的学生成绩信息管理系统数据库时,其具体功能应包括:可提供课程成绩数据的添加、插入、删除、更新、查询,以及学生基本信息查询的功能等。首先讲解数据库设计的基本阶段。

1. 数据库设计

按照规范设计的方法,考虑数据库及其应用系统开发全过程,可以将数据库设计分为六个阶段,即需求分析、概念结构设计、逻辑结构设计、物理结构设计、数据库实施、数据库运行和维护,如图 1-2-1 所示。

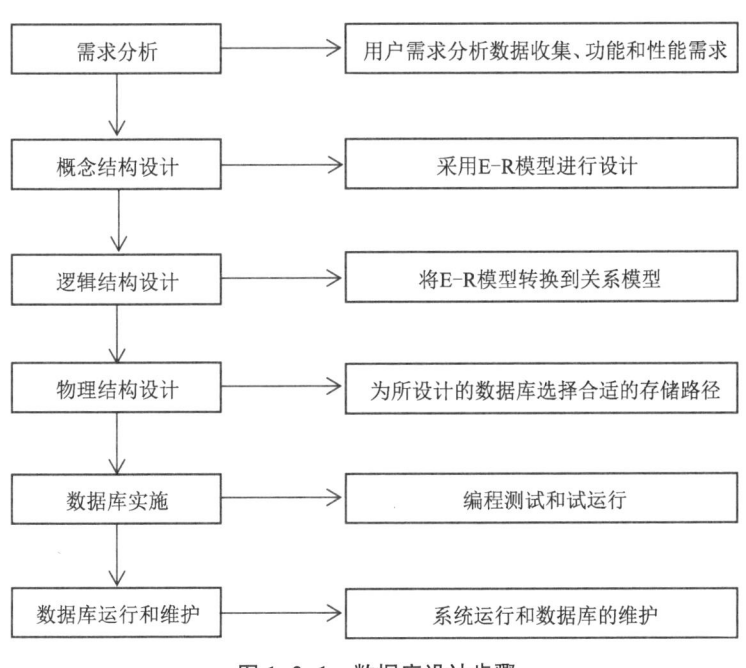

图 1-2-1 数据库设计步骤

在数据库设计过程中,需求分析和概念设计可以独立于任何数据库管理系统,逻辑结构设计和物理结构设计与选用的 DBMS 密切相关。

1) 需求分析阶段

在需求分析阶段,根据用户的需求收集数据,是设计数据库的开始。需求分析的结果不正确(对用户的实际需求反映不够直接)将直接影响后面的设计阶段,并影响数据库的设计是否合理与实用。需求分析阶段包括数据收集、系统功能和性能需求。

2）概念结构设计

概念结构设计是整个数据库设计的关键，它通过对用户的需求进行综合、归纳与抽象，形成一个独立于具体 DBMS 的概念模型，主要采用 E-R 模型进行设计，包括画 E-R 图。

3）逻辑结构设计

逻辑结构设计是指将概念模型转换成某个 DBMS 所支持的数据模型，并对其进行优化，将概念结构设计中的 E-R 图转换成表，实现从 E-R 模型到关系模型的转换。

4）物理结构设计

物理结构设计是指为逻辑数据模型选取一个适合应用环境的物理结构，主要是为所设计的数据库选择合适的平台和存取路径。

5）数据库实施

数据库的实施是指在数据库实施阶段，设计人员运用 DBMS 提供的数据语言及其宿主语言，根据逻辑设计和物理设计的结果创建数据库（此项工作在项目 3 中体现），编制与调试应用程序，组织数据入库，并进行试运行。此阶段包括编程、测试和试运行。

6）数据库运行与维护

数据库运行与维护是指数据库应用系统正式投入运行后，在数据库系统运行过程中必须不断地对其进行评价、调整和修改。

在数据库设计中，前两个阶段是面向用户的应用需求和具体需分析与解决的问题，中间两个阶段是面向数据库管理系统，最后两个阶段是面向具体的实施方法。前四个阶段可统称为"分析和设计阶段"，后两个统称为"实现和运行阶段"。

2. 数据模型的三要素

模型（Model）是对现实世界的抽象，反映了客观事物及事物之间的关系形式。数据库中的数据是按一定的数据结构组织存放的，这种数据结构反映数据间的相互联系，又称数据模型（Data Model）。在关系数据库技术中，数据模型是数据特征的抽象，是数据库管理的教学形式框架，用来描述数据库的结构和语义。数据模型体现的是实体与实体之间的关系。

数据模型严格定义来说是一组概念的集合，这些概念精确地描述系统静态特征（数据结构）、动态特征（数据操作）和数据约束条件，这是数据模型的三要素。这三个要素也是关系数据库之父 E. F. Codd 提出的，他认为一个数据模型是一组向用户提供的规则，这些规则定义了数据如何组织及允许何种操作。

1）数据结构

数据结构的研究对象是数据集合，用于描述系统的静态特征。数据库中每个对象都不是独立存在的，而是有着某种联系，这些对象是数据库的组成部分。它一方面描述与数据内容、类型和性质有关的对象，也就是数据本身（如关系模型中的域、属性、关系等）；另一方面又描述数据之间是如何相互关联的，这是数据之间的联系（如关系模型中的主码、外码等）。数据结构是数据模型的基础，数据操作和数据约束都建立在数据结构之上，不同的数据结构具有不同的操作和约束。

2) 数据操作

数据操作主要是数据库中每个数据对象是否允许执行的操作的集合,用于描述系统的动态特征。它描述了在相应数据结构上的操作类型和操作方式,具有相应的操作规则,主要有检索和更新(包括插入、删除和修改)两类。

3) 数据约束条件

数据约束条件是用来描述数据结构内数据间完整性规则的集合。完整性规则是给定数据模型中的数据及其关系所具有的制约和存储规则,用以限定符合数据模型的数据库的语法关系和它们之间的制约与依存及数据动态的限制,以保证数据的正确、有效和相容。

3. 数据模型的分类

数据模型从面向用户到物理实现,从不同的用户视角来看,数据模型可分为概念模型、逻辑模型和物理模型。

1) 概念模型

概念模型是对信息世界的建模,它应当能够全面、准确地描述信息世界,是信息世界的基本概念。它是面向用户的数据模型,是用户容易理解的现实世界特征的数据抽象,它表示实体类型及实体间的联系,是独立于计算机系统的模型。概念模型用于建立信息世界的数据模型,强调其语义表达功能,要求概念简单、清晰,易于用户理解,它是现实世界的第一层抽象,常作为业务人员和技术人员之间沟通的桥梁。作为现实世界的概念化结构,这种数据模型使得数据库的设计人员在最初的数据库设计阶段将精力集中在数据之间的联系上,而不用同时关注数据的底层细节(如所用的计算机系统的特性以及数据库管理系统DBMS的特性)。最常用的概念模型是实体—关系模型(Entity-Relationship Model,E-R模型)。

(1) 概念模型的操作对象

概念模型主要操作对象如下。

① 实体(Entity)

它是客观存在的可以相互区分的事物,可以是抽象的事件,也可以是具体的事物,如一件衣服、一名教师、一名学生、一件商品等属于具体的事物,教师授课、比赛、学生选课、借阅图书等属于抽象的事件。

② 属性(Attribute)

每个实体都拥有一系列的特性,每个特性可以看作是实体的一个属性。一个实体可以用若干个属性来描述,如学生实体由学号、姓名、性别、出生日期等若干个属性组成。实体的属性用型(Type)和值(Value)来表示。举例说明,学生是一个实体,其姓名、性别等是属性的型,也称属性名;而具体的学生姓名如"梁温馨",具体描述学生性别的"男"或"女"等是属性的值。

③ 标识符(Identifier)或码(Key)

它是能够唯一标识实体的属性或属性的集,如学生的学号、教师的编号等。

④ 域(Domain)

它是指属性的取值范围。如学号的域为 10 位整数,姓名的域为字符串集合,性别的域为男或女等。

⑤ 实体型(Entity Type)

它是指具有相同属性的实体必然具有共同的特征和性质,用实体名及其属性名的集合来抽象和刻画同类实体。如学生(学号,姓名,性别,出生日期,班级,学院)就是实体型。

⑥ 实体集(Entity Set)

它是指具有相同属性实体的所有实例的集合。例如,"学生成绩"是学生综合管理系统中的一个实体集,通过对学生信息实体的课程号、课程名、成绩等属性的描述,当属性值越多时,所描述的实体越清晰。

一个实体集中有多个实例,数据库中存储的每名学生的信息都是学生信息实体集中的实例。如表 1-2-1 所示。

表 1-2-1 实体集和实例

学生实体	学号	姓名	性别	年龄	出生日期	班级	学院
实例 1	2028033101	梁温馨	女	17 岁	2003.10.4	网络 201	信息学院
实例 2	2028033102	李玉竹	男	18 岁	2002.3.5	动漫 202	信息学院

实体通过一组属性来表示,属性是实体集中每个成员所拥有的特性,不同的实体属性值不同。在 E-R 图 (Entity-Relationship Diagram,ERD)中,实体用矩形表示,实体属性用椭圆表示,实体属性和实体之间用实线相连,如图 1-2-2 所示。

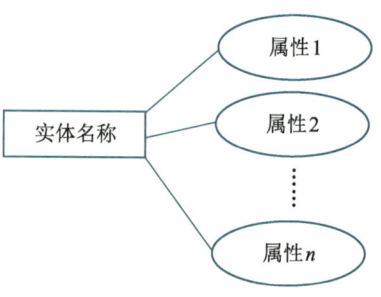

图 1-2-2 实体属性表示

概念模型主要的贡献在于分析数据之间的关系,它是用户对数据存储的一种高度抽象,反应的是用户的一种业务层面的综合信息需求。在这个阶段一般会形成整个数据模型或者是软件系统中的实体的概念以及实体之间的关系,为构建逻辑模型奠定基础。图 1-2-3 中描述了现实世界和信息世界以及最终转换成计算机世界信息的转换流程。

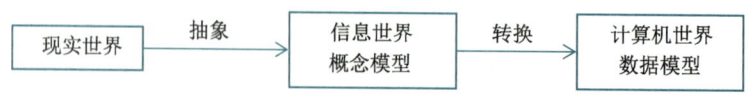

图 1-2-3 数据的抽象以及转换流程图

⑦ 关系(Relationship)

在现实世界中,事物内部以及事物之间是有关系的,这些关系在世界中反映为实体

(型)内部的关系和实体(型)之间的关系。实体内部的关系通常是指组成实体的各属性之间的关系,实体之间的关系通常是指不同实体集之间的关系。两个实体型之间的对应关系有三种,即一对一、一对多、多对多。

一对一关系(One-to-One Relationship)。如果对于实体集 A 中的每一个实体,实体集 B 中至多存在一个实体与之对应,或者,对于实体集 B 中的每一个实体,实体集 A 中至多存在一个实体与之对应,则称实体集 A 与实体集 B 之间存在一对一的关系,记作 1∶1。例如一个学校中一个班级只有一个班长,而一个班长只能在一个班级任职,所以班级与班长之间是一对一关系,如图 1-2-4 所示。

一对多关系(One-to-Many Relationship)。如果对于实体集 A 中的每一个实体,实体集 B 中存在多个实体与之对应,或者,对于实体集 B 中的每一个实体,实体集 A 中至多存在一个实体与之对应,则称实体集 A 与实体集 B 之间存在一对多的关系,记作 1∶n。例如一个学校中的一个班级中有很多学生,一个学生只能在一个班级注册,则班级与学生之间存在一对多关系,如图 1-2-5 所示。

多对多关系(Many-to-Many Relationship)。如果对于实体集 A 中的第一个实体,实体集 B 中存在多个实体与之关系,或者,对于实体集 B 中的每一个实体,实体集 A 中也存在多个实体与之联系,则称实体集 A 与实体集 B 之间存在多对多的关系,记作 m∶n。例如一个学生可选多门课程,一门课程也可同时被多个学生选修,则课程和学生之间存在多对多联系,如图 1-2-6 所示。

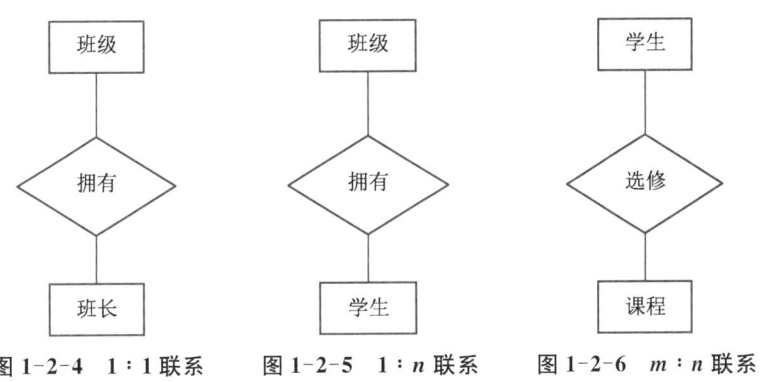

图 1-2-4　1∶1 联系　　图 1-2-5　1∶n 联系　　图 1-2-6　m∶n 联系

两个以上的实体集之间也存在着一对一、一对多、多对多的关系。

⑧ 数据表

在关系数据库中,关系就是一张二维表。表的每一行对应于一个元组,表的每一列称为属性,属性的名字必须是唯一的。属性的取值范围称为值或域。

⑨ 关键字

在数据库中,一个表或一个文件中可能存储着很多记录,为了能唯一地标识一个记录,必须在一个记录的各个数据项中确定一个或几个数据项。这些数据项的集合称为关键字。

⑩ 候选关键字

一个属性的值能唯一标识一个关系的元组而又不包含多余的属性,这组属性的集合称为候选关键字。

⑪ 主关键字

如果一个关系有多个候选关键字,可选择其中一个作为关系的主关键字,也称主键或主码。每个关系必须有且只有一个主关键字。

概念模型在对上述对象进行设置时要注意关系完整性。关系完整性是为保证数据库中数据的正确性和相容性,即对关系模型提出的某种约束条件或规则。完整性通常包括实体完整性、参照完整性和用户定义完整性(又称域完整性),其中实体完整性和参照完整性是关系模型必须满足的完整性约束条件。

实体完整性是指主关键字的值不能为空。例如学生基本情况(学号,姓名,性别,年龄,班级,学院),其中学号是主码,它的值绝对不能为空。

参照完整性是指如果关系中存在外部关键字,则它的值要么与主关系中的某个无级的主关键字取值相匹配,要么值为空。例如,学生(学号,姓名,性别,年龄,专业代码)和专业(专业代码,专业名称)两个关系中,学生表中的"专业代码"的取值要么与专业表中的主关键字"专业代码"中某个元组相匹配,要么取值为空。

用户定义完整性是指用户定义的完整性规则由用户根据实际情况对数据库中的内容进行规定,也称为域完整性规则。例如,学生关系中的"性别"取值只能是"男"或"女",年龄取值范围可定义在 16～30 岁。

(2) 概念模型设计的方法

概念模型设计(概念结构设计)是整个数据库设计的关键,它通过对用户需求进行综合、归纳与抽象,形成了一个独立于具体 DBMS 的概念模型。通常设计概念模型有自顶向下、自底向上、逐步扩张和混合策略四类方法。

① 自顶向下

即首先定义全局概念结构的框架,再逐步细化,如图1-2-7所示。

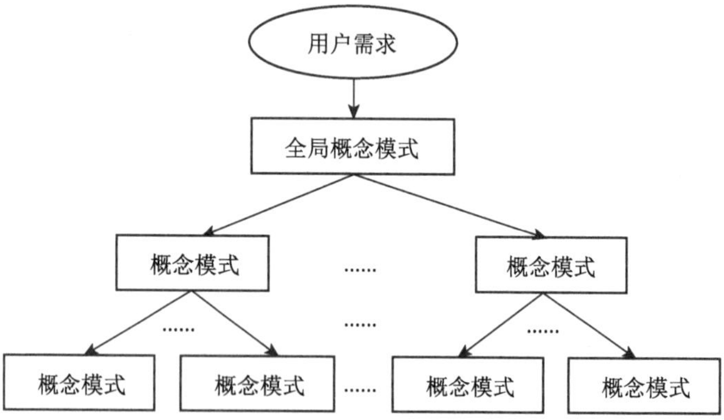

图 1-2-7 自顶向下策略

② 自底向上

即首先定义各局部应用的概念结构,然后再将他们集成起来,最后得到全局概念结构,如图 1-2-8 所示。

③ 逐步扩张

即首先定义最重要的核心概念结构,然后向外扩张,以滚雪球的方式逐步生成其他的

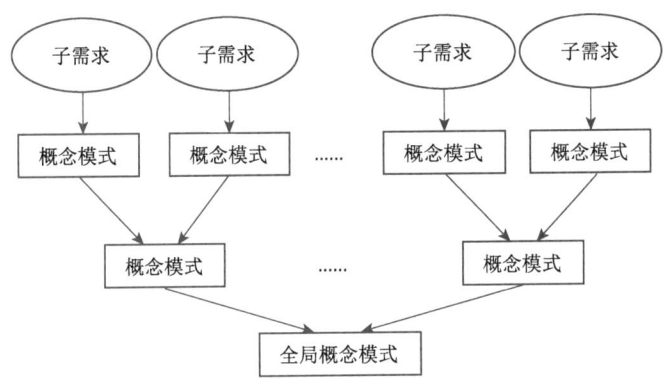

图 1-2-8　自底向上策略

概念结构,直至得到总体概念结构,如图 1-2-9 所示。

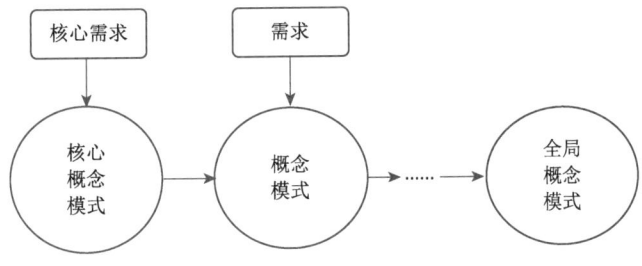

图 1-2-9　逐步扩张策略

④ 混合策略

即自顶向下和自底向上相结合。首先自顶向下进行需求分析,再自底向上设计概念模型,如图 1-2-10 所示。

以上四种策略中,通常采用的是混合策略。

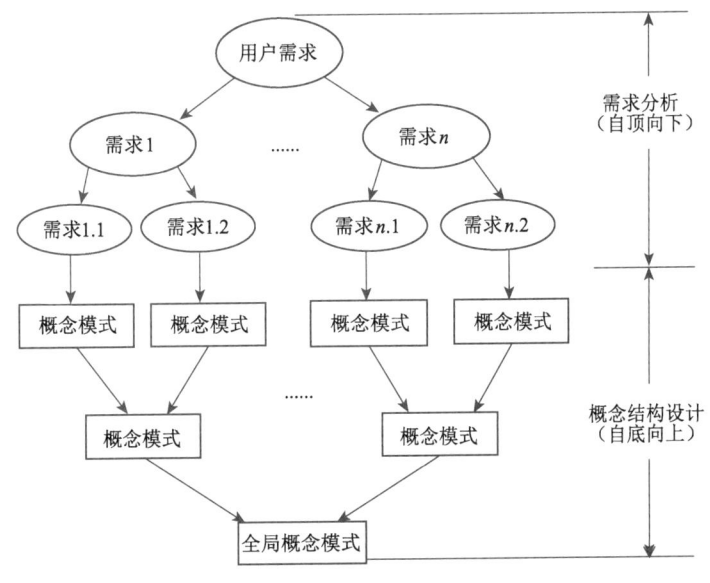

图 1-2-10　混合策略概念模式设计

2）逻辑模型

逻辑模型是直接面向数据库的结构，是现实世界的第二层抽象。它通常由概念模型转换得到，是面向计算机系统的，是所使用的数据库管理系统所支持的数据类型，是用户从数据库的角度能够看到的数据的模型，这种数据模型架起了用户和系统之间的桥梁，既要面向用户，同时也考虑到了所用的DBMS所支持的特性。

逻辑模型反映了系统分析设计人员针对数据在特定的存储系统的观点，是对概念模型的进一步细化和划分。逻辑模型是根据业务之间的规则产生的，是关于业务对象、业务对象数据以及业务对象彼此之间关系的蓝图。

逻辑模型的内容包括所有的实体、实体的属性、实体之间的关系以及每个实体的主键、实体的外键（用于维护数据完整性）。其主要目标是尽可能详细地描述数据，但是并不涉及这些数据的具体物理实现。逻辑模型不仅会影响数据库的设计方向，而且会影响数据库的性能（如主键设计、外键等都会最终影响数据库的查询性能）。

逻辑模型是开发物理数据库的完整文档，逻辑模型主要采用的是层次模型、网状模型、关系模型，其中最常用的是关系模型，关系模型对应的数据库称之为关系数据库，如MySQL。

逻辑模型包括以下四个部分。

① 表(Table)：相同结构的记录集合构成一个数据表，每个数据表对应于概念模型中的实体集。

② 关键字(Key)：能够唯一标识记录集中每个记录的字段或字段集，对应于概念模型中的实体标识符。

③ 记录(Record)：用来表示概念模型中的一个实体。

④ 字段(Field)：用来表示概念模型中实体的属性，它是数据库中可以命名的最小信息单位。每个属性对应一个字段。

3）物理模型

物理模型描述数据在物理存储介质上的组织结构，它与具体的DBMS相关，也与操作系统和硬件相关，是物理层次上的数据模型。每种逻辑模型在实现时都有其对应的物理模型。

在数据库应用系统中，概念模型、逻辑模型和物理模型之间的关系如图1-2-11所示。

物理设计是为逻辑模型选取一个最适合应用环境的物理结构（包括存储结构和存取方法）。首先要对运行的事务进行详细分析，获取并选择物理数据库设计所需要的参数；其次，要充分了解所用的DBMS的内部特征，特别是系统提供的存取方法和存储结构。

常用的存取方法有索引方法、聚簇方法(Clustering)方法、

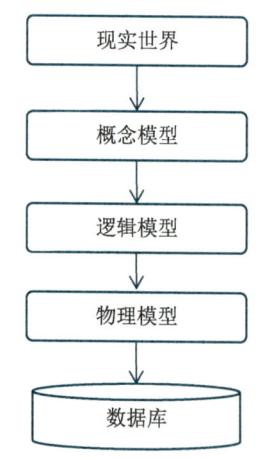

图1-2-11　数据模型的关系

Hash 方法三类。

4. 常见的数据模型

任何一个数据库管理系统都是基于某种数据模型的,数据库管理系统所支持的数据模型有层次模型、网状模型、关系模型。使用支持某种数据模型的数据库管理系统开发出来的应用系统相应地称为层次数据库、网状数据库、关系数据库。层次模型和网状模型代表产品是 1969 年 IBM 公司研发的层次模型数据库管理系统 IMS,目前大部分数据库采用的是关系数据库。1970 年,时任 IBM 公司研究员的 E. F. Codd 提出了关系模型,其代表产品为 Sysem R 和 Inges,支持面向对象,具有开放性,能够在多个平台上使用。

数据模型是数据库系统的关键概念,不同的数据模型,其数据库系统是完全不同的,所有的数据库管理系统都是基于某种数据模型设计的。随着数据库技术的发展和面向对象程序设计概念的提出以及关系模型结构的改进,又提出了关系对象模型。

1) 层次模型

层次模型用"树"结构来表示数据之间的关系,在数据库中定义满足下面两个条件的记录以及它们之间联系的集合为层次模型:有且只有一个结点没有双亲结点,这个结点称为根结点;根以外的其他结点有且只有一个双亲结点。其基本结构如图 1-2-12 所示(以教师与学生为例)。

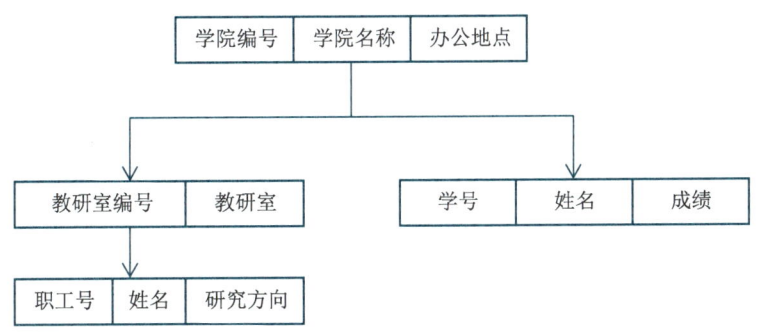

图 1-2-12　教师与学生层次数据库模型

层次模型优点:数据模型比较简单,且操作简单;对于实体间联系是固定的,且预先定义好的应用系统,性能较高,有良好的完整性支持。

层次模型缺点:不适合非层次性的联系;对插入和删除操作限制比较多,查询子女结点必须通过双亲结点;由于结构严密,层次命令比较程序化。

2) 网状模型

网状模型是用网络结构表示实体类型及其实体之间联系的模型,是一种可以灵活地描述事物及其之间关系的数据库模型。

网状模型的数据结构具有"允许一个以上的结点无双亲"和"一个结点可以有多于一个的双亲"两个特征。数据结构如图 1-2-13 所示(以学生、选课为例)。

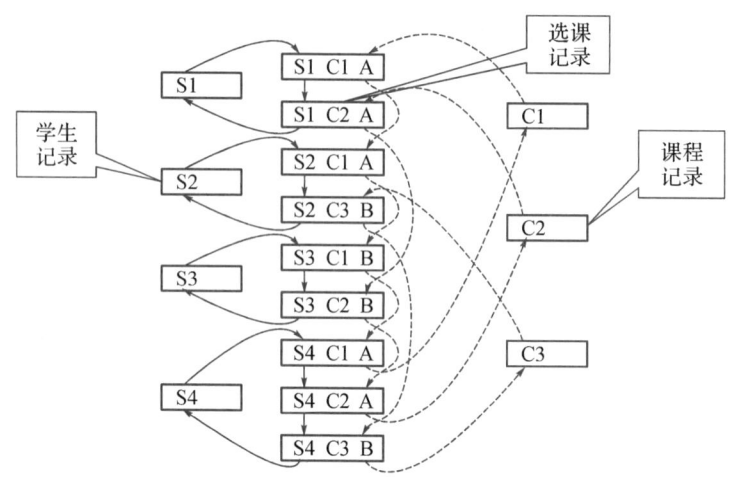

图 1-2-13　学生选课网状数据库模型

网状数据模型的优点：能够更直接地描述现实世界，如一个结点可以有多个双亲，结点之间可以有多种联系；具有良好的性能，存取效率较高。

网状数据模型的缺点：结构比较复杂，而且应用环境范围越大，数据库的结构就变得越复杂，不利于终端用户掌握；数据定义语言（DDL）、数据操作语言（DML）复杂，用户不容易使用。

3）关系模型

关系模型是用二维表的形式表示实体和实体间联系的数据模型，由行和列组成。它提供了关系操作的特点和功能要求，但不对 DBMS 的语言给出具体的语法要求。对关系数据库的操作是高度非过程化的，用户不需要指出特殊的存取路径，路径的选择由 DBMS 的优化机制来完成。

关系代数

关系模型是以集合论中的关系概念为基础发展起来的，其中无论是实体还是实体间的联系均由单一的结构类型——关系来表示，在实际的关系数据库中关系也称为表，一个关系数据库由若干个表组成，如表 1-2-2 所示（以学生信息表为例）。

表 1-2-2　学生信息表

学号	姓名	性别	年龄	出生日期	班级	学院
2028033101	梁温馨	女	17 岁	2003.10.4	网络 201	信息学院
2028033102	李玉竹	男	18 岁	2002.3.5	动漫 202	信息学院

关系模型主要有以下三个优点。

（1）数据结构单一。关系模型中，不管是实体还是实体之间的联系，都用关系来表示，而关系都对应一张二维数据表，数据结构简单、清晰。

（2）关系规范化，并建立在严格的理论基础上。构成关系的基本规范要求关系中每个属性不可再分割，同时关系建立在具有坚实理论基础的严格数学概念的基础上。

(3) 概念简单,操作方便。关系模型最大的优点就是简单,用户容易理解和掌握,一个关系就是一张二维表格,用户只需用简单的查询语言就能对数据库进行操作。

4) 关系对象模型

关系对象模型一方面对数据结构方面的关系结构进行改进,如 Oracle 8 就提供了关系对象模型的数据结构描述;另一方面,人们对数据操作引入了对象操作的概念和手段,现今的数据库管理系统基本上都提供了这方面的功能。

总的来说,在层次模型、网状模型、关系模型三种数据模型中,关系模型结构简单,数据之间的关系容易实现,因此关系模型是目前广泛使用的数据模型,并且关系数据库也是目前流行的数据库。

1.2.3 任务总结

使用计算机对成绩信息的管理,具有手工管理所无法比拟的优点。比如,信息存储及时、检索迅速、查找方便、可靠性高、存储量大、保密性好、寿命长、成本低等。这些优点能够极大地提高学生成绩管理的效率,也是高校成绩正规化管理的重要途径。

本任务主要讲解了关系数据库的概念以及数据库的设计、数据模型中的知识点。通过本任务的学习,读者应具备运用混合策略方法设计概念模型的能力。

任务 3
设 计 E-R 图

1.3.1 任务描述

需求分析阶段解决了客户的业务和数据处理需求后,就进入了概念结构设计阶段。我们需要将收集到的与学生信息管理系统相关的数据和团队的其他成员及客户沟通,讨论数据库的设计是否满足客户的业务员和数据处理需求,找出它们之间的关系,并进行概念模型设计(也称 E-R 图设计)。

E-R 图也称实体—关系图,用于描述现实世界的事物,以及事物与事物之间的关系。其中,E 表示实体,R 表示关系。它提供了表示实体类型、属性和关系的方法。

本任务以学生管理系统为例,主要讲解实体、属性、关系,以及 E-R 图的设计原则。

1.3.2 知识准备

E-R 图最早由美籍华裔计算机科学家陈品山提出,是概念数据模型的高层描述所使用的数据模型或模式图。它为表述实体—关系模式的数据模型提供了图形符号。这种数据模型用在信息系统设计的第一阶段,在基于数据库的信息系统设计的情况下,在后面的逻辑设计阶段,概念模型要映射到逻辑模型(如关系模型)上。

E-R图是一种概念模型,该模型方便转化为数据库管理系统实际支持的数据模型(比如关系模型)。E-R图最重要的三个部分分别是实体、属性、关系。

E-R图设计(概念结构设计)是整个数据库设计的关键,它是各种数据模型的共同基础,比数据模型更独立于计算机、更抽象,从而也更加稳定,依据常用概念结构设计的方法,E-R图设计的步骤包括:局部E-R图设计;集成各局部E-R图,形成全局E-R图并优化全局E-R图,形成最终E-R图。如图1-3-1所示。

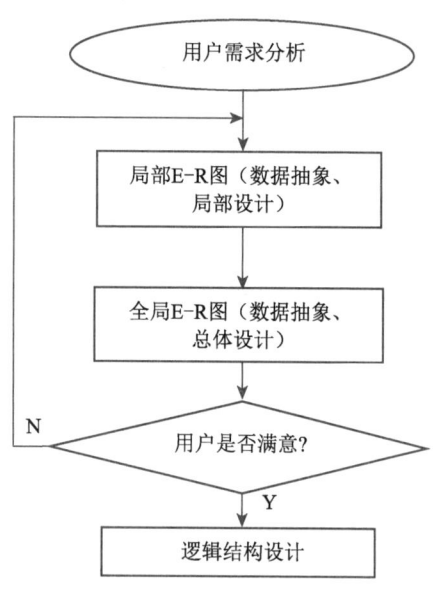

图1-3-1　E-R图设计步骤

1. 局部E-R图设计

局部E-R图设计第一步需要根据系统的具体情况,在多层数据流图中选择一个适当层次的数据流图,让这组图中的一部分对应一个局部应用。第二步,以这一层次的数据流图为出发点,设计局部E-R图。最后,分别将每个局部应用所涉及的数据从总数据流图中抽取出来,参照数据流图,确定各局部应用中的实体、标识实体的码、实体之间的联系及其类型,即确定实体型间的关系($1:1,1:n,m:n$)。

在实际应用中,实体和属性是相对而言的。一个事物在一种应用环境中称为"属性",而在另一种应用环境中又可能称为"实体"。

【例1-3-1】　辽宁林业职业学院下的某个"学院",在某种情况下只是作为"学生"实体的一个属性,表示某一个学生属于哪个学院。而在另一种环境中,"学院"下面应该有学院名称、院长、教师人数、学生人数、办公地点等,这时"学院"就需要作为实体。如图1-3-2所示。

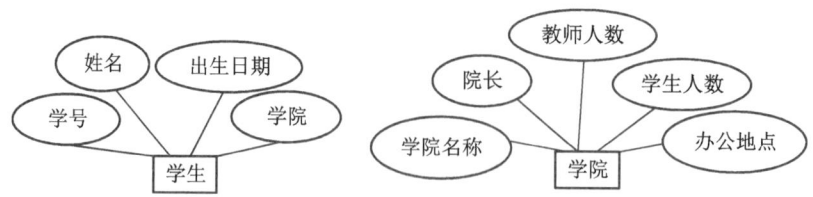

图1-3-2　学院作为属性和实体的关系

因此,为了解决上述问题,在设计局部E-R图时应遵循以下基本原则。

① 属性必须是不可分的数据项,不能再由另一些属性组成,即属性中不能包含具有需要描述的性质。

② 关系只发生在实体之间,属性不能与实体有关系。

为了简化E-R图的设计,现实世界中的事物只要能作为属性的,都应尽量作为属性。

【例 1-3-2】 在学生选课和教师授课局部 E-R 图中设有学生、课程、教师和单位实体。

学生：学号、姓名、性别、选修课程号、成绩。

课程：课程号、课程名称、教师编号。

教师：教师编号、姓名、性别、职称、教授课程号。

单位：单位名称、单位电话、教师编号、教师姓名。

这些实体间有如下关系：

① 一个学生可以选修多门课程，一门课程可以被多个学生选修；

② 一名教师可以讲授多门课程，一门课程可以被多个教师讲授；

③ 一个单位可以有多名教师，一名教师只能属于一个学院；

④ 一个学院可以有多名学生，一名学生只能属于一个学院。

根据上述分析，可以得到学生选课局部 E-R 图和教师授课局部 E-R 图，如图 1-3-3 和图 1-3-4 所示。

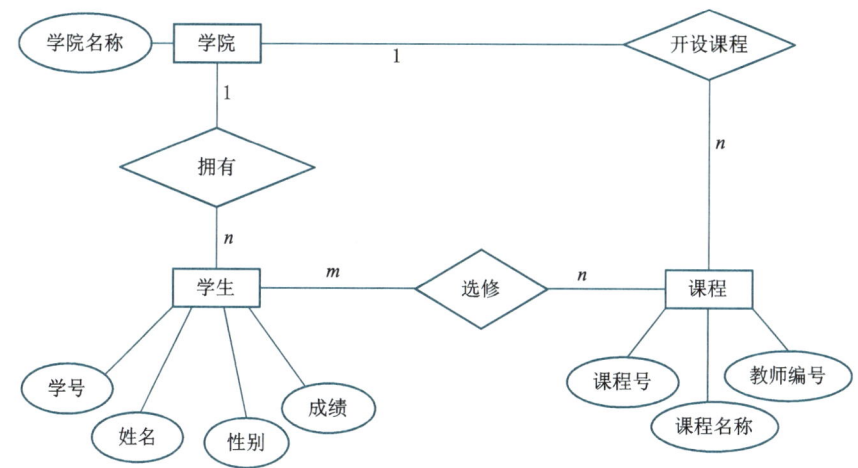

图 1-3-3　学生选课局部 E-R 图

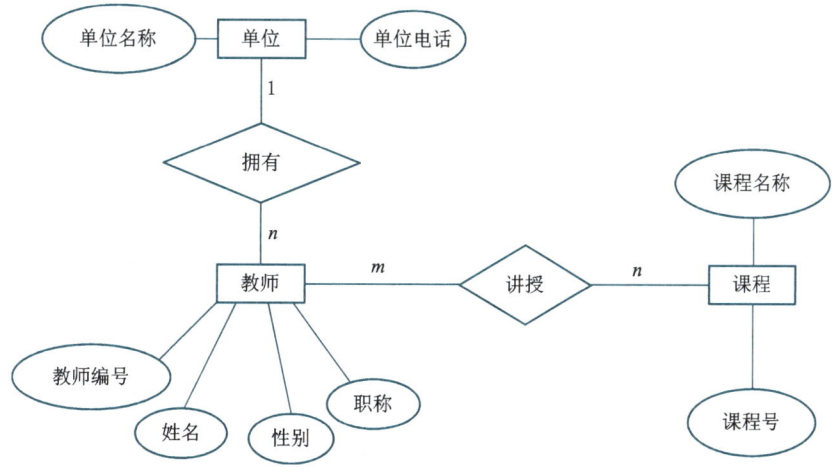

图 1-3-4　教师授课局部 E-R 图

2. 集成全局 E-R 图

各个局部 E-R 图建立好以后,还需要对它们进行合并,集成一个整体的概念数据结构,即全局 E-R 图。

1) 局部 E-R 图的集成方法

局部 E-R 图的集成方法有两种。

(1) 多元集成法(图 1-3-5)

即一次将多个局部 E-R 图合成一个全局 E-R 图,也称一次集成法。

(2) 二元集成法(图 1-3-6)

即先集成两个重要的局部 E-R 图,然后用累加的方法一步一步地加进新 E-R 图,也叫逐步集成法。在实际应用中,一般采用逐步集成法,即每次综合两个图,这样可以降低难度,如果局部图比较简单,可以采用一次集成法。

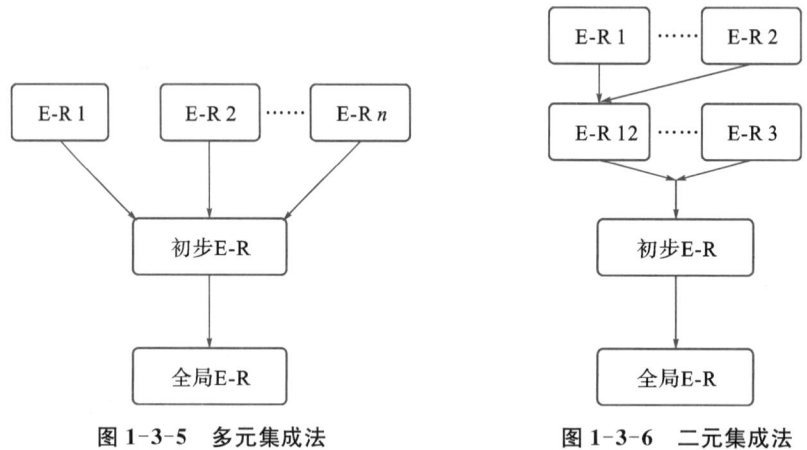

图 1-3-5 多元集成法　　　　图 1-3-6 二元集成法

2) 局部 E-R 图的集成步骤

在实际应用中,可以根据数据库系统的复杂程度来选择集成的方法。如果局部 E-R 图比较简单,可以采用一次集成法。在一般情况下,采用逐步集成法,即每次只综合两个图,这样可以降低难度。实际上,无论采用哪一种方法来集成全局 E-R 图,它的步骤只有两步:第一步就是合并,消除局部 E-R 图之间的冲突,生成初步 E-R 图;第二步就是优化,在优化初步 E-R 图的过程中要消除数据间不必要的冗余,生成基本 E-R 图。

(1) 合并局部 E-R 图,生成初步 E-R 图

这是生成基本 E-R 图的第一步,这个步骤是将所有设计的局部 E-R 图合成全局概念结构。全局概念结构不仅要支持所有的局部 E-R 模型,还要合理地表示一个完整、一致的数据库概念结构。

由于各个局部应用所面向的问题不同,并且通常由不同的设计人员进行局部 E-R 图设计,因此,各局部 E-R 图不可避免地会有许多不一致的地方,通常把这种现象称为冲突。因此当合并局部 E-R 图时,并不是简单地将各个局部 E-R 图集合在一起,而是必须消除各个

局部 E-R 图中的冲突,这样做就是让合并后的全局概念结构不仅能支持所有的局部 E-R 模型,而且必须是一个能为全系统中所有用户共同理解和接受的统一的概念模型。那么合理消除各局部 E-R 图中的冲突就是我们要解决的问题。

E-R 图中的冲突一共有三种,即命名冲突、属性冲突和结构冲突。

① 命名冲突:命名冲突可能发生在实体名、属性名或关系之间,在 E-R 图设计中最为常见。命名冲突有两种,即同名异义和异名同义。

同名异义就是同一名称的对象在不同的局部具有不同的意义。例如"单位",在某些部门它表示人员所在的部门,而在某些部门可能表示物品的重量、长度等。

异名同义就是同一意义的对象在不同的局部应用中具有不同的名称。例如系部和学院,它们都表示学校下面设置的各个分院。

② 属性冲突:属性冲突又分为属性值域冲突和属性取值单位冲突。

属性值域冲突:属性值的类型、取值范围或取值集合不同。例如,在给学生的学号设定取值范围时,有的设计人员用数字表示,而有些设计人员将其设置为字符型,这种就是属性值域冲突。

属性取值单位冲突:比如定义长度的单位,有的以 m 为单位,有的以 cm 为单位。

属性冲突属于用户业务的约定,必须与用户沟通协商来解决。

③ 结构冲突:结构冲突就好比一根橡皮筋把该物体拉向目标,而另一根橡皮筋要把该物体拉回原点,这种系统中的力量互相冲突就是结构冲突。

a. 同一对象在不同应用中有不同的抽象,可能为实体也可能为属性。例如,学校中的系部,在有些局部应用中可能作为实体,而在另一种局部应用中又可能作为属性出现,解决的方法是要么将实体转换为属性,要么将属性转换为实体。

b. 同一实体在不同局部应用中的属性组成不同,可能是属性个数或属性的排列次序不同,解决方法是合并后的实体的属性组成为各局部 E-R 图中的同名实体属性的并集,再适当调整顺序。

c. 实体之间的联系在不同应用中呈现不同的类型。例如,局部应用 X 中 b1 与 b2 可能是一对一联系,而在另一局部应用 Y 中可能是一对多或多对多联系,也可能是在 b1、b2、b3 三者之间有联系,解决方法是根据应用语义对实体联系的类型进行综合调整。

下面以图 1-3-3 和图 1-3-4 的局部 E-R 图为例来说明如何消除各局部 E-R 图之间的冲突,并进行局部 E-R 模型的合并,从而生成初步 E-R 图。

首先,这两个局部 E-R 图中存在命名冲突,学生选课局部 E-R 图中的实体"学院"与教师授课局部 E-R 图中的实体"单位"都是指学校下设的二级院部,即所谓的异名同义,合并后统一改为"学院",这样属性"名称"和"单位名称"即可统一为"学院"。

其次,还存在结构冲突,实体"学院"和实体"单位"在两个局部 E-R 图中的属性组成不同,合并后这两个实体的属性组成为各局部 E-R 图中同名实体发生的并集。

解决以上两个冲突后,合并两个局部 E-R 图,就能生成初步 E-R 图。

（2）消除不必要的冗余，生成基本 E-R 图

在初步的 E-R 图中，可能存在冗余的数据和冗余的实体之间的关系。冗余的数据是指可由基本数据导出的数据，冗余的联系是指可由其他联系导出的关系。冗余的存在容易破坏数据库的完整性，给数据库的维护增加困难，应该消除。当然，不是所有的冗余数据和冗余联系都必须消除，有时为了提高某些应用的效率，不得不以冗余信息作为代价。设计数据库概念模型时，需要根据用户的整体需求来确定哪些冗余信息必须消除，哪些冗余信息允许存在。消除了冗余的初步 E-R 图称为基本 E-R 图。

通常采用分析的方法消除冗余。数据字典是分析冗余数据的依据，还可以通过数据流图分析出冗余的关系。

在如图 1-3-3 和图 1-3-4 所示的局部 E-R 图中，"课程"实体中的属性"教师编号"可由"讲授"这个教师与课程之间的关系导出，而学生的成绩可由"选修"联系中的属性"成绩"计算出来，所以"课程"实体中"教师编号"与"学生"实体中的"成绩"均属于冗余数据；"学院"和"课程"之间的关系"开设课程"，可以由"单位"和"教师"之间的"拥有"关系与"教师"和"课程"之间的"讲授"关系推导出来，所以"开设课程"属于冗余关系。

这样，图 1-3-3 和图 1-3-4 所示的局部 E-R 图在消除冗余数据和冗余联系后，便可得到基本 E-R 图，如图 1-3-7 所示。

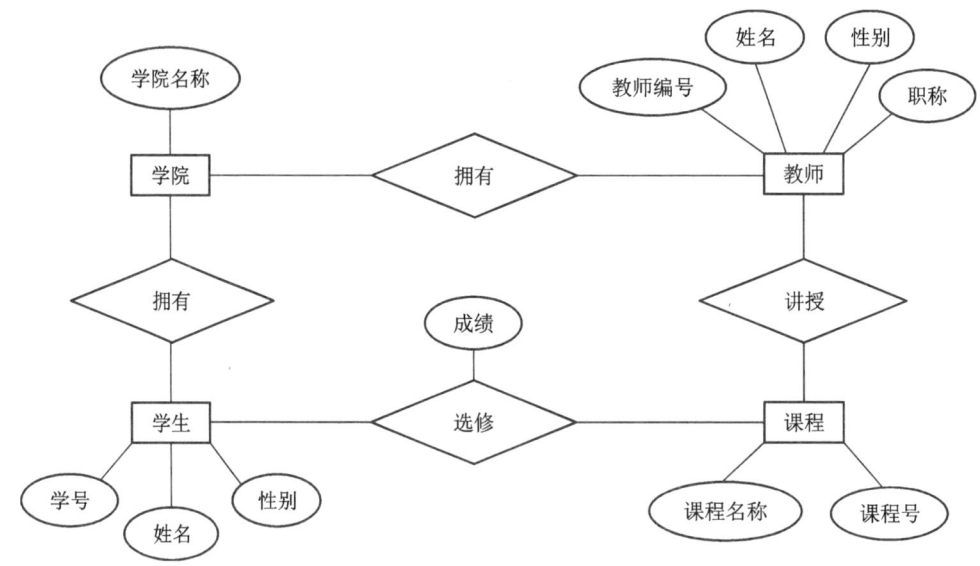

图 1-3-7　优化后的基本 E-R 图

通过解决冲突和消除不必要的冗余就得到基本 E-R 模型，也就是概念模型，它代表了用户的数据要求，是沟通"要求"和"设计"的桥梁，它决定数据库的总体逻辑结构，是成功创建数据库的关键。如果设计不当，就不能充分发挥数据库的功能，无法满足用户的处理要求。

注意：用户和数据库人员必须对这一模型反复讨论，在用户确认这一模型已正确无误地反映了他们的要求之后，才能进入下一阶段的设计工作。

1.3.3 任务总结

在数据库领域中，客观世界中的万事万物都被称为实体，实体是客观存在并相互区别的事物。实体的特征（外在表现）称为属性，通过属性可以区分同类实体。在通常情况下，开发人员在设计 E-R 图时，使用矩形表示实体，在矩形内框中写明实体名，使用椭圆表示属性，并且使用无向边将其与实体连接起来。

本任务主要讲解 E-R 图设计的概念、特点以及设计方法。通过本任务的学习，读者应熟练掌握设计 E-R 图的方法，能够合理地设计 E-R 图。

任务 4
设 计 数 据 表

1.4.1 任务描述

设计人员用 E-R 图表示数据和数据之间的关系，这种表示方法不能直接在计算机上实现。为了创建用户所要求的数据库，需要将概念模型转换为某个具体的 DBMS 所支持的数据模型（逻辑结构）。逻辑结构设计（数据表）是将概念结构设计阶段完成的概念模型，转换成能被选定的 DBMS 支持的数据模型。本任务主要将 E-R 模型转换为关系模型（数据表）。

将原始数据进行分解、合并后重新组织起来的数据库全局逻辑结构，包括所确定的关键字和属性、重新确定的记录结构和文件结构、所建立的各个文件之间的相互关系，形成本数据库的数据库管理员视图。

1.4.2 知识准备

概念结构设计阶段得到的 E-R 模型是用户的模型，它独立于任何一种数据模型，独立于任何一个具体的 DBMS。为了创建用户所要求的数据库，需要把上述概念模型转换为某个具体的 DBMS 所支持的数据模型。数据库逻辑设计的过程是将概念结构转换成特定 DBMS 所支持的数据模型的过程。从此开始便进入了"实现设计"阶段，需要考虑具体的 DBMS 的性能、具体的数据模型特点。

E-R 图所表示的概念模型可以转换成任何一种具体的 DBMS 所支持的数据模型，如网状模型、层次模型和关系模型。这里只讨论关系数据库的逻辑设计问题，因此只介绍 E-R 图如何向关系模型进行转换。

1. E-R 图转换数据表

概念模型设计中得到的 E-R 图是由实体、属性和联系组成的,而关系数据库逻辑设计的结果是一组关系模式的集合。因此将 E-R 图转换为关系模型实际上就是将实体、属性和联系转换成关系模式。转换时需遵循以下原则。

1) 实体类型的转换

一个实体转换为一个数据表(关系模式),实体的属性即为关系的属性,实体的标识符即关系模式的码。

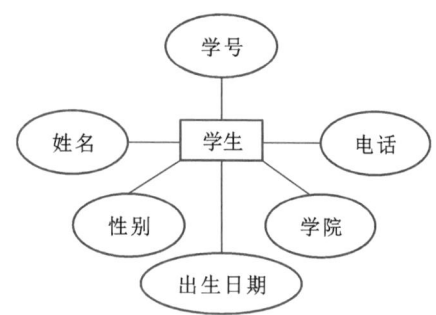

图 1-4-1 学生实体 E-R 图

【例 1-4-1】 将图 1-4-1 所示的学生实体转换为关系模式。

与其对应的关系模式为:

学生(学号,姓名,性别,出生日期,学院,电话)

2) 关系的转换

关系的转换需要根据不同的关系类型做不同的处理。

(1) 1∶1 的转换可以在两个实体类型转换成的两个关系模式中的任意一个当中加入另一个关系模式的主关键字和关系类型的属性。

① 转换为一个独立的关系模式:转换后关系模式中关系的属性包括与该关系相连的各实体的主关键字以及关系本身的属性,关系的主关键字为两个实体的主关键字的组合。

【例 1-4-2】 将图 1-4-2 的 E-R 图按方法①转换主关系模式。

转换后的关系模式如下:

班级(班号,班级名称,入学时间,专业代码,教师编号)

学生(学号,姓名,性别,出生日期,学院,住址,电话)

班级-班长(班号,学号,任期)

② 与某一端对应的关系模式合并。合并后关系模式的属性包括自身关系模式的属性和另一关系模式的主关键字及联系本身的属性,合并后主关键字不变。

【例 1-4-3】 将图 1-4-2 的 E-R 图按方法②转换主关系模式。

转换后的关系模式如下:

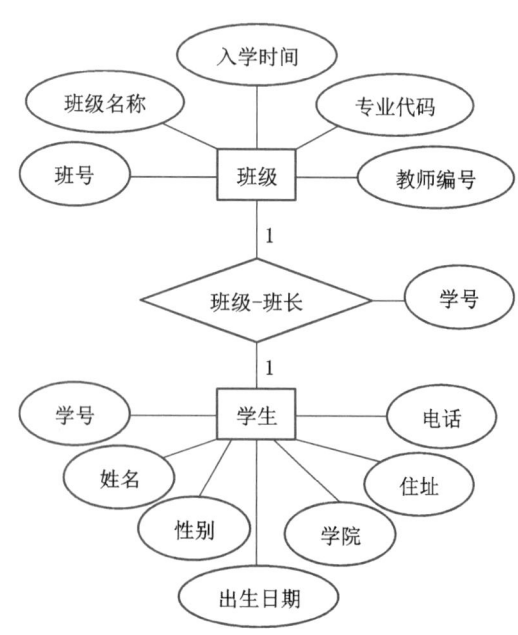

图 1-4-2 1∶1 关系 E-R 图

班级(班号,班级名称,入学时间,专业代码,班长学号,教师编号)

学生(学号,姓名,性别,出生日期,学院,住址,电话)

(2) 1∶n 关系可以转换为一个独立的关系模式(数据表),也可以与 n 端对应的关系模式合并。

① 转换为一个独立的关系模式。关系的属性包括与该关系相连的各实体的主关键字以及关系本身的属性,关系的主关键字为 n 端实体的主关键字。

【例 1-4-4】 将图 1-4-3 的 E-R 图按方法①转换为关系模式。

转换后的关系模式如下:

专业(专业代码,专业名称,学院代码,教师编号,课程名称)

学生(学号,姓名,性别,学院,住址,出生日期)

学习(学号,专业代码,入学时间)

② 与 n 端对应的关系模式合并。合并后关系的属性包括在 n 端关系中加入 1 端关系的主关键字和联系本身的属性,合并后关系的主关键字不变。

【例 1-4-5】 将图 1-4-3 的 E-R 图按方法②转换为关系模式。

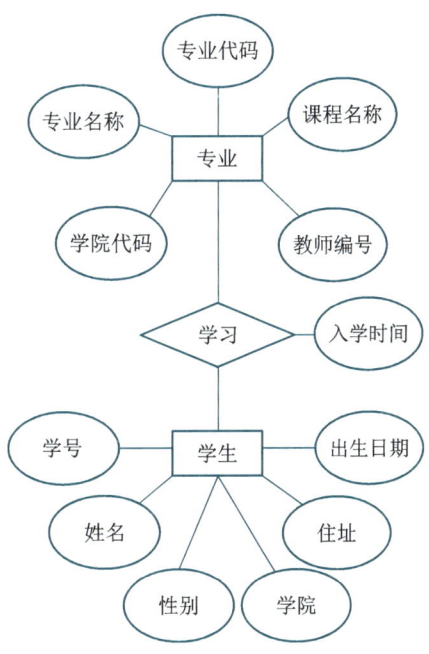

图 1-4-3　1∶n 关系 E-R 图

转换后的关系模式如下:

专业(专业代码,专业名称,学院代码,教师编号,课程名称)

学生(学号,姓名,性别,学院,住址,出生日期,专业代码,入学时间)

(3) 一个 m∶n 关系转换为一个关系模式。

可以将关系类型也转换成关系模式,其属性为两端实体类型的主关键字加上关系类型的属性,而主关键字为两端各实体主关键字的组合。如可转换为如下关系模式:

学生(学号,姓名,性别,学院,住址,出生日期);

课程(课程号,课程名,学分,教师编号);

选课(学号,课程号,成绩)。

2. 数据表优化

数据库逻辑设计的结果不是唯一的。为了进一步提高数据库应用系统的性能,需要对逻辑模型进行适当修改和调整,即数据模型的优化。关系数据模型的优化以规范化理论为指导,关系模式设计的好坏将直接影响数据库设计的成败。将关系模式规范化,使之达到较高的范式是设计好关系模式的唯一途径;否则,设计的数据库会产生一系列的问题。

1) 存在的问题及解决方法

(1) 存在的问题

下面以一个实例说明如果一个关系没有经过规范化可能会出现的问题。

我们要设计一个教学管理数据库,该数据库中要有学生的学号、姓名、性别、出生日期、学院、院长姓名、学生学习的课程名和该课程的成绩信息。若要将这些信息设计为一个关系,则关系的模式如下:

Student(sno,sname,ssex,sbir,sdept,mname,cname,score)

在这个关系中,各个属性的关系为:一个学院有若干学生,但一个学生只属于一个学院;一个学院只有一名院长,但一个院长可以同时兼任几个学院的院长;一个学生可以选修多门课程,每门课程可被若干个学生选修;每个学生学习的每门课程都有一个成绩。

可以看出,此关系模式的主关键字为(sno,cname)。仅从关系模式来看,该关系模式已经包括了需要的信息,如果按此关系模式建立关系,并对它进行分析,就会发现其中的问题。关系模式 Student 的实例见表 1-4-1。

表 1-4-1 关系模式 Student 的实例

sno	sname	ssex	sbir	sdept	mname	cname	score
20200101	孙红	女	2000-06-12	林学院	李明哲	树木学	67
20200101	孙红	女	2000-06-12	林学院	李明哲	林下经济	89
20200101	孙红	女	2000-06-12	林学院	李明哲	计算机基础	96
20200201	张雷	男	2001-05-23	信息工程学院	雷鸣	数据库原理及应用	68
20200201	张雷	男	2001-05-23	信息工程学院	雷鸣	三维设计	87
20200201	张雷	男	2001-05-23	信息工程学院	雷鸣	二维动画	98
20200201	张雷	男	2001-05-23	信息工程学院	雷鸣	C语言程序设计	90
20200202	林雨	男	2001-12-25	信息工程学院	雷鸣	数据库原理及应用	78
20200202	林雨	男	2001-12-25	信息工程学院	雷鸣	三维设计	87
20200202	林雨	男	2001-12-25	信息工程学院	雷鸣	二维动画	89
20200202	林雨	男	2001-12-25	信息工程学院	雷鸣	C语言程序设计	62

从表 1-4-1 中的数据情况可以看出,存在以下问题。

① 数据冗余大。其中的学院名称和院长姓名的存储次数等于该学院学生人数乘以每名学生选修的课程门数,数据重复量太大,存在冗余。

② 插入异常。一个新的学院在没有招生时,或学院里有学生但没有选修课程,学院名称和院长姓名无法插入到数据库中。因为在这个关系模式中主关键字是(sno,cname),这时没有学生而使得学号无值,或学生没有选课而使得课程名无值。但在一个关系中,主关键字属性不能为空值,因此,关系数据库无法操作,导致插入异常。

③ 删除异常。当某个学院的学生全部毕业且没有录入新生信息时,删除学生信息的同时,学院及院长姓名的信息也随之删除,但这个学院依然存在,这时在数据库中却不能找到该学院的信息,这就出现了删除异常。

④ 更新异常。若某个学院更换院长,数据库中该学院的学生记录应全部修改。如果稍有不慎,某些记录漏改了,则造成数据的不一致,即出现了更新异常。

那出现这些插入和删除异常是为什么呢?原因就是在这个关系模式中,属性与属性之间存在数据依赖不好。一个"好"的关系模式不应当发生这些异常情况,冗余度要尽量减少。

(2) 解决方法

对于存在问题的关系模式,可以通过模式分解的方法使之规范化。

可以将表 1-4-1(Student)关系模式分解成三个关系模式(即学生表、成绩表和院系表)。

```
Student(sno,sname,ssex,sbir,sdept)
Sc(sno,cname,score)
Dept(sdept,mname)
```

这样分解后,三个关系模式都不会发生插入、删除异常问题,数据冗余也得到了控制,数据的更新也变得简单。

"分解"是解决冗余的主要方法,也是规范化的一条原则,"关系模式有冗余问题,就分解它"。

注意:以上关系模式的分解方案是否就是最佳的,这也不是绝对的。如果要查询某位学生所在学院的院长姓名,就要对两个关系做连接操作,而连接的代价也是很大的。一个关系模式的数据依赖会有哪些不好的性质?如何改造一个模式?这就是规范化理论所讨论的问题。

2) 数据表优化的方法

① 确定数据依赖;

② 对各个模式之间的数据依赖进行极小化处理,消除冗余的联系;

③ 根据数据依赖理论对关系模式进行分析,考察是否存在部分函数依赖、传递函数依赖、多值依赖等,确定关系模式属于第几范式;

④ 按照需求分析阶段得到各种要求,分析关系模式是否合适,是否需要合并或分解;

⑤ 对关系模式进行必要的分解。

数据库的规范化理论为判断关系模式的优劣提供了理论依据,数据库设计人员可依据规范化理论和方法对数据库的关系模型进行优化。

注意:属性间的函数依赖不是指关系模式 R 的某个或某些关系满足上述限定条件,而是指 R 的一切关系都要满足定义中的限定。只要有一个具体关系 r 违反了定义中的条件,就破坏了函数依赖,使函数依赖不成立。

3) 范式

利用规范化理论,使关系模式的函数依赖集满足特定的要求,满足特定要求的关系模式称为范式。

范式

关系按其规范化程度从低到高可分为五级范式(Normal Form),分别为第一范式(1NF)、第二范式(2NF)、第三范式(3NF、BCNF)、第四范式(4NF)、第五范式(5NF)。规范化程度较高者必是较低者的子集。

5NF⊂ 4NF⊂ BCNF⊂ 3NF⊂ 2NF⊂ 1NF

一个低一级范式的关系模式,通过模式分解可以转换成若干个高一级范式的关系模式的集合,这个过程称为规范化。

由于数据库中最低要求是满足第三范式,因此本书只讲解到第三范式。

在任何一个关系数据库中,第一范式是对关系模式的最低要求,不满足第一范式的数据库就不是关系数据库。

(1) 1NF

第一范式是指数据库表的每一列都是不可分割的基本数据项,同一列中不能有多个值,即实体中的某个属性不能有多个值或者不能有重复的属性。第一范式就是无重复的列。

(2) 2NF

第二范式是在第一范式的基础上建立起来的,即满足第二范式必须先满足第一范式。第二范式(2NF)要求实体的属性完全依赖于主关键字。所谓完全依赖是指不能存在仅依赖主关键字一部分的属性,如果存在,那么这个属性和主关键字的这一部分应该分离出来形成一个新的实体,新实体与原实体之间是一对多的关系。为实现区分,通常需要为表加上一个列,以存储各个实例的唯一标识。简而言之,第二范式就是主属性只依赖于主关键字。

(3) 3NF

第三范式是在第二范式的基础上建立起来的,即满足第三范式必须先满足第二范式。第三范式既要求关系表中不存在非关键字对任一候选关键字的传递函数依赖,也要求关系表不包含其他表中已包含的非主关键字段信息。

(4) BCNF

BCNF(Boyce Codd Normal Form)是由 Boyce 和 Codd 提出的,通常认为 BCNF 是修正的第三范式,有时也称为扩充的第三范式。一个满足 BCNF 的关系模式有以下特点:

① 所有非主属性对每一码都是完全函数依赖。
② 所有的主属性对每一个不包含它的码也是完全函数依赖。
③ 没有任何属性完全函数依赖于非码的任何一组属性。

范式具有避免数据冗余、减少数据库占用空间、减轻维护数据完整性的工作量等优点,但是随着范式的级别升高,其操作难度越来越大,同时性能也随之降低。因此在数据库设计中,寻求数据库可操作性和可维护性之间的平衡,对数据库设计人员是比较困难的。

3. 用户子模式设计

用户子模式又称外模式。我们根据用户需求设计了局部 E-R 图,将 E-R 图转换为数据表后,即生成了整个应用系统的模式后,还应该根据局部应用需求,结合具体 DBMS 的特点来设计用户子模式。

定义数据库模式主要是从系统的时间效率、空间效率、易维护等角度出发。用户子模式与模式是独立的,因此在定义用户子模式时应该更注重考虑用户的习惯与方便程度。

1) 使用更符合用户习惯的别名

在合并各局部 E-R 模型时,应该消除命名冲突,以使数据库系统中同一关系或属性具有相同的名称。但有时命名统一后,会使用户感觉不方便,这时,可以对子模式中的关系和属性名重新命名,使其与用户习惯一致。必要时,可以对子模式中的关系和属性名重新命名,使其与用户习惯一致。

2) 对不同级别的用户定义不同的子模式

针对不同级别的用户定义不同的子模式,可以满足系统对安全性的要求。

3) 简化用户对系统的使用

在实际应用中,经常会使用比较复杂的查询,包括多表连接、分组、统计等。为了方便用户,可以将这些复杂查询定义为视图,用户每次只需对定义好的视图进行查询,避免了每次查询都要对其进行重复描述,大大简化了应用。

1.4.3 任务总结

在创建数据库表之前,首先需要设计数据库表。设计数据库表的主要工具是 E-R 图。E-R 图给出了数据库表的图形化描述,但还缺乏建立数据库表必要的字段名称、字段类型等信息,所以需要把 E-R 图归纳为满足一定约束条件的二维表形式,创建数据库表。

本任务主要讲解如何将 E-R 图转换成数据表,以及数据表的优化和用户子模式设计三个知识点。通过本任务的学习,读者应掌握将 E-R 图转换成数据表的方法,了解用户子模式设计的规则,具备数据表优化的能力。

项目 2　MySQL 的安装与配置

　　MySQL 数据库是一种关系数据库管理系统,所使用的 SQL 语言是用于访问数据库的最常用的标准化语言,其特点为体积小、速度快、成本低、开放源码等,在 Web 应用方面,MySQL 是目前最好的关系数据库管理系统应用软件之一。

　　本项目主要讲解 MySQL 的安装与配置,并以 Windows 平台为例,详细讲解安装和配置 MySQL 数据库的方法。

> **学习目标**
> - 掌握 MySQL 数据库的安装方法;
> - 能够 Windows 平台下安装 MySQL 数据库;
> - 学会配置 MySQL 数据库;
> - 能够启动、登录 MySQL 系统。

任务 1
MySQL 的安装

MySQL 简介

2.1.1 任务描述

设计人员接下来的工作是安装与配置 MySQL。由于 MySQL 适用于多平台,故不同平台下的安装和配置过程也不相同。本任务重点讲解 Windows 平台下 MySQL 的安装与配置的具体步骤。

2.1.2 知识准备

MySQL 是目前应用最广泛的开源关系数据库。MySQL 最早是由瑞典的 MySQL AB 公司开发的,该公司在 2008 年被 SUN 公司收购,紧接着,SUN 公司在 2009 年被 Oracle 公司收购,因此 MySQL 最终就变成了 Oracle 旗下的产品。

要使用 MySQL 数据库,需要提前在本地电脑上安装 MySQL 数据库,下面详细讲解 Windows 平台下 MySQL 的安装与配置过程。

MySQL 的安装一般可以分为下载、安装、配置、运行四个步骤,具体过程如下。

1. 下载

进入 MySQL 官方下载页面 https://www.mysql.com/downloads/,如图 2-1-1 所示。

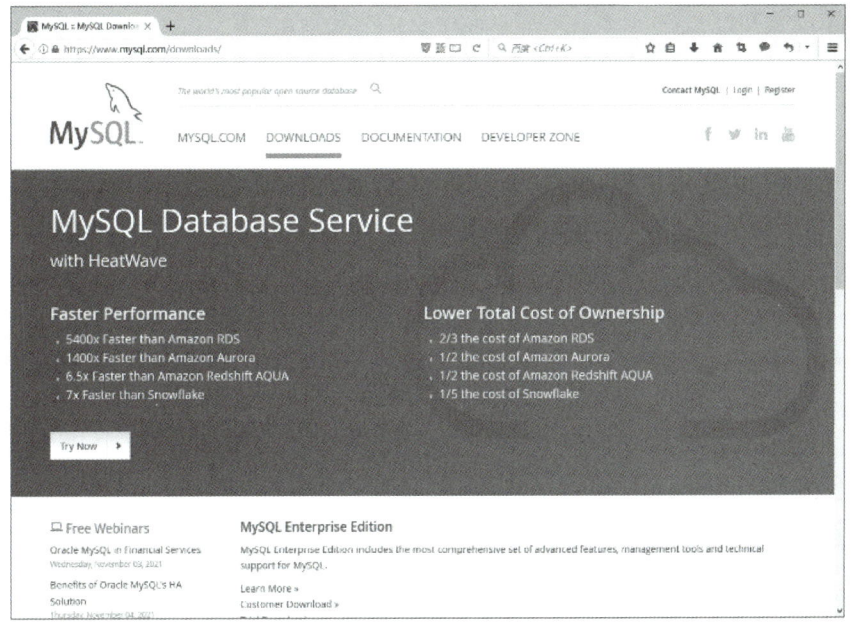

图 2-1-1　MySQL 官方下载界面

可以看到,页面中有多个 MySQL 版本可供选择。

① MySQL Enterprise Edition:企业版本。该版本拥有丰富的功能,需付费,适合对数据库可靠性和安全性要求较高的企业用户。

② MySQL Cluster CGE:高级集群版本,需付费。

③ MySQL Community (GPL):社区版本。该版本开源且免费,是开发者的首选。

滚动至该界面底部,选择"MySQL Community (GPL) Downloads",如图 2-1-2 所示。

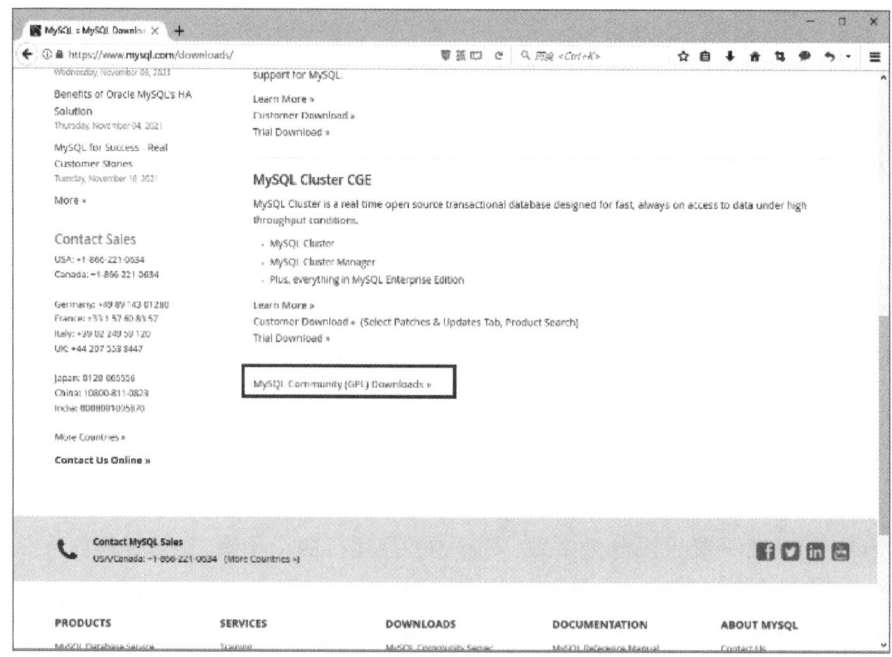

图 2-1-2　选择适合的 MySQL 版本

进入 MySQL 社区版本的下载页面,如图 2-1-3 所示。

图 2-1-3　MySQL 社区版下载页面

选择"MySQL Installer for Windows",进入 MySQL Community Downloads 页面,如图 2-1-4 所示。

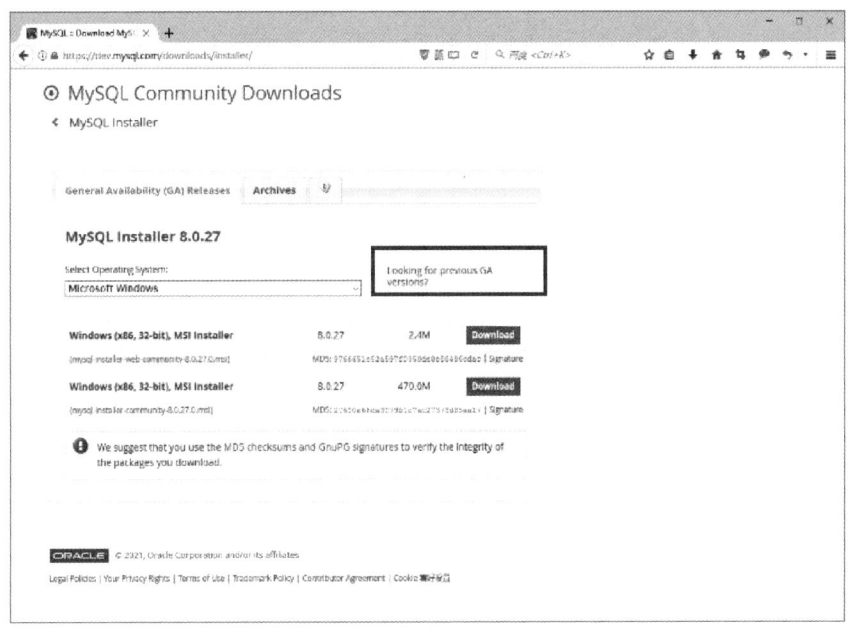

图 2-1-4　MySQL Community Downloads 页面

这里默认提供的是 MySQL Installer 8.0.27,选择"Looking for previous GA versions",进入 MySQL 5.7.x 版本下载页面,如图 2-1-5 所示。

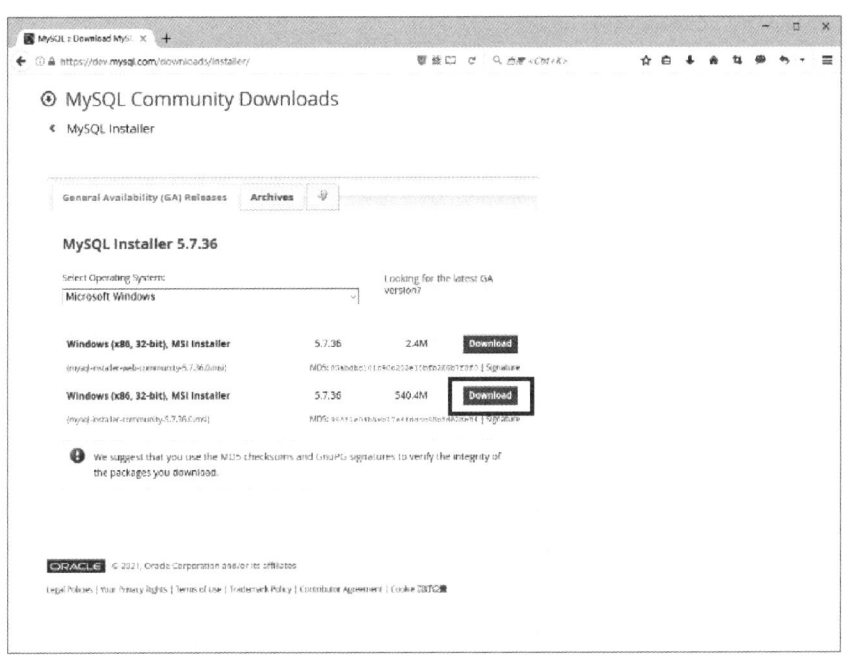

图 2-1-5　MySQL 5.7.x 版本下载页面

页面上提供了两个选项,第一个 Windows（x86,32－bit）,MSI Installer 表示基于 web 的在线安装包,第二个 Windows（x86,32－bit）,MSI Installer 表示完整的 msi 安装包,这里选择第二个,点击"Download"按钮,出现界面如图 2-1-6 所示。

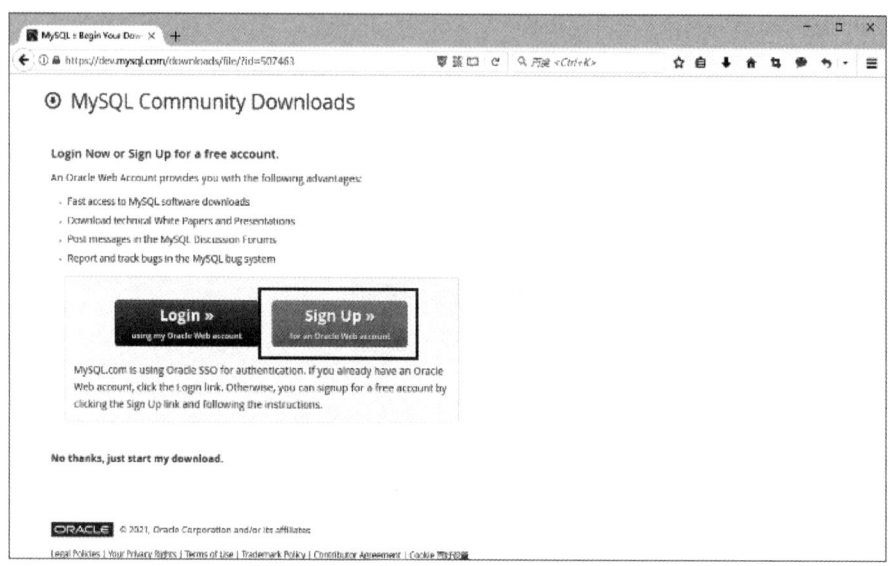

图 2-1-6　点击"Download"按钮后出现界面

无须登录或注册,选择直接下载,如图 2-1-7 所示。

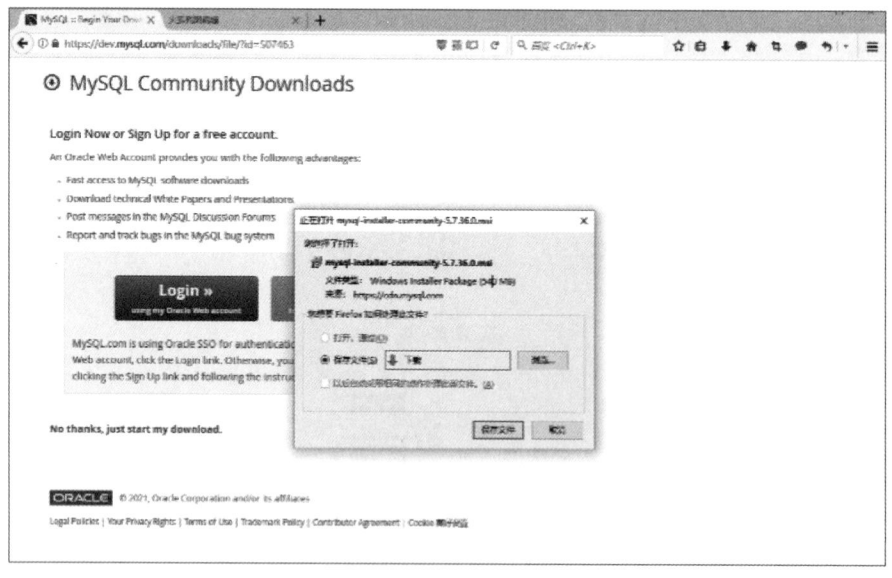

图 2-1-7　下载安装界面

选择保存文件,下载完成后,就得到了一个 mysql-installer-community-5.7.36.0.msi 的文件,如图 2-1-8 所示。

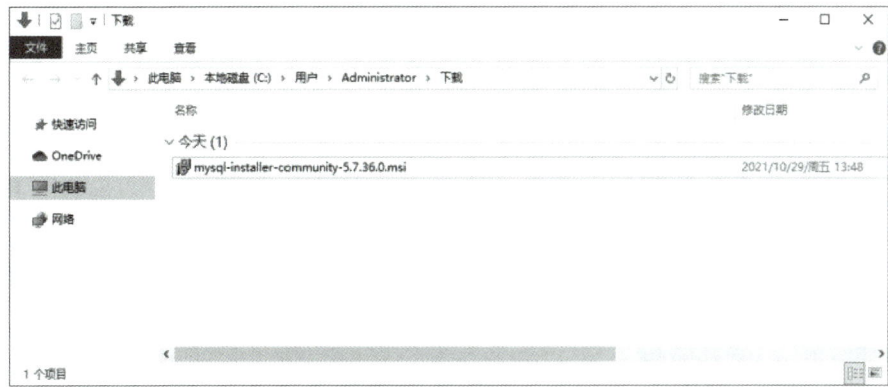

图 2-1-8　完成文件下载

2. 安装

双击 mysql-installer-community-5.7.36.0.msi 文件,提示安全警告,如图 2-1-9 所示。

图 2-1-9　MySQL 安装文件运行界面

点击"运行",开始准备安装,如图 2-1-10 所示。

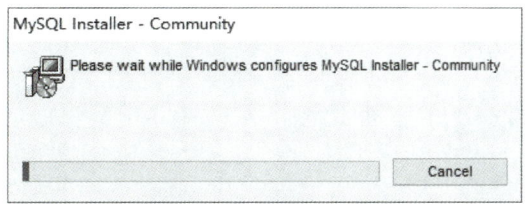

图 2-1-10　MySQL 安装界面

准备完成进入 MySQL Installer 界面,如图 2-1-11 所示。

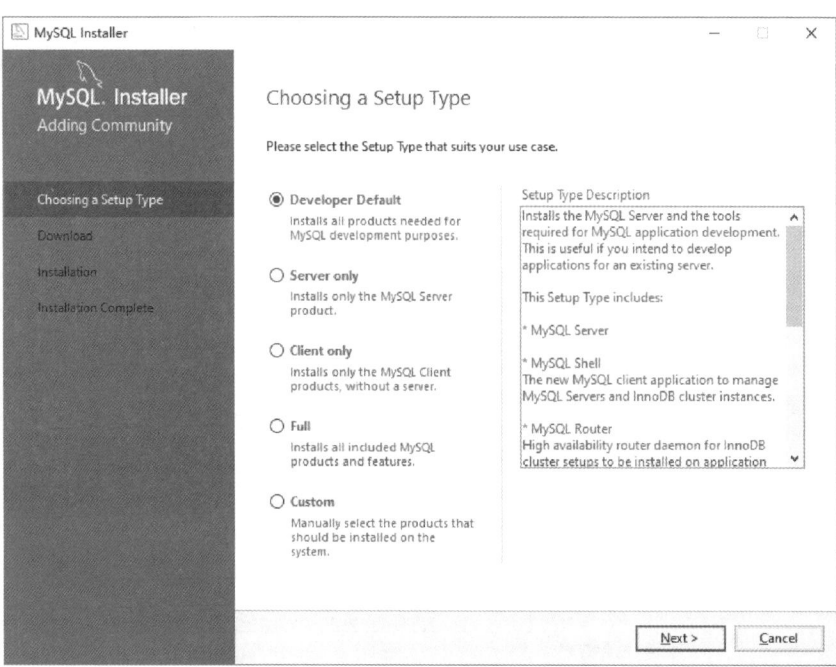

图 2-1-11　MySQL 选择安装类型界面

选择安装类型,这里一共有五个选择,建议选择默认的"Developer Default",它包含了 MySQL Server、MySQL Shell、MySQL Workbench 等开发者所需要的各个组件。选择"Next",进入 Check Requirements 界面,如图 2-1-12 所示。

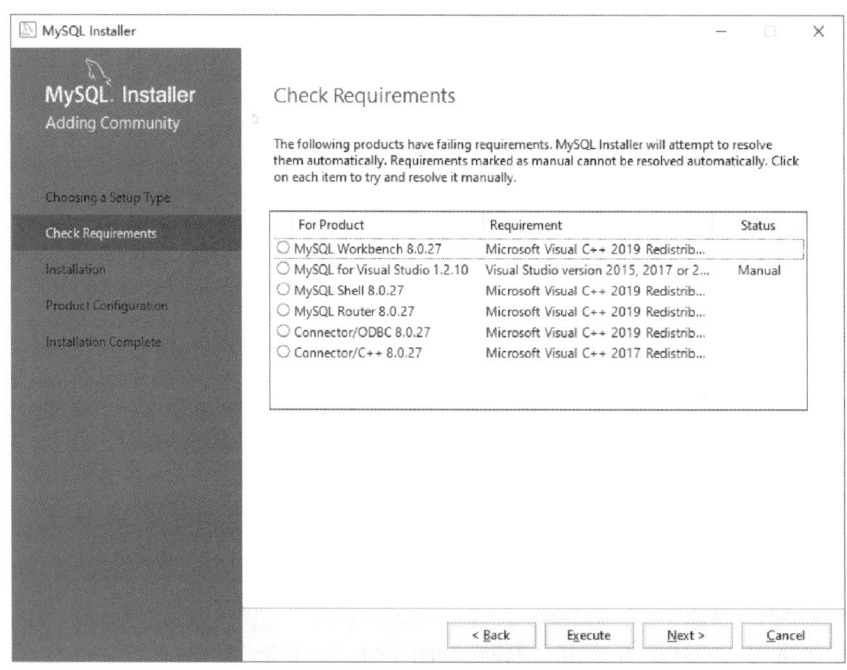

图 2-1-12　Check Requirements 界面

这里会进行环境依赖的检查，一般是 Microsoft Visual C++ 的一些组件，这里点击 "Execute"执行，进行组件的自动下载和安装，如图 2-1-13 所示。

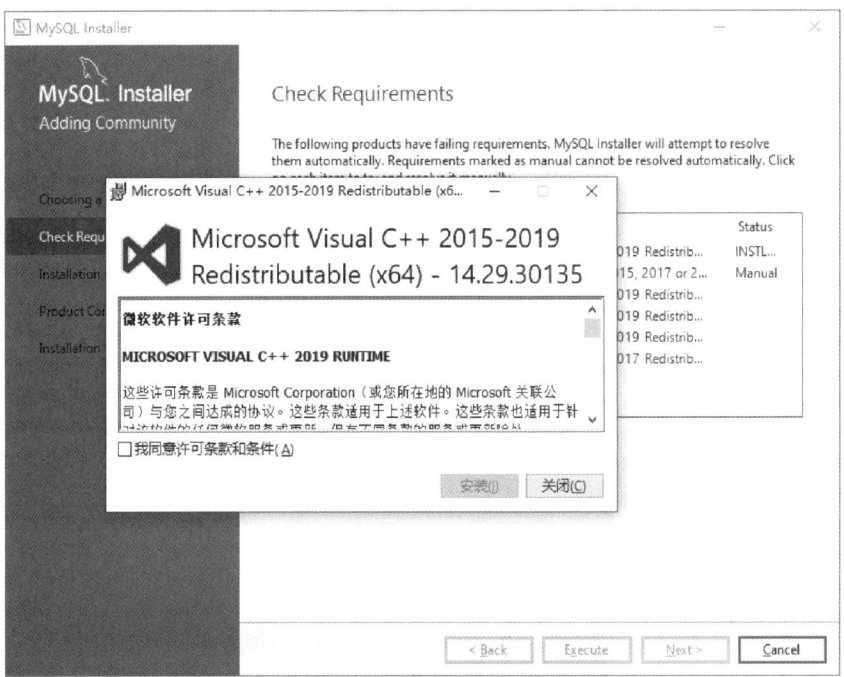

图 2-1-13　组件自动下载界面

同意许可条款并安装，安装完成后，界面如图 2-1-14 所示。

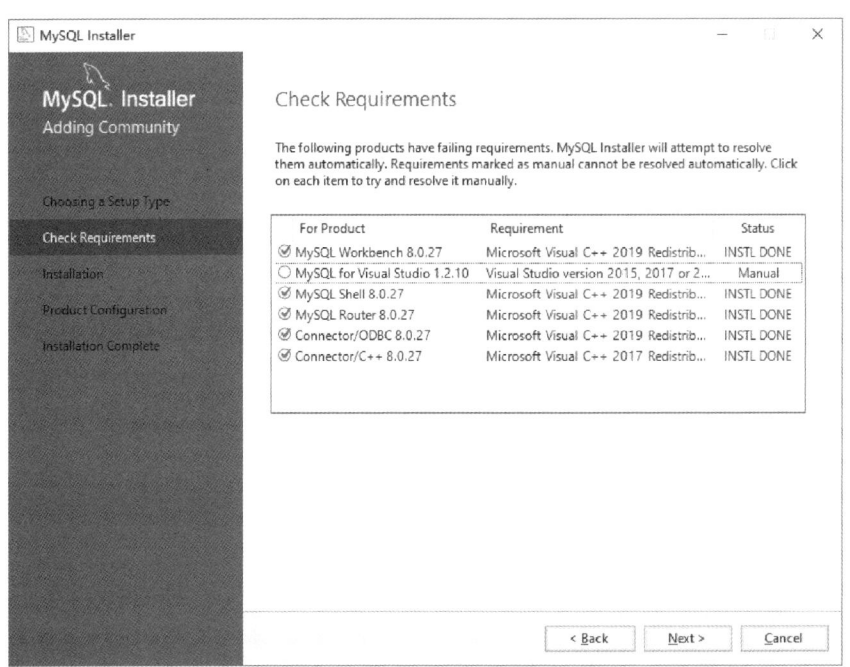

图 2-1-14　安装完成界面

可以看到基本的组件都安装完成，安装成功的组件会显示 INSTL DONE，这里有一个显示 Manual，意思是如果需要的话，可以自行选择手动安装。接下来选择"Next"，如图 2-1-15 所示。

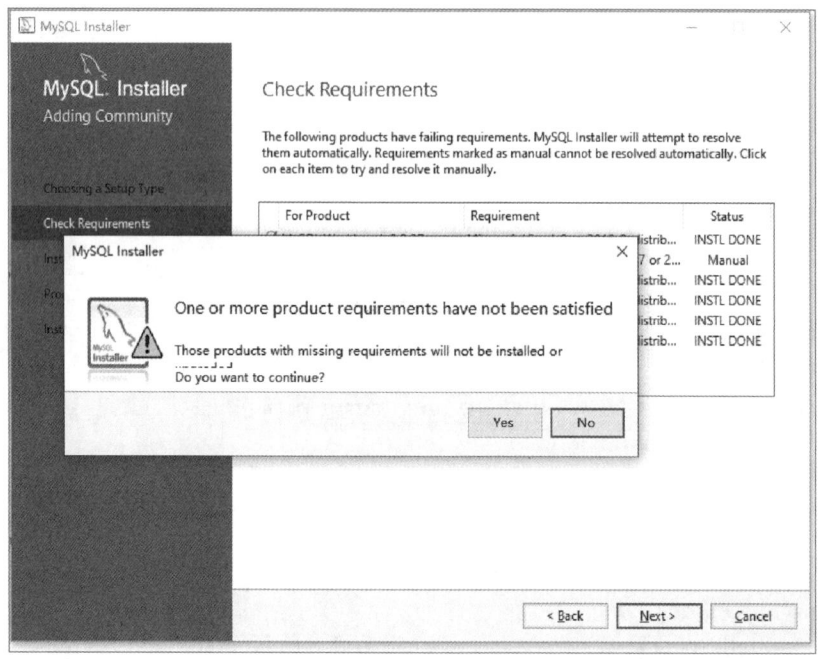

图 2-1-15　INSTL DONE 界面

这里提示有某个需要的组件未安装，是否继续。选择"Yes"，继续安装，进入 Installation 界面，如图 2-1-16 所示。

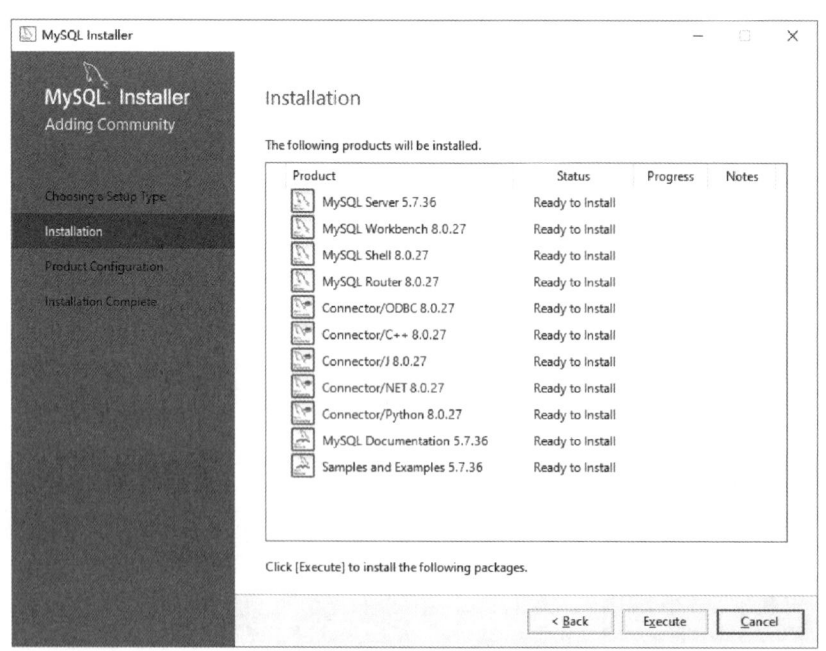

图 2-1-16　Installation 界面

这里显示的是接下来会安装的 MySQL 的组件，包括 MySQL Server、MySQL Workbench、MySQL Shell 等，点击"Execute"执行，会进行顺序安装，如图 2-1-17 所示。

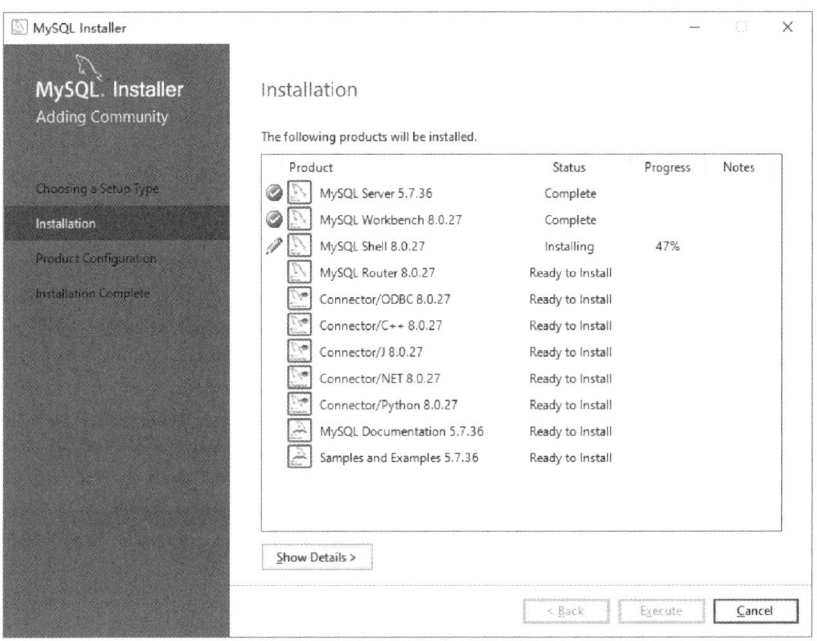

图 2-1-17　进入 MySQL 组件安装界面

这里需要花几分钟时间，等待安装完成即可。所有组件安装完成后，界面如图 2-1-18 所示。

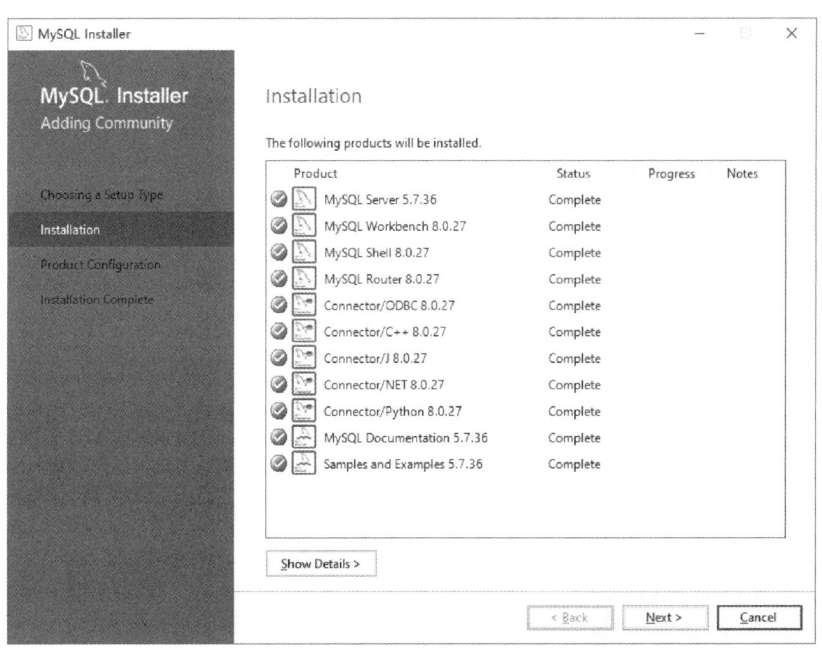

图 2-1-18　MySQL 安装完成界面

3. 配置

在上述安装过程完成后,点击"Next"进入配置界面,如图 2-1-19 所示。

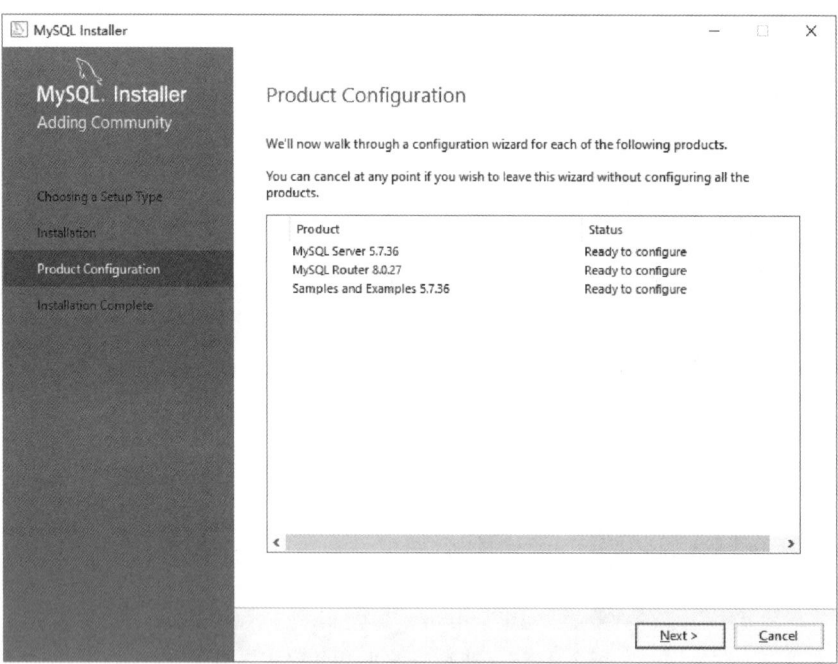

图 2-1-19　MySQL 配置界面

点击"Next",进入服务器类型和网络配置界面,如图 2-1-20 所示。

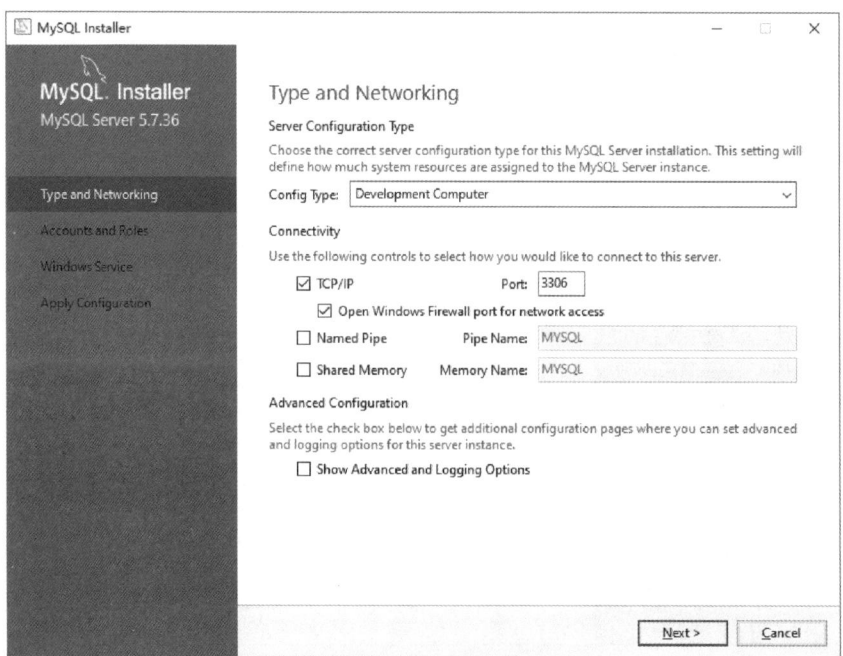

图 2-1-20　网络配置界面

这里 Config Type 设置默认的 Development Computer，端口号 Port，默认是 3306，也无须修改，直接点击"Next"，进入账户设置界面，如图 2-1-21 所示。

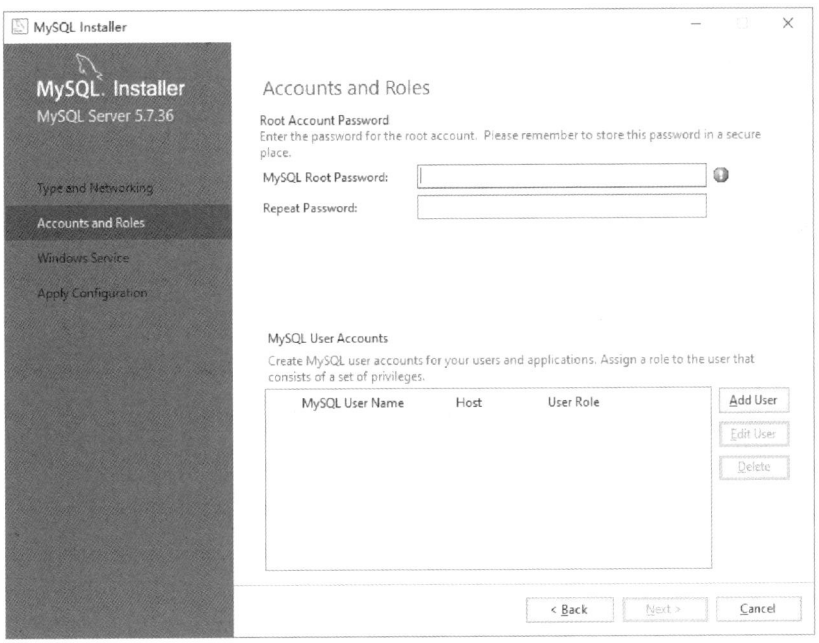

图 2-1-21　设置端口号界面

MySQL 会自动创建一个 root 用户，这里要求设置 Root Account Password，自行设置一个密码即可，比如 123456，如图 2-1-22 所示。

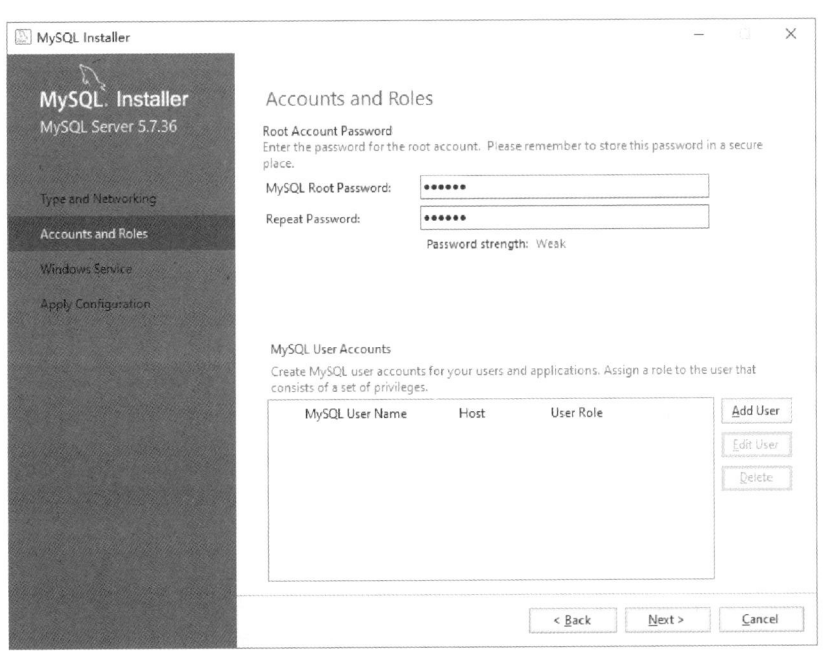

图 2-1-22　设置账户界面

点击"Next",进入 Windows Service 界面,如图 2-1-23 所示。

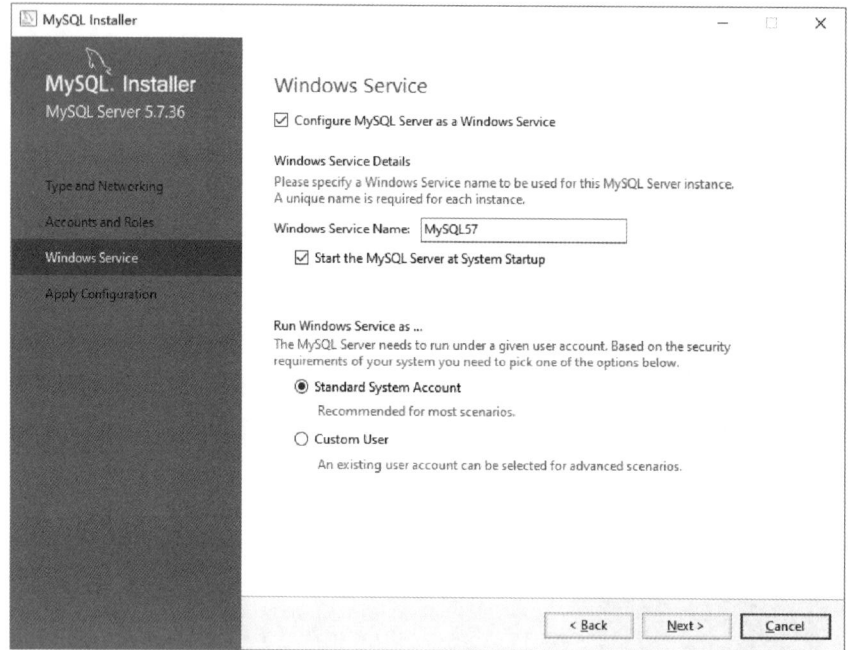

图 2-1-23　Windows Service 界面

这里进行 Window Service 的一些配置,比如开机时要不要启动 MySQL 服务,保持默认即可。点击"Next",进入 Apply Configuration 界面,如图 2-1-24 所示。

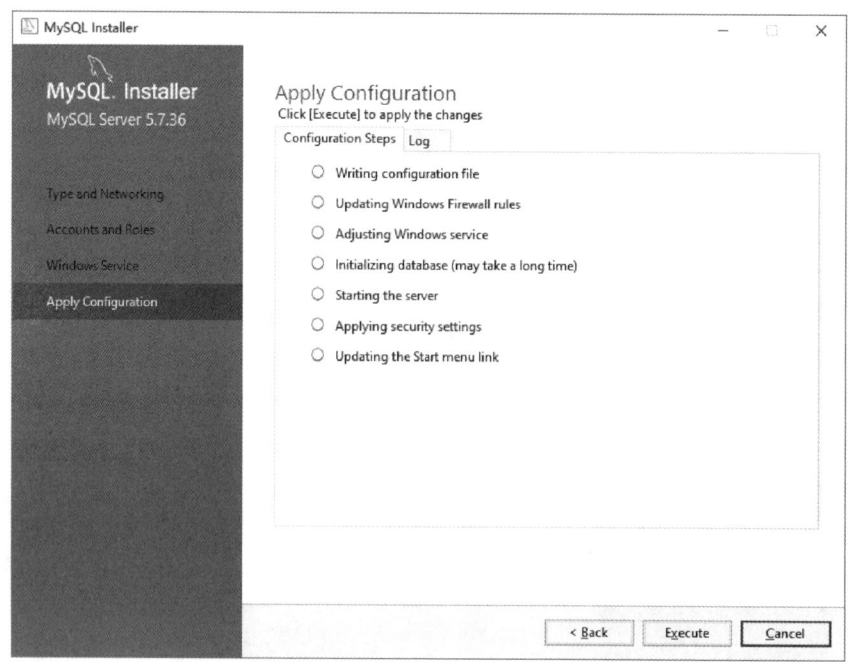

图 2-1-24　Windows Service 配置界面

点击"Execute",将刚才的配置应用生效,配置完成后,界面如图 2-1-25 所示。

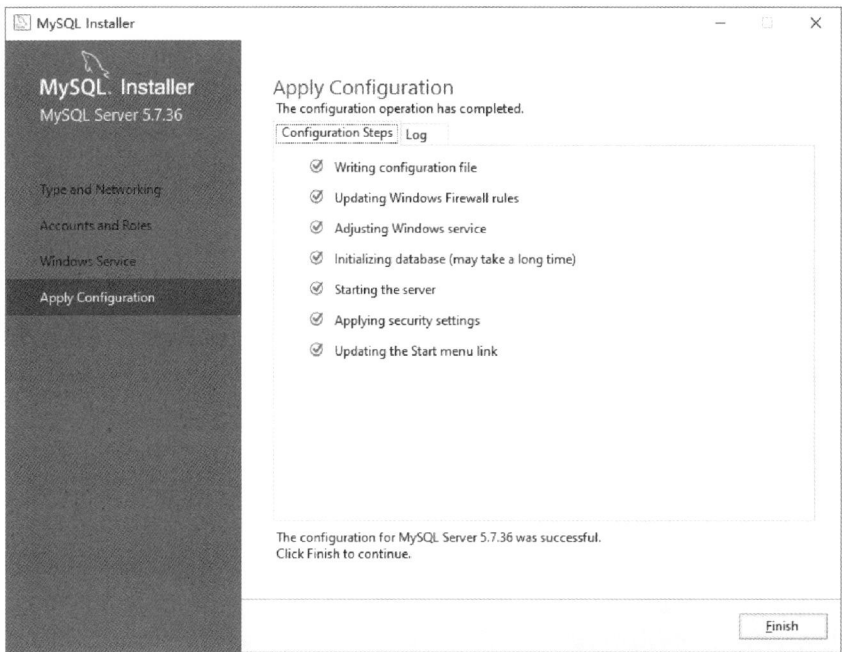

图 2-1-25　Windows Service 配置完成界面

表示 MySQL Server 配置完成,点击"Finish",如图 2-1-26 所示。

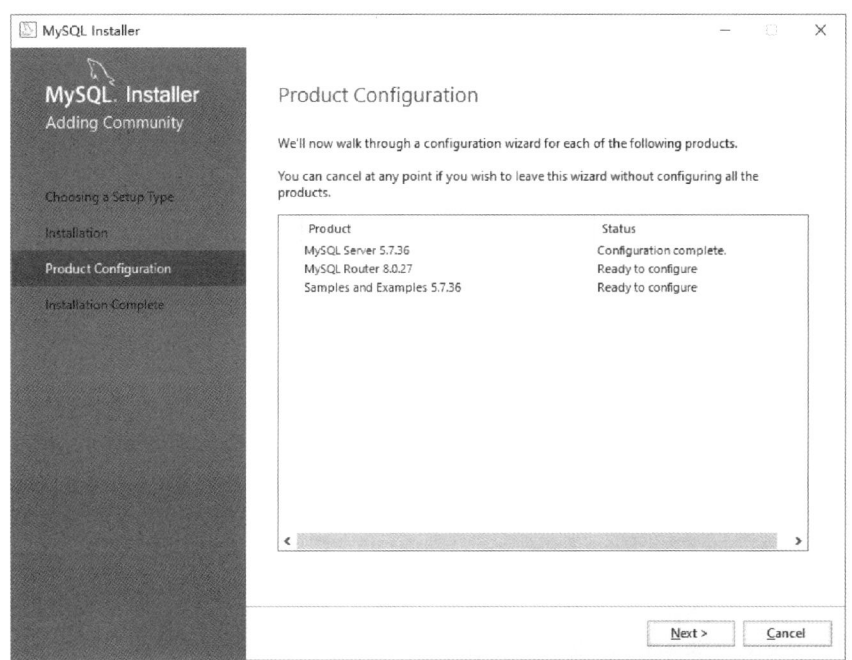

图 2-1-26　MySQL 服务器配置完成界面

接下来进行 MySQL Router 配置,直接点击"Next",如图 2-1-27 所示。

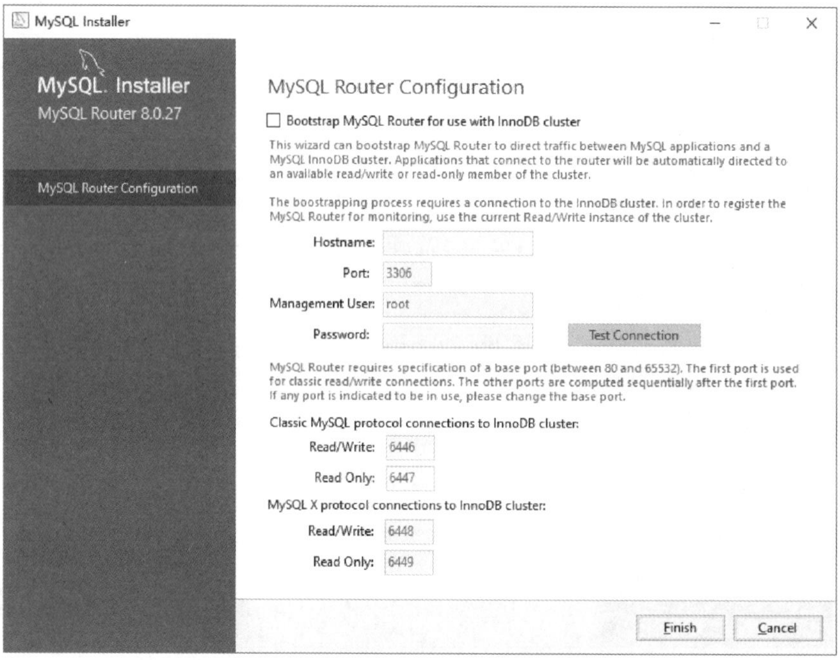

图 2-1-27　MySQL Router 配置界面

无须修改,直接点击"Finish",表示 MySQL Router 配置完成,如图 2-1-28 所示。

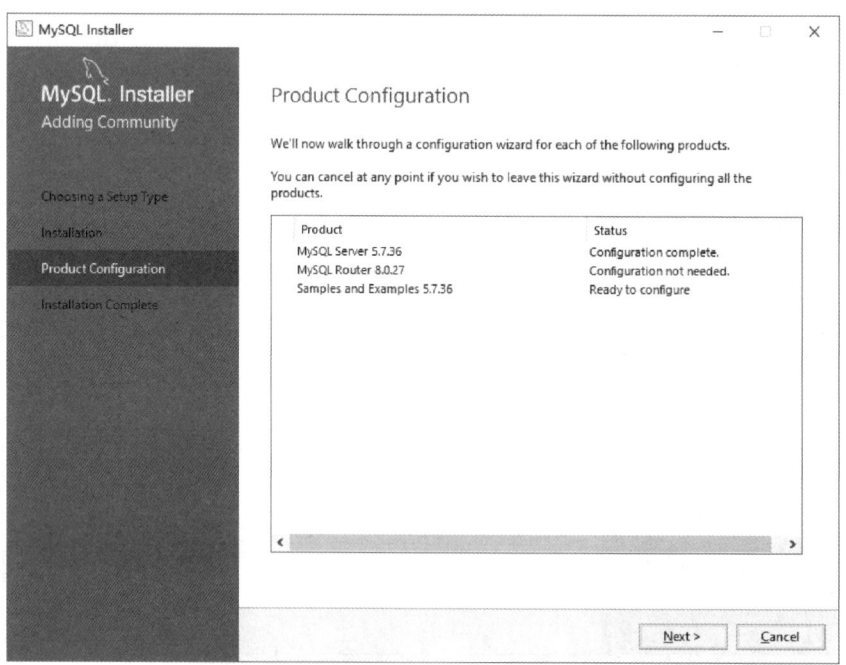

图 2-1-28　MySQL Router 配置完成界面

接下来进行示例的安装和配置,点击"Next",如图 2-1-29 所示。

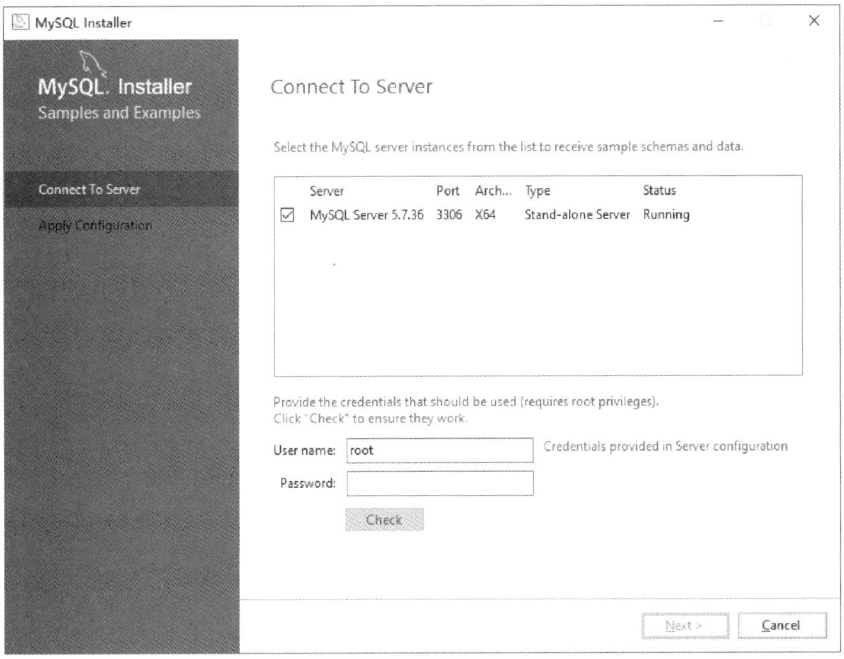

图 2-1-29　MySQL 示例安装和配置界面

这里需要连接 MySQL 服务器,使用 root 用户登录,输入密码为之前设置的 123456,点击"Check",如图 2-1-30 所示。

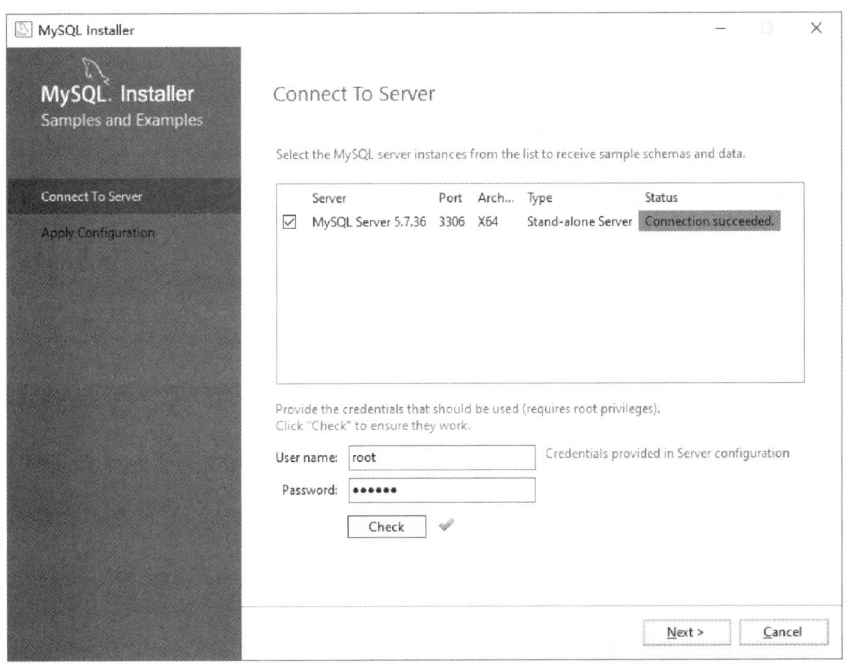

图 2-1-30　连接 MySQL 服务器界面

绿色的对号表示连接成功,选择"Next",进入 Apply Configuration 界面,如图 2-1-31 所示。

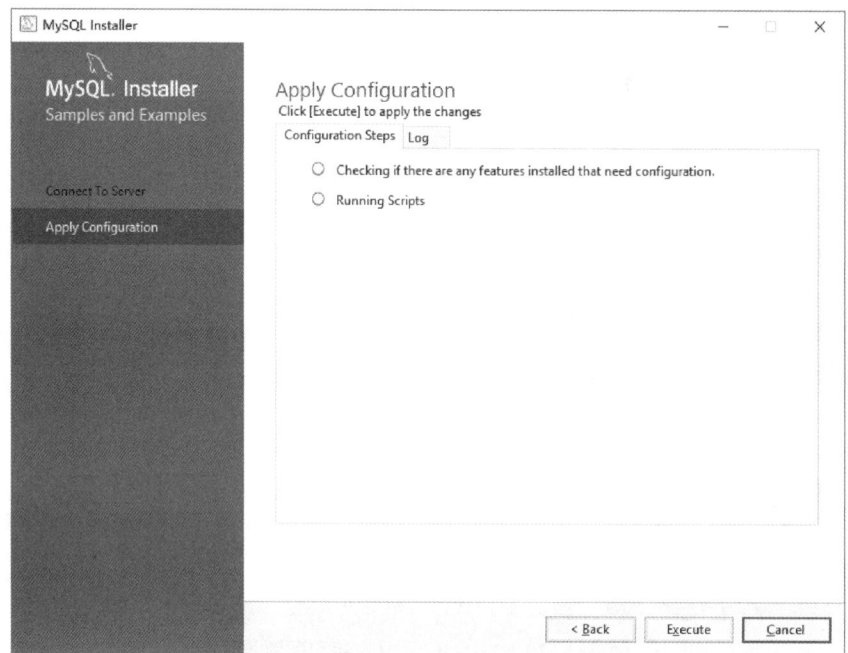

图 2-1-31　连接 MySQL 服务器成功界面

点击"Execute",将刚才的配置应用生效,配置完成后,界面如图 2-1-32 所示。

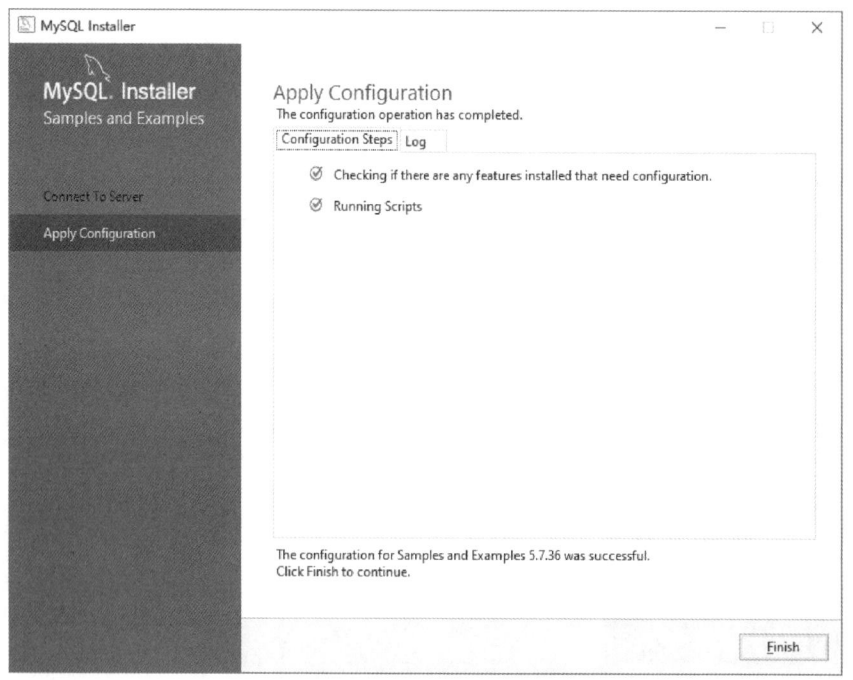

图 2-1-32　MySQL 配置完成界面

表示 Product 配置完成,点击"Finish",如图 2-1-33 所示。

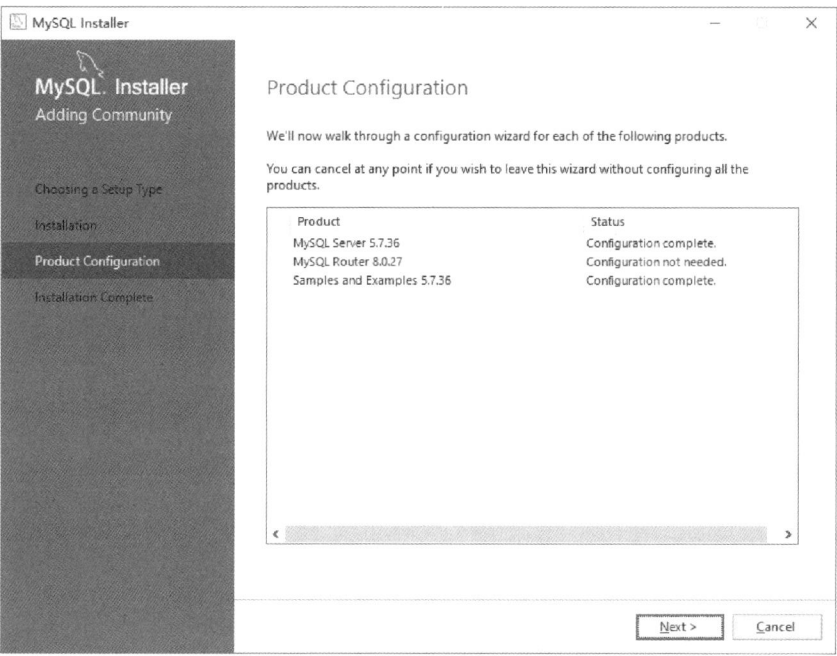

图 2-1-33　Product 配置完成界面

点击"Next",进入 Installation Complete 界面,如图 2-1-34 所示。

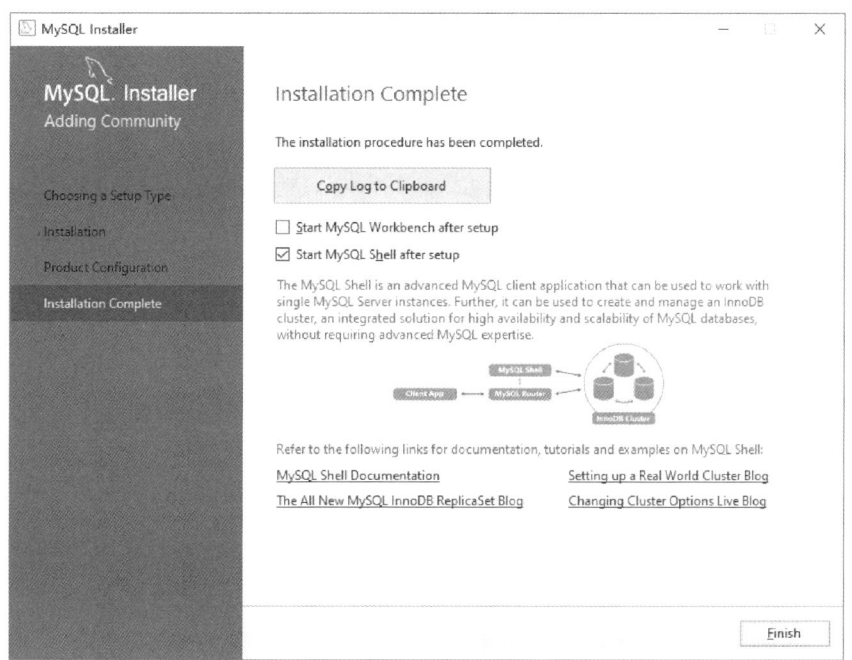

图 2-1-34　Installation Complete 界面

点击"Finish",表示 MySQL 配置完成。

4. 运行

安装完成后,打开开始菜单,选择 MySQL 5.7 Command Line Client,如图 2-1-35 所示。

图 2-1-35 打开 MySQL 5.7 界面

这里需要输入 root 用户的密码 123456,效果如图 2-1-36 所示。

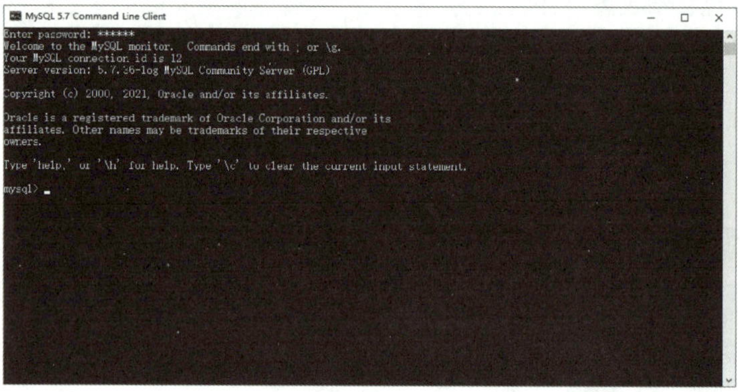

图 2-1-36 进入 root 用户界面

试着输入命令"show databases;",可以看到效果如图 2-1-37 所示。

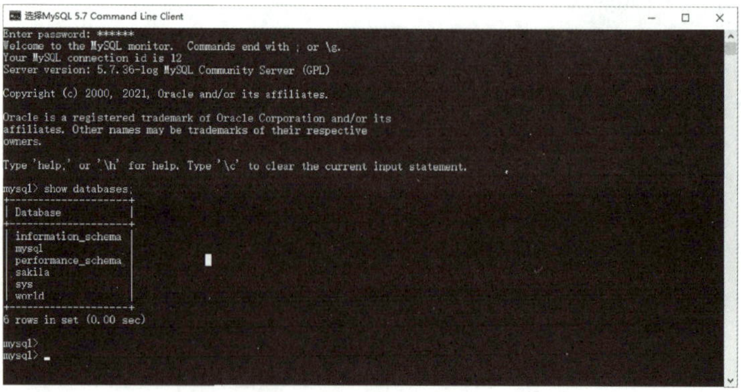

图 2-1-37 查看数据库界面

至此，MySQL 下载、安装、配置、运行成功。

2.1.3　任务总结

MySQL 是一个关系数据库管理系统，由瑞典 MySQL AB 公司开发，目前属于 Oracle 公司。关系数据库将数据保存在不同的表中，而不是将所有数据放在一个大仓库内，这样就提高了速度和灵活性。

本任务主要讲解了 Window 平台下 MySQL 的安装与配置。通过本任务的学习，使读者掌握 MySQL 在 Windows 平台的安装与配置，为后续课程搭好环境。

任务 2　MySQL 的使用

2.2.1　任务描述

MySQL 为关系数据库，这种所谓的"关系"可以理解为"表格"的概念，一个关系数据库由一个或数个表格组成。接下来我们需要了解 MySQL 的基本知识，掌握使用 MySQL 数据库的方法。

本任务主要讲解 MySQL 的特点、MySQL 的工作原理、MySQL 的存储引擎、MySQL 服务的启动与登录，以及 MySQL 的相关命令等内容。

2.2.2　知识准备

MySQL 作为关系数据库，由于其体积小、开放源码、成本低等优点，当前被广泛地应用于 Internet 的中小型网站。

1. MySQL 的特点

1）可移植性好

MySQL 支持至少 20 种的开发平台，包括 Linux、Windows 等。这使得在任何平台下编写的程序都可以进行移植，而不需要对程序做任何修改。

2）强大的数据存储功能

MySQL 数据库的最大有效容量通常是由操作系统对文件大小的限制决定的，而不是由 MySQL 内部限制决定的。InnoDB 存储引擎将 InnoDB 表保存在一个表空间内，该表空间可由数个文件创建，表空间的最大容量为 64 TB，可以轻松处理拥有上万条记录的大型数据库。

3）能提供多种存储器引擎

MySQL 支持的存储引擎有 InnoDB 引擎、MyISAM 引擎、MEMORY 引擎、CSV 引擎、

ARCHIVE 引擎和 BLACKHOLE 引擎。

4）功能强大

MySQL 中提供了多种数据库存储引擎，各个引擎各有所长，适用于不同的应用场合。用户可以选择最合适的引擎以得到最高性能，这些引擎甚至可以应用处理每天访问量数亿次的高强度 Web 搜索站点。MySQL 支持事务、视图、存储过程和触发器等。

5）安全性高

灵活安全的权限和密码系统允许主机的基本验证。连接到服务器时，所有的密码传输均采用加密形式，从而保证了密码的安全。

6）运行速度快

高速是 MySQL 的显著特性。在 MySQL 中，使用了极快的 B 书磁盘表（MyISAM）和索引压缩；通过使用优化的单扫描多连接，能够极快地实现连接；SQL 函数使用高度优化的类库实现，运行速度极快。

2. MySQL 的工作原理

为了理解 MySQL 的工作原理，我们用一张表来理解它，如表 2-2-1 所示。

MySQL 的内部架构由以下几个部分组成。

表 2-2-1　MySQL 的内部架构

编程语言交互接口 Native CAP、JDBC、ODBC、.NET、PHP、Perl、Python、Ruby、Cobol				
系统管理 和控制工具集合	连接池			
^	SQL 接口	解析器	查询优化器	查询缓存
存储引擎 MyISAM、InnoDB、Archive、Federated、Memory、Partner、Community、Custom				
文件系统				

1）编程语言交互接口（Connectors）

编程语言交互接口指的是不同语言与 SQL 的交互接口，比如说 Java 的 JDBC、.Netframework 的 ODBC。

2）系统管理和控制工具集合（Management Services & Utilities）

系统管理和控制工具集合提供管理配置服务、备份还原、安全复制等功能。

3）连接池（Connection Pool）

连接池用于接受客户端的请求，缓存请求，检查内存可利用的情况，如果没有可用线程,就创建纯种执行任务,有可用线程就重复利用。

4)解析器(Parser)

解析器的功能是解析验证 SQL 语法,分解 SQL 成相应的数据结构,以备后面处理。

5)查询优化器(Optimizer)

查询优化器对 SQL 语句进行优化处理,优化执行路径,生成执行树,最终数据库会选择认为最优的方案执行并返回结果。

6)SQL 接口(SQL Interface)

SQL 接口接受用户的 SQL 命令,并返回结果。

7)查询缓存(Cache & Buffer)

查询缓存用于缓存查询结果。若 SQL 查询中命中查询结果,则将直接从缓存中返回结果,不再执行 SQL 分析等操作性;若没有命中,则会进行后续的解析、查询优化、执行 SQL 接口,返回结果,同时将结果加入缓存中。

8)存储引擎(Pluggable Storage Engines)

存储引擎是 MySQL 中具体的与文件打交道的子系统,可以看到它是以插件形式存在的,意味着可以自定义存储引擎,这是很特别的地方。MySQL 提供了很多存储引擎,其优势各不一样,有的查询效率高,有的支持事务等,最常用的有 MyISAM、InnoDB、BDB。

9)文件系统(File System)

文件系统是存放数据库表数据以及相关配置的地方。

3. MySQL 的存储引擎

MySQL 中的数据可以采用不同的技术存储在文件(或内存)中。这些技术都使用不同的存储机制、索引技巧、锁定水平,并且最终提供广泛的、不同的功能和能力。通过选择不同的技术,能够获得额外的速度或者功能,从而改善应用的整体功能,每种技术以及配套的相关功能就可以看成是一种数据库存储引擎,MySQL 默认配置了许多不同的存储引擎,这些存储引擎可以预先设置或者在 MySQL 服务器中启用。

比如,银行系统中客户进行转账交易时就需要一个支持事务处理的数据库,以确保事务处理不成功时数据的回退能力,这时就可以采用 InnoDB 存储引擎,而不能选用 MyISAM 存储引擎,这是因为 MyISAM 是非事务性存储引擎,只有 InnoDB 才能保证银行系统客户转账交易的正常进行。

1)InnoDB 引擎

作为 5.5 版本以上的默认存储引擎,InnoDB 具备以下主要优势。

① 数据操纵语言(Data Manipulation Language,DML)操作遵循事务的 4 个特征——原子性(Atomicity)、一致性(Consistency)、隔离性(Isolation)、持久性(Durability),并通过 commit、rollback、crash-recovery 保障数据的安全。具体来说,crash-recovery 就是指如果服务器因为硬件或软件问题而崩溃,不管当时数据是怎样的状态,在重启 MySQL 后,InnoDB 会自动恢复到发生崩溃之前的状态,并回到用户离开的地方。另外,如果数据在磁

盘或者内存中损坏,校验机制会提醒当前数据为虚假数据。

② 具有行级锁和 Oracle 风格的读一致性,通过一种更改缓存机制对新增、更新和删除进行优化,增加了对用户并发读写操作和性能。

③ 对表进行基于主键的优化查询,每张表都有一个基于主键的聚焦索引,以此减少磁盘 I/O,进而提高了搜索效率和性能。

④ 支持外键约束,检查外键、插入、更新和删除,以确保数据的完整性。

⑤ InnoDB 存储引擎提供了专门的缓存池,在内存中缓存了表和索引的数据,常用的数据可以直接从内存中处理,比直接从磁盘获取数据处理速度快。

⑥ 可以压缩表和相关索引,创建和删除索引以达到提高性能的目的。

⑦ 快速压缩表空间,并能释放磁盘空间,保证系统能够重用,而不仅仅是腾出空间给 InnoDB 复用。

⑧ 表的存储分为两个文件:frm 文件——存储表结构;ibd 文件——存储表的数据和索引。

⑨ 创建表时可以通过以下语句显示指定表引擎。

```
CREATE TABLE TABLE_NAME(I INT)ENGINE = INNODB;
```

2) MyISAM 引擎

MyISAM 引擎不支持事务、外键,但它访问速度非常快,表的存储分为以下 3 个文件。

① frm 文件:存储表定义。

② MYD(MYData)文件:存储数据。

③ MYI(MYIndex)文件:存储索引。

MyISAM 有以下特点:

① 所有数值类型键值都是以高字节存储的,以便于更好地索引压缩。

② 每张 MyISAM 最多支持$(2^{32})^2$($\sim$1.844E+19)行。

③ 每张 MyISAM 表支持的最大索引数是 64,每个索引最多 16 列。

④ 当表字段是通过自增长(auto_increment)生成的,索引树节点只会包含一个键,这样可以提高索引的空间利用率。

⑤ 在 INSERT、UPDATE 时,MyISAM 有内部处理会自动更新 auto_increment 字段,这使 auto_increment 处理更快(至少 10%)。

⑥ 当进行混合操作(删除、更新、插入同时进行)时,MyISAM 通过自动合并和扩展删除块,减少了行碎片。

⑦ MyISAM 支持并发插入数据:如果一个表中的数据文件中没有空闲块,则可以在插入数据的同时通过其他线程读取表数据。空闲块是由删除或更新操作时数据长度超过当前行内容长度引起的。

⑧ 可以通过将数据文件(MYD)和索引文件(MYI)放在不同物理设备的不同目录上来

更快地创建表。

⑨ BLOB 和 TEXT 可以被索引，索引列中允许 NULL 值，不过需要占 0～1 个字节。

⑩ 创建表时可以通过以下语句显示指定表引擎。

```
CREATE TABLE TABLE_NAME(I INT)ENGINE = MyISAM;
```

3）MEMORY 引擎

MEMORY 引擎又称 HEAP 引擎，用来创建特殊用途的表，且内容存储在内存中。将数据存储在内存中，能够实现快速访问和低延迟。

因此，使用 MEMORY 引擎，在出现数据崩溃、硬件故障等问题时，数据极易丢失；它适合临时态和非关键数据的操作。

创建表时可以通过以下语句显示指定表引擎。

```
CREATE TABLE TABLE_NAME(I INT)ENGINE = MEMORY;
```

4）CSV 引擎

当用户创建一个 CAV 引擎的表时，服务器会在数据库目录中创建一个"表名.frm"的格式文件，同时还会创建一个"表名.csv"的数据文件，该数据文件中的数据是以逗号分隔保存的，它主要用于 CSV 报表格式的数据存储，应用面比较窄。

缺点：CSV 引擎不支持索引，也不支持分区，并且所有列必须指明为 NOT NULL。

创建表时可以通过以下语句显示指定表引擎。

```
CREATE TABLE TABLE_NAME(I INT)ENGINE = CSV;
```

5）ARCHIVE 引擎

ARCHIVE 引擎用于数据归档，它的压缩比例非常高，适合存储历史数据（前提是不做查询操作），所占的存储空间不到 InnoDB 引擎的 1/10；它支持行级锁定实现并发插入操作，却不支持事务，其设计目的在于提供高速插入和压缩功能；另外，它不支持索引。

创建表时可以通过以下语句显示指定表引擎。

```
CREATE TABLE TABLE_NAME(I INT)ENGINE = ARCHIVE;
```

6）BLACKHOLE 引擎

BLACKHOLE 引擎是一种很特别的引擎，它的表不存储任何数据，就像是"黑洞"一样。它主要用于充当伪服务器、日志服务器、增量备份服务器等。

创建表时可以通过以下语句显示指定表引擎。

```
CREATE TABLE TABLE_NAME(I INT)ENGINE = BLACKHOLE;
```

除此，MySQL 还支持其他引擎，如 MERGE 引擎、FEDERATED 引擎、EXAMPLE 引擎等，若有需要，可以查阅相关文档，这里就不一一列举。

7) 存储引擎特点对比

MySQL 在 5.5 版本之后默认的存储引擎是 InnoDB 存储引擎，下面对比一下 MySQL 5.7 版本存储引擎的特点，如表 2-2-2 所示。

表 2-2-2 MySQL 5.7 版本存储引擎的特点

特点	InnoDB	MyISAM	MEMORY	ARCHIVE
存储限制	64 TB	没有	有	没有
事务安全	支持			
锁机制	行锁	表锁	表锁	行锁
B 树索引	支持	支持	支持	
哈希索引	支持		支持	
全文索引	支持	支持		
集群索引	支持			
数据缓存	支持		支持	
数据可压缩		支持		支持
空间使用	高	低		非常低
内存使用	高	低	中等	低
批量插入速度	低	高	高	非常高
支持外键	支持			
复制支持	支持	支持	支持	支持
查询缓存	支持	支持	支持	支持
备份恢复	支持	支持	支持	支持

选择存储引擎的建议如下：

① MySQL 的存储引擎很多，不同的库、不同的表可以选择不同的存储引擎，推荐同一个库用同一种存储引擎，因为不同存储引擎的表之间 join 操作比较慢。

② InnoDB 存储引擎提供了具有提交、回滚和崩溃恢复能力的事务安全表，如果需要事务处理、ACID 事务支持，则选择 InnoDB 存储引擎。

③ Memory 将所有数据保存在缓存 RAM 中，可以提供极快的访问速度。

④ 尽量不要选择 MyISAM 存储引擎，它只能用单个 CPU，内存只能用到 4GB，内存里只有索引，而且并发能力差。

4. MySQL 服务的启动与登录

MySQL 数据库管理系统分为服务器端(Server)和客户端(Client)两部分。只有在服务器端的服务开启后，才可以通过客户端登录 MySQL。

1) 启动 MySQL 的服务

① 通过 Windows 服务管理器启动 MySQL 的服务。打开 Windows 运行对话框，输入

"services.msc"单击"确定"按钮,打开 Windows 的【服务】管理器(图 2-2-1),在其中可以看到服务名为"MySQL"的服务项,其启动类型为自动,如果没有"已启动"字样,说明 MySQL 服务未启动。这时可以双击"MySQL"服务,打开"MySQL 的属性"对话框(如图 2-2-2 所示),单击"启动""停止""暂停"或"恢复"按钮来更改服务状态。也可以用鼠标右键弹出下拉列表来选择启动或停止。

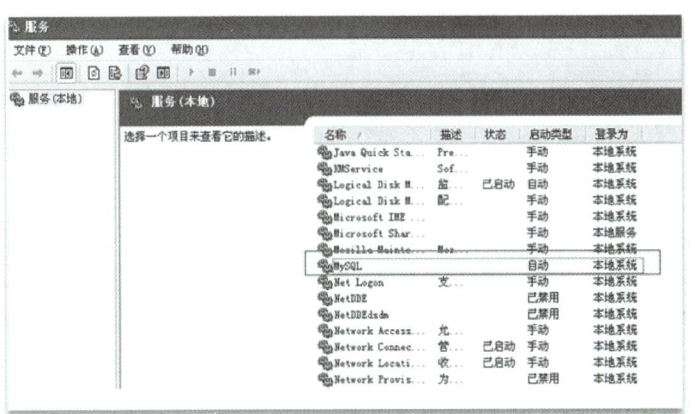 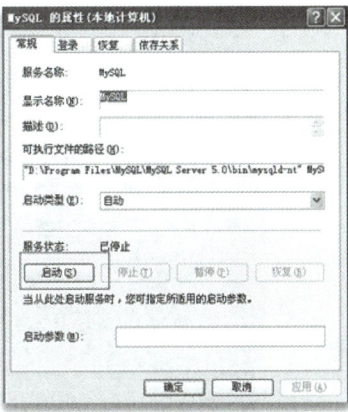

图 2-2-1　服务对话框　　　　　　　　图 2-2-2　MySQL 服务对话框

② 通过行命令启动 MySQL 的服务。通过行命令启动 MySQL 的服务的方法是:【开始】→【运行】对话框中输入"cmd",回车后弹出命令提示符窗口,然后输入"net start mysql"启动服务,输入"net stop mysql"停止服务(图 2-2-3)。

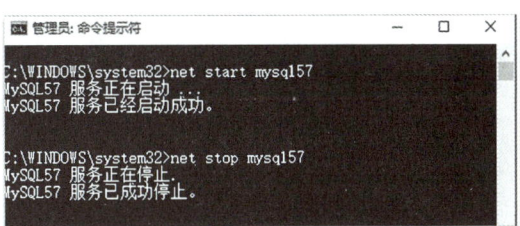

图 2-2-3　DOS 命令启动或停止 MySQL 的服务器

2) 登录 MySQL

① 使用 MySQL 控制台登录。MySQL 的登录可以利用 MySQL 的控制台进行登录,在开始菜单中找到 MySQL 5.7 Command Line Client,如图 2-2-4 所示。

② 使用命令登录 MySQL。具体命令如下:

`Mysql -uroot -pitcast`

上述命令中"-u"后面用于输入用户名,"-p"后面用于输入用户的登录密码。此命令执行结果如图 2-2-5 所示。

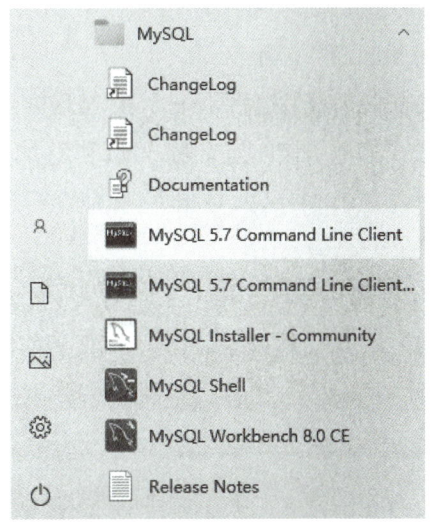

图 2-2-4　MySQL 控制台启动服务器

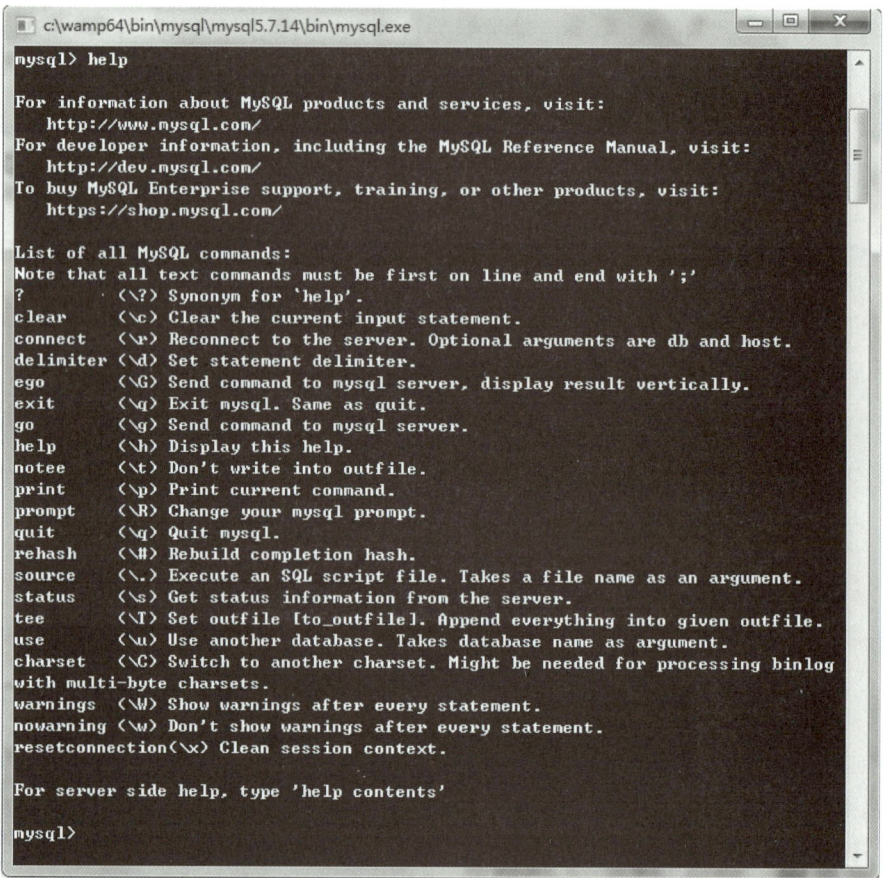

图 2-2-5　帮助命令

5. MySQL 的相关命令

1) Help 和\h 命令

在安装、管理和使用 MySQL 过程中,需要记忆很多的 MySQL 命令。有时很多的命令不知道或忘记了该如何应用,可以利用 MySQL 的 Help 或\h 命令。

首先登录 MySQL 数据库,在命令行窗口中输入"help"或"\h"命令,此时就会显示帮助信息(图 2-2-5)。

2) \s 命令

输入\s 命令后,显示了 MySQL 当前的版本、字符集编码以及问号等信息(图 2-2-6)。

2.2.3　任务总结

可以使用命令行工具管理 MySQL 数据库,也可以从 MySQL 的网站下载图形管理工具 MySQL Workbench 进行管理。

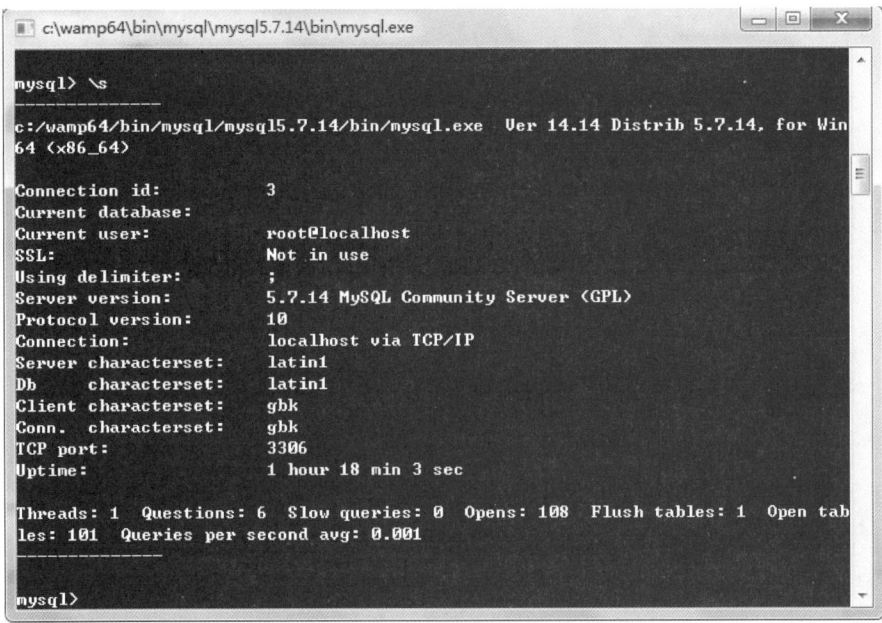

图 2-2-6 /s 命令

本任务主要讲解了 MySQL 服务的启动和登录、MySQL 的相关命令与配置等知识点，通过本任务的学习，读者应重点掌握 MySQL 服务的启动和登录的方法，并能简单地应用 MySQL 相关命令进行数据库的管理。

项目 3　数据库的基本操作

关于数据库等对象的操作,可以通过图形管理工具和 SQL 语句两种方式来完成,使用图形化操作工具可以完成的操作,以命令行的方式都可以完成,但是通过命令行方式可以完成的操作,图形化工具并不一定可以实现。当前常用的图形管理工具主要有 MySQL Workbench、phpMyAdmin、Navicat、MySQLDumper、SQLyog、MySQL ODBC Connector。MySQL 图形化管理工具可以让用户通过鼠标拖拽、点击的方式来完成数据库的操作与管理,这种方式虽然方便有效,但是需要人机交互,并不能实现在代码中创建数据库及其对象,为此本书主要讲解如何使用 SQL 语句通过命令行的方式创建并管理数据库及其对象。

在 phpMyAdmin 中建立数据库和数据表

学习目标

- 了解数据库的概念;
- 掌握建立、修改、删除、使用数据库的方法;
- 学会设计数据表的结构;
- 掌握建立、修改、删除数据表的方法;
- 理解数据的意义,掌握增加、修改、删除数据的方法;
- 学会使用 SQL 语句创建表;
- 能够使用 SQL 语句修改表结构;
- 掌握建立表的约束的语法格式;
- 熟练掌握向表中添加数据的方法和技巧;
- 掌握修改表中数据的方法;
- 能够根据需求删除表中数据。

任务 1
数据库操作

3.1.1 任务描述

要利用数据库管理系统进行数据管理工作,首先需要程序设计人员创建符合系统要求的数据库。在本任务中,以学生成绩管理系统数据库 StudentScore 为例,使用 SQL 语句创建数据库,并对该数据库进行使用、修改、删除操作。

通过本任务的学习,能够掌握建立数据库操作的语法格式、功能,理解参数的意义。能够使用 SQL 语句熟练地进行建立、修改、使用、删除数据库等操作。

3.1.2 知识准备

1. SQL 介绍

SQL 是一种专门用来与数据库通信的语言,它是结构化查询语言(Structured Query Language)的缩写,是集数据定义、数据查询、数据操纵和数据控制于一体的语言,功能丰富、简单易学,已经成为关系数据库的标准数据语言。

SQL 是一种数据库查询和程序设计语言,主要是完成数据存取、查询、数据库管理等操作。它是高级的非过程化编程语言,对于数据的存放方法没有要求,不需要用户了解数据的存放方式,所以可以兼容不同底层结构的数据库系统。SQL 语言是目前广泛使用的关系数据库标准语言,是各种数据库交互方式的基础,不同的数据库对 SQL 语言的支持与标准只存在细微的不同。

2. SQL 特点

SQL 语言主要有以下 4 个特点。

1) 一体化

SQL 语言集合了数据定义语言(DDL)、数据操纵语言(DML)、数据控制语言(DCL)于一体,可以完成数据库管理中的全部工作。

2) 使用方式灵活

SQL 不仅可以通过控制台命令行交互方式操作数据库,而且可以嵌入高级语言中,在开发中操作数据库。

SQL 语言

3) 非过程化

SQL 一次处理一条记录,对数据提供自动导航,允许用户在更高层次的数据结构中工作、使用,用户不需要知道数据的存储方法,可以将精力集中在数据上。解决具体问题时,只需提出要求,告诉系统要做什么,不需要告诉它怎么做,存储路径的选择和操作的执行由数据库管理系统自动完成。

4) 语言简洁、语法简单

SQL 功能很强,但完成核心功能的动词不多,而且语法简单,都是由描述性很强的英语单词组成的,容易学习和掌握。

3. SQL 语言分类

根据 SQL 语言的功能,主要将 SQL 语言分为 3 种类型:数据定义语言、数据操纵语言、数据控制语言。

1) 数据定义语言(Data Definition Language,DDL)

DDL 主要功能是创建数据库,以及创建数据库中的各种数据库对象。这是进行数据库操作的基础,因为只有创建了数据库和数据库中各种对象后,才可以对它们进行管理和操作。这相当于我们一般计算机语言中的声明定义变量和实例化对象,需要先声明后使用。在数据库中也是同样的,数据表、视图、存储过程、索引、触发器、函数、用户等这些数据库对象都需要先定义后使用。DDL 语句主要由 CREATE、DROP、ALTER 组成。

① DROP:删除数据库和表等对象。

② CREATE:创建数据库和表等对象。

③ ALTER:修改数据库和表等对象的结构。

2) 数据操纵语言(Data Manipulation Language,DML)

DML 主要完成数据的查询、插入、更新、删除,主要由 SELECT、INSERT、UPDATE、DELETE 组成。

① SELECT:查询表中的数据。

② INSERT:向表中插入数据。

③ UPDATE:更新表中的数据。

④ DELETE:删除表中的数据。

3) 数据控制语言(Data Control Language,DCL)

DCL 负责处理数据库中用户的角色和权限,主要由 GRANT、REVOKE、COMMIT、ROLLBACK 语句组成。

① GRANT:赋予用户操作权限。

② REVOKE:取消用户的操作权限。

③ COMMIT:确认对数据库的数据进行变更。

④ ROLLBACK:取消对数据库中的数据进行的变更。

4. MySQL 数据库文件特点

1) MySQL 数据库文件

数据库管理系统的主要工作是作为程序员创建、管理数据库的工具。在 MySQL 中,每创建一个数据库,就会在指定路径下创建一个与数据库同名的文件夹。在文件夹中主要存放描述数据库表结构的文件、表的数据文件以及数据文件中的索引文件,以上三个文件分别以.FRM、.MYD、.MYI 后缀名命名。数据库创建的默认位置是数据库管理系统安装路径下的 Data 文件夹下。

2) MySQL 自动安装时自动建立的系统数据库

我们在使用 MySQL 数据库管理系统时会发现,系统已经自动建立了 4 个数据库,分别为 information_schema、mysql、performance_schema、test。

information_schema 数据库是 MySQL 自带的,它提供了访问数据库元数据的方式。mysql 数据库是 MySQL 的核心数据库,类似于 SQL Server 中的 master 表,主要负责存储数据库的用户、权限设置、关键字等 MySQL 自己需要使用的控制和管理信息,不可以删除;如果对 MySQL 不是很了解,也不要轻易修改这个数据库里面的表信息。performance_schema 数据库是 MySQL 5.5 版本,新增了一个性能优化的引擎。test 数据库是安装时创建的一个测试数据库,是一个空数据库,没有表,可以删除。

5. 创建数据库

创建数据库的语法格式如下:

```
CREATE DATABASE [IF NOT EXISTS] 〈数据库名〉
[[DEFAULT] CHARACTER SET 〈字符集名〉] [[DEFAULT] COLLATE 〈校对规则名〉];
```

最简单的建库语句为:

```
CREATE DATABASE database_name;
```

功能:创建一个数据库。

说明:

① CREATE 是创建命令关键字。

② DATABASE 表示创建一个数据库。

③ database_name 为要创建的数据库名称,另外 MySQL 的数据存储区将以目录方式表示 MySQL 数据库,因此数据库名称必须符合操作系统的文件夹命名规则,不能以数字开头,尽量见名知意。并且在 MySQL 中不区分大小写。

④ IF NOT EXISTS 用来判断数据库是否存在,只有该数据库不存在时才能执行操作。此选项可以用来避免数据库已经存在而重复创建的错误。

⑤ CHARACTER SET 用来指定字符集;指定字符集的目的是避免在数据库中出现乱码的情况。如果在创建数据库时不指定字符集,那么就使用系统默认的字符集。

⑥ COLLATE 用来指定校对规则,MySQL 的字符集和校对规则是两个不同的概念。字符集是用来定义 MySQL 存储字符串的方式,校对规则定义了比较字符串的方式。

一个数据库创建成功之后,可以使用 SHOW 命令来进行查看。数据库可以看作是一个专门存储数据对象的容器,每一个容器都有唯一的名字。

6. 查看数据库

数据库创建成功之后,我们可以继续查看该数据库的信息,语法格式如下:

```
SHOW  CREATE  DATABASE  〈数据库名〉;
```

功能:查看指定的数据库信息。

也可以使用 SHOW 命令查看 MySQL 服务器当前都有哪些数据库,语法格式如下:

```
SHOW  DATABASES;
```

功能:显示当前路径下的所有数据。

7. 修改数据库

当创建了数据库之后,可以通过 SHOW CREATE DATABASE database_name;语句查看该数据库的定义声明。

例如:查看 StudentScore 数据库的定义声明。

```
mysql> show create database studentscore;
+--------------+--------------------------------------------------------------------------+
| Database     | CreateDatabase                                                           |
+--------------+--------------------------------------------------------------------------+
| studentscore | CREATE DATABASE 'studentscore'/ * ! 40100 DEFAULT CHARACTER SET latin1 * /|
+--------------+--------------------------------------------------------------------------+
1 row in set (0.02 sec)
```

如果我们对于建立的数据库不满意,想进行修改,那么要如何操作呢?

在 MySQL 中,可以使用 ALTER DATABASE 语句来修改已经被创建或者存在的数据库的相关参数。在 MySQL 中,对于已经建立的数据库,我们只能对这些数据库的字符集和校对规则进行修改,数据库的名称是不能够更改的。

使用 ALERT DATABASE 来修改数据库管理系统中已经存在的数据库的字符集和校对规则的语法格式如下:

```
ALTER DATABASE   database CHARSET SET 〈字符集名〉COLLATE 〈校对规则名〉;
```

功能:修改数据库的字符集、校对规则。

说明:

① ALTER DATABASE 用于更改数据库的全局特性。

② 使用 ALTER DATABASE 需要获得数据库 ALTER 权限。

③ 数据库名称可以忽略,此时语句对应默认数据库。

④ CHARACTER SET 子句用于更改默认的数据库字符集。

8. 使用数据库

若想对数据库中的数据库对象进行操作,首先应该把该数据库设置为当前数据库,或者说是进入数据库,或者叫打开数据库。

在 MySQL 中可以用 USE 语句来完成一个数据库到另一个数据库的跳转。当使用 CREATE DATABASE 语句创建数据库之后,该数据库不会自动成为当前数据库,需要用 USE 语句来指定当前数据库。

使用 SQL 语句打开数据库的语法格式为:

```
USE database_name;
```

功能:选择打开指定的数据库。

说明:database_name 是要使用的数据库名称。

该语句可以通知 MySQL 把 database_name 所指定的数据库作为当前数据库。该数据库保持为默认数据库,直到语段的结尾,或者直到遇见一个不同的 USE 语句。只有使用 USE 语句来指定某个数据库作为当前数据库之后,才能对该数据库及其存储的数据对象执行操作。

9. 删除数据库

数据库管理系统是数据库管理员与数据库之间的接口,是一个系统软件,如果管理的数据库过多,就会占用大量的系统资源,导致系统速度变慢,甚至导致系统崩溃。为此,对那些无用的数据库应该及时删除,以确保数据库存储空间中存放的是有效数据。删除数据库是将已经存在的数据库从磁盘空间上清除,清除之后,数据库中的所有数据也将一同被删除。在 MySQL 中,当需要删除已经创建的数据库时,可以使用 DROP DATABASE 语句。

使用 SQL 删除数据库的语法格式如下:

```
DROP DATABASE database_name;
```

功能:删除指定的数据库。

说明:

① database_name 为要删除的数据库名称。

② DROP DATABASE 删除数据库中的所有表格并同时删除数据库。使用此语句要非常小心,应该提前做好数据备份,以免错误删除。如果要使用 DROP DATABASE,需要获得数据库 DROP 权限。

3.1.3 任务实现

【例 3-1-1】 创建学生成绩管理系统数据库 studentscore。

```
mysql> CREATE DATABASE studentscore;
Query OK, 1 row affected (0.00 sec)
```

其中,"Query OK"表示上面的命令执行成功,"1 row affected"表示操作只影响了数据库中一行记录,"0.00sec"则记录了操作执行的时间,时间为 0.00 s 并不代表没有花费时间,而是时间非常短,小于 0.01 s。

使用 SHOW DATABASES 命令查看所有数据库,从显示结果可以发现 studentscore 数据库已经创建成功。

```
mysql> SHOW DATABASES;
+--------------------+
| Database           |
+--------------------+
| information_schema |
| mysql              |
| performance_schema |
| studentscore       |
| studentscore1      |
| studentscore2      |
| sys                |
+--------------------+
7 rows in set (0.00 sec)
```

通过结果能看到在 MySQL 数据库中存在系统数据库和自定义数据库,系统数据库是在安装 MySQL 后系统自带的数据库,自定义数据库是由用户自定义创建的数据库。

若再次输入 CREATE DATABASE studentscore,则系统给出错误提示。

```
mysql> CREATE DATABASE studentscore;
ERROR 1007 (HY000): Can't create database 'studentscore'; database exists
```

提示不能创建"studentscore"数据库,数据库已经存在。MySQL 中不允许在同一系统下创建两个相同名称的数据库。

可以加上 IF NOT EXISTS 从句,就可以避免类似问题。

```
mysql> CREATE DATABASE IF NOT EXISTS studentscore;
Query OK, 1 row affected, 1 warning (0.00 sec)
```

【例 3-1-2】 创建一个使用 UTF-8 字符集的 studentscore1 数据库。

```
mysql> CREATE DATABASE studentscore1 charset=utf8;
Query OK, 1 row affected (0.00 sec)
```

使用 SHOW DATABASE 命令查看所有数据库,发现 studentscore1 数据库已经创建成功。

```
mysql> SHOW DATABASES;
+--------------------+
| Database           |
+--------------------+
| information_schema |
| mysql              |
| performance_schema |
```

```
| studentscore  |
| studentscore1 |
| studentscore2 |
| sys           |
+---------------+
7 rows in set (0.00 sec)
```

在上面创建数据库的命令语句中添加了 charset=utf8 子句,这是用来指定数据库的字符集。因为数据库也是由字符编码的,在创建数据库时,设置数据库的编码方式与客户端语句的字符集一致,可以避免后期从数据库中读取的数据在客户端显示为乱码。MySQL 默认 latin1(其实就是 ISO-8859-1)字符集。这显然不能满足我们的需要,因此我们把它调整为 UTF-8 字符集,以兼容大多数字符集。字符集影响数据在传输和存储过程中的处理方式。

使用 SHOW CREATE DATABASE 来查看 studentscore1 数据库的定义声明,可以发现该数据库的指定字符集为 UTF-8。

```
mysql> SHOW CREATE DATABASE studentscore1;
+---------------+-----------------------------------------------------------------------------+
| Database      | Create Database                                                             |
+---------------+-----------------------------------------------------------------------------+
| studentscore1 | CREATE DATABASE 'studentscore1' /*! 40100 DEFAULT CHARACTER SET utf8 */     |
+---------------+-----------------------------------------------------------------------------+
1 row in set (0.02 sec)
```

可以使用下面的语句查看字符集编码设置。

```
mysql> show variables like '%character%';
+--------------------------+----------------------------------------------+
| Variable_name            | Value                                        |
+--------------------------+----------------------------------------------+
| character_set_client     | gbk                                          |
| character_set_connection | gbk                                          |
| character_set_database   | latin1                                       |
| character_set_filesystem | binary                                       |
| character_set_results    | gbk                                          |
| character_set_server     | latin1                                       |
| character_set_system     | utf8                                         |
| character_sets_dir       | d:\wamp64\bin\mysql\mysql5.7.14\share\charsets\ |
+--------------------------+----------------------------------------------+
8 rows in set, 1 warning (0.08 sec)
```

【例 3-1-3】 创建一个使用 UTF-8 字符集,并带比较规则的 studentscore2 数据库。

```
mysql> CREATE DATABASE studentscore2 CHARACTER SET utf8 COLLATE utf8_general_ci;
Query OK, 1 row affected (0.00 sec)
```

使用 SHOW DATABASE 命令查看所有数据库发现 studentscore2 数据库已经创建成功。

```
mysql> SHOW DATABASE;
+--------------------+
| Database           |
+--------------------+
| information_schema |
| mysql              |
| performance_schema |
| studentscore       |
| studentscore1      |
| studentscore2      |
| sys                |
+--------------------+
7 rows in set (0.00 sec)
```

校对规则是在字符集内用于字符比较和排序的一套规则,比如有的规则区分大小写,有的则无视。

我们可以使用 SHOW COLLATION 命令来查看数据库支持的校对规则。

```
mysql> show collation;
```

Collation	Charset	Id	Default	Compiled	Sortlen
big5_chinese_ci	big5	1	Yes	Yes	1
big5_bin	big5	84		Yes	1
dec8_swedish_ci	dec8	3	Yes	Yes	1
dec8_bin	dec8	69		Yes	1
cp850_general_ci	cp850	4	Yes	Yes	1
cp850_bin	cp850	80		Yes	1
hp8_english_ci	hp8	6	Yes	Yes	1
hp8_bin	hp8	72		Yes	1
koi8r_general_ci	koi8r	7	Yes	Yes	1
koi8r_bin	koi8r	74		Yes	1
latin1_german1_ci	latin1	5		Yes	1
latin1_swedish_ci	latin1	8	Yes	Yes	1
latin1_danish_ci	latin1	15		Yes	1
latin1_german2_ci	latin1	31		Yes	2
latin1_bin	latin1	47		Yes	1
latin1_general_ci	latin1	48		Yes	1
latin1_general_cs	latin1	49		Yes	1
latin1_spanish_ci	latin1	94		Yes	1
latin2_czech_cs	latin2	2		Yes	4
latin2_general_ci	latin2	9	Yes	Yes	1
latin2_hungarian_ci	latin2	21		Yes	1

上面只显示其中一部分。

【例 3-1-4】 把 studenscore 数据库的字符集改为 gb2312,默认校对规则修改为 gb2312_unicode_ci。

```
mysql> ALTER DATABASE studentscore DEFAULT CHARACTER SET gb2312 DEFAULT COLLATE gb2312_unicode_ci;
Query OK, 1 row affected (0.12 sec)
```

查看结果。

```
mysql>SHOW CREATE DATABASE studentscore;
+--------------+----------------------------------------------------------------------------+
| Database     | Create Database                                                            |
+--------------+----------------------------------------------------------------------------+
| studentscore | CREATE DATABASE 'studentscore' /*！40100 DEFAULT CHARACTER SET gb2312 */ |
+--------------+----------------------------------------------------------------------------+
1 row in set (0.00 sec)
```

【例 3-1-5】 打开数据库 studentscore。

```
mysql> USE studentscore;
Database changed;
```

看到 Database changed 提示,说明选择数据库成功。

【例 3-1-6】 删除数据库 studentscore。

```
mysql> DROP DATABASE studentscore;
Query OK, 5 rows affected (0.51 sec);
```

此时数据库 studentscore 已经不存在,如果再次执行相同的命令,系统会报错。

```
mysql> DROP DATABASE studentscore;
ERROR 1008 (HY000): Can't drop database 'studentscore'; database doesn't exist;
```

如果在命令中使用 IF EXISTS 子句,可以防止系统报此类错误。

```
mysql> DROP DATABASE IF EXISTS studentscore;
Query OK, 0 rows affected, 1 warning (0.00 sec);
```

这里要特别注意:使用 DROP DATABASE 命令时要非常谨慎,在执行该命令后 MySQL 不会给出任何提示确认信息。数据库删除是一个非常危险的工作,删除前一定做好数据库的迁移和备份,因为数据库删除语句执行后,数据库中的所有数据将全部删除并且不能恢复。并且在安装 MySQL 数据库管理系统时,自动安装的系统数据库保存了系统的关键数据,一定不要删除,否则将导致 MySQL 数据库管理系统崩溃。

3.1.4 训练任务

随着信息技术的发展,计算机作为今天使用最广泛的现代化工具已经深入各个领域。在这样的大背景下,现代图书馆的管理方式发生了翻天覆地的变化,主要表现在图书馆工作、管理、服务平台发生了变化。图书馆不再是传统的手工操作、人工管理,而是全面地实行计算机管理。一个简单的图书管理数据库包括图书的信息、学生的信息以及学生的借阅信息,下面我们来逐步完成。首先完成数据库的管理。

(1) 学院 E 时代社团有 2 000 余册专业书籍,需要一个图书管理系统管理。请建立图书管理数据库 Ebook。

（2）请把图书管理系统数据库 Ebook 的字符集改为 gb2312，默认校对规则修改为 gb2312_unicode_ci。

（3）请把 Ebook 图书数据库设置为当前数据库。

（4）请使用 MySQL 语句删除 Ebook 图书管理系统数据库。

3.1.5 任务总结

在本任务中，我们学习了使用 MySQL 语句创建、修改、使用、删除数据库的方法，了解了 MySQL 语言的基本概念和特点。在 MySQL 数据库环境下，如果要创建一个数据库，可以使用 CREATE DATABASE 命令，使用方法为：CREATE DATABASE〈数据库名〉，这种创建方法并没有指定此数据库采用的字符集，也就是数据的编码格式，MySQL 的版本不同，默认采用的是字符集也不同，默认采用的校验规则也不同。因为我们可能会创建多个数据库来管理不同的数据，但只能同时操作一个数据库，所以这时就需要选择一个数据库先进行使用，使用方法为 use〈数据库名〉语句。在修改数据库时，一般修改的都是数据库的字符集，当然也可以修改校验规则，修改数据库时的命令是 ALTER DATABASE，修改的方法是 ALTER DATABASE 数据库名字 charset＝字符集名字。在删除数据库时，可以使用 DROP DATABASE 命令，使用方法就是 DROP DATABASE〈数据库名〉。在以后的章节中会逐步使用 MySQL 语句完成相应的任务。

任务 2

数 据 表 操 作

表的概述

3.2.1 任务描述

数据库中的数据是存放在表中的，所以建立了数据库之后，第一项任务就是创建存放数据的表。我们可以把表看成容器，数据看成放在容器中的物体，那么容器就要根据所放物体的大小尺寸、形状特点进行设计。首先看看表，也就是所谓的容器的结构特点，然后考察数据，也就是所谓的物体的类型属性，从而根据需要在已经创建的数据库中建立新表。

3.2.2 知识准备

数据库可以简单地理解为存放数据的仓库。那么数据库中的数据具体存放在哪里呢？或者说是数据库中的哪个对象用来存放数据呢？在物理上数据库中的数据是存放在表上的。表是数据库中最重要的数据库对象。数据表是数据库的重要组成部分，每一个数据库都是由若干个数据表组成的。没有数据表就无法在数据库中存放数据。对数据表的操作主要包括创建表、修改表、删除表和对表中数据的操作。创建表主要是设计表的结构，根据项目存放数据类型的需要，设计相应的、符合项目中数据类型的字段来存放数据。对数据表中数据的操作主要是对数据的增、删、改、查操作。

在本书项目学生成绩管理系统中，每位学生都会上多门新课，这就需要向课程表添加新的记录，在学生的基本信息发生变化时需要修改学生表。当学生查询考试成绩时，需要查找成绩表中的记录；当有教师离职时，需要删除教师表中的记录。本项目根据对学生成绩管理的需要，在 studentscore 数据库中建立 students 表、teachers 表、courses 表、scores 表，并对表中的数据进行增、删、改、查操作。

1. 表结构

在数据库中表是存放数据的基本单位。表是由行和列组成的，一行叫作一条记录（元组），一列叫作一个字段（属性）。同一个数据库中，不同的表不能同名，也就是表名不能重复。表中的列也有名字，列名叫字段名。表名和列名的命名一定要符合 SQL 语言标识符的命名规则，并且不能是保留字。不同的数据库可以有相同名称的表。同一个表不能有同名字段，不同的表可以有同名字段。实际上创建一个表就是设计表中有多少个字段，每个字段的字段名称，每个字段可以存放什么数据类型的数据，字段可以存放该数据的最大长度，数据的精度，数据可以显示的小数位数，字段是否可以为空，是否设置默认值等信息。

2. 数据类型

数据类型是数据库系统中的数据特征，它决定数据的存储格式，代表不同的信息类型。为什么要使用数据类型来描述数据的特点呢？这是由于在数据库中存放的数据多种多样，各种数据所占内存空间大小、数据可以进行的运算、数据的取值范围、数据的表达精度都不一样，因此我们用数据类型进行分类。数据类型实际上是数据在内存中存储方式的一种约定。我们已经知道创建表的过程实际上是创建表的结构，也就是设计表中都有哪些字段，设计字段的名称，指定字段中可以存放哪种数据类型的数据。

3. 字段属性

在 MySQL 数据库的表中，一个字段中只能存放一种数据类型的数据，不同数据类型的数据不能同时存放在一个字段中。例如：学生表中学号字段用来存放学生的学号信息，字段数据类型应该定义为字符型，学生表的出生日期字段用来存放学生的出生日期信息，则字段数据类型应该定义为日期型。

MySQL 数据库提供了多种数据类型，设计数据库可以方便地描述所存储数据的特点。主要有数值类型、字符类型、日期和时间类型等，设计表结构过程中，可以指定某一列可以存放数据，也可以为空，也就是在向容器中放置物体时可以放也可以不放。有时指定某一列在存放数据时不能为空，也就是在向容器中放置物体时必须放。向表中添加数据时，也可以指定某一列的默认值，在用户没有给出时自动添加默认值，这样可以保证数据的完整性。或者指定表的某列值自动增加，当有新数据插入时该列值自动增加。这个设置主要是针对主键字段，主键字段是表中唯一可以确定一条记录的字段。当主键字段设置为自动增加后,每插入一条新记录,该字段就会自动赋予一个新值。自动增加默认情况下是自动+1递增的。具体数据类型描述及表的约束内容在后面的项目4中详细讲解。

4. 设计学生成绩管理系统数据表的结构

在前面分析项目实体之间的关系后,确定了学生成绩管理系统由学生表、教师表、课程表、成绩表组成,分析各个实体的属性后可以确定各个表的结构,如表 3-2-1~表 3-2-4 所示。

表 3-2-1 学生表(students)

列名	数据类型	长度	是否为空	键	说明
sid	int	10	否	主键,自增	学号
sname	varchar	20	否		姓名
sclass	int	10	是		班级
sgender	varchar	10	否		男或女
smajor	varchar	20	是		专业
sbirthday	datetime		是		出生日期
credit_points	int	5	是		学分

表 3-2-2 教师表(teachers)

列名	数据类型	长度	是否为空	键	说明
tid	int	10	否	主键/自增	教师编号
tname	varchar	20	是		教师姓名
tschool	varchar	20	是		所在院系

表 3-2-3 课程表(courses)

列名	数据类型	长度	是否为空	键	说明
cid	int	10	否	主键/自增	课程编号
tid	int	10	否	外码	教师标号
cname	varchar	20	是		课程名称
credit_point	int	5	是		学分

表 3-2-4 成绩表(scores)

列名	数据类型	长度	是否为空	键	说明
sid	int	10	否	主键,外键	学生编号
cid	int	10	否	主键,外键	课程编号
score	decimal	5.2	是		小数后两位

5. 建立数据表的结构

要存储数据，首先要在数据库中创建数据表，创建 MySQL 数据表主要是确定表名、字段名、定义每个字段属性。在 SQL 中，使用 CREATE TABLE 语句创建表，语法格式如下：

```
CREATE TABLE table_name (column_name column_type) [表选项];
```

功能：建立数据表的结构。

说明：

① CREATE TABLE 用于创建给定名称的表，必须拥有表 CREATE 的权限。

② Table_name 是指定要创建表的名称，在 CREATE TABLE 之后给出，必须符合标识符的命名规则。可以在表名前标准指定的数据库如 db_name.table_name 的形式在特定的数据库中创建表。在当前数据库中创建表时，可以省略数据库名称。

③ 创建表时要定义列名、列的相关属性，字段之间用逗号隔开。

④ 默认情况下表被创建在当前数据库中。若表已经存在或者没有当前数据库则会出现错误。

⑤ 表选项用于设置表的存储引擎、字符集、校对规则等相关特性。

⑥ 创建表的名称不区分大小写，不能使用 SQL 语言中的保留字。

6. 查看数据表的结构

数据表建立成功之后，可以使用 SHOW 命令来查看数据库的所有表、数据表的结构和相关信息。

1）查看所有表

查看所有表的语法格式如下：

```
SHOW TABLES;
```

功能：显示当前数据库中的所有表。

2）查看指定数据表的信息

查看指定数据表信息的语法格式如下：

```
SHOW CREATE TABLE 〈数据表名〉;
```

功能：显示当前数据库中的指定数据表的信息。

```
mysql> show create table students;
+----------+----------+
| Table    | Create Table |
+----------+----------+
| students | CREATE TABLE 'students' (
  'sid' int(10) unsigned NOT NULL AUTO_INCREMENT,
  'sname' varchar(20) COLLATE utf8_bin DEFAULT NULL,
```

```
'sclass' int(10) DEFAULT NULL,
'sgender' varchar(10) COLLATE utf8_bin DEFAULT NULL,
'smajor' varchar(20) COLLATE utf8_bin DEFAULT NULL,
'sbirth' date DEFAULT NULL,
'credit_points' smallint(5) DEFAULT NULL,
PRIMARY KEY ('sid')
) ENGINE=MyISAM AUTO_INCREMENT=2 DEFAULT CHARSET=utf8 COLLATE=utf8_bin|
+------------------+
1 row in set (0.00 sec)
```

3）查看指定数据表的结构（字段）

查看指定数据表结构的语法格式如下：

```
DESCRIBE 〈数据表名〉；
```

功能：显示当前数据库中的指定数据表的结构。

例如要查看学生表的结构，可以使用下面的语句完成。

```
mysql> DESCRIBE students;
+---------------+--------------+------+-----+---------+----------------+
| Field         | Type         | Null | Key | Default | Extra          |
+---------------+--------------+------+-----+---------+----------------+
| sid           | int(10) unsigned | NO   | PRI | NULL    | auto_increment |
| sname         | varchar(20)  | YES  |     | NULL    |                |
| sclass        | int(10)      | YES  |     | NULL    |                |
| sgender       | varchar(10)  | YES  |     | NULL    |                |
| smajor        | varchar(20)  | YES  |     | NULL    |                |
| sbirthday     | date         | YES  |     | NULL    |                |
| credit_points | int(5)       | YES  |     | NULL    |                |
+---------------+--------------+------+-----+---------+----------------+
7 rows in set (0.09 sec)
```

上面的2）和3）两种操作方法功能相近，都能查看用户建立的数据表的结构，但有一定的区别，使用格式2）除了能查看数据表的字段信息之外，还能查看数据表的存储引擎、外键等信息。使用格式3）是以表格的形式来显示字段名、字段类型、是否允许空值、默认值等，更加直观。

通过这些信息可以查看创建的表结构是否与设计得一致。

7. 修改数据表结构

在创建表结构之后，也可以对表的结构进行调整、修改。创建表过程中定义了表的结构，包括定义表名、定义字段名、字段数据类型设置表的完整性约束条件等，这些都可以进行修改。除此之外还可以根据需要增加删除字段。

在MySQL中修改表的结构使用ALTER TABLE命令完成。修改数据表的前提是数据库中已经存在该表。修改表主要有增加或删除列、更改原有列类型、重命名列或表名称等。修改表的语法格式如下：

```
ALTER TABLE〈数据表名〉[修改选项]；
```

功能:修改数据表的结构。

说明:

① ALTER TABLE:修改数据表的命令动词;

②[修改选项]:根据具体要求使用不同的动词来完成各种修改操作。

在数据表结构修改过程中,可以利用[修改选项]来进行数据表名称、字段属性(名称、类型、宽度)、增减字段、字符集、校对规则等内容的修改。[修改选项]的具体使用方法如下。

1) 修改数据表名

在数据库中,不同的数据表是通过数据表名来区分的。修改数据表名的语法格式如下:

```
ALTER TABLE 〈原表名〉 RENAME [to] 〈新表名〉;
```

功能:修改指定数据表的名称。

说明:

① 原表名必须是已经存在的数据表。

② 新表名原则上不与原表名相同,否则 MySQL 会报错。

③ to 是可选项,可以省略,不影响命令语句的执行。

2) 修改字段名

在数据表中,不同的字段是通过字段名来区分的。修改字段名的语法格式如下:

```
ALTER TABLE 〈数据表名〉 CHANGE 〈原字段名〉 〈新字段名〉 〈新数据类型〉;
```

功能:修改数据表中指定的字段名称和数据类型。

说明:

①〈原字段名〉必须是数据表中已经存在的字段名。

②〈新字段名〉原则上不与〈原字段名〉相同,否则无法修改字段名。

③〈新数据类型〉是修改后该字段的数据,且不能为空,即使修改前后的数据类型相同也要重新设置。

3) 修改字段的数据类型

修改字段的数据类型,也就是将字段的数据类型更改为一种新的数据类型,保持字段的名称不变。修改字段名的语法格式如下:

```
ALTER TABLE 〈数据表名〉 MODIFY 〈字段名〉 〈数据类型〉;
```

功能:修改数据表中指定字段的数据类型。

说明:

①〈字段名〉必须是数据表中已经存在的字段名。

②〈数据类型〉是修改后该字段的数据类型。

4) 修改字段的排列顺序

在创建数据表的时候，每个字段在数据表中的位置已经按照录入的顺序确定下来了，但是如果有不同的需求，也可以对表中各个字段的位置进行调整，只要在 CHANGE 或 MODIFY 命令中加入 FIRST|AFTER 短语就行了，修改字段排列顺序的语法格式如下：

```
ALTER TABLE〈数据表名〉MODIFY〈字段名1〉〈数据类型〉FIRST|AFTER〈字段名2〉;
或者
ALTER TABLE〈数据表名〉CHANGE〈原字段名〉〈新字段名〉〈数据类型〉
    FIRST|AFTER〈字段名2〉;
```

功能：修改数据表中指定字段的位置。

说明：

① FIRST 是将语句中指定的字段设置为第一个字段。

② AFTER 是将语句中指定的字段插入到〈字段名2〉指定的字段之后。

5) 添加新字段

在数据表创建之后，用户的需求可能随时都会发生变化，如果想在已经建好的数据表中添加新字段，可以使用 ALTER TABLE 语句中 ADD 子句来实现。添加字段的语法格式如下：

利用函数快速修改数据

```
ALTER TABLE 〈数据表名〉 ADD 〈新字段名〉〈新数据类型〉[约束条件]
            [FIRST|AFTER〈已存在的字段名〉];
```

功能：在指定的数据表中添加一个新字段。

说明：

①〈新字段名〉不能与数据表中已经存在的字段重名。

② [FIRST|AFTER〈已存在的字段名〉]，该短语用于指定新字段的位置。

6) 删除字段

数据表创建之后，根据用户的需求，有时要添加新字段，有时也会删某些字段。删除字段可以使用 ALTER TABLE 语句中 DROP 子句来实现。删除字段的语法格式如下：

```
ALTER TABLE 〈数据表名〉 DROP 〈字段名〉;
```

功能：在指定的数据表中删除指定的字段。

说明：〈字段名〉必须是数据表中已经存在的字段名，否则会报错。

7) 修改字符集

一个数据表创建之后，它的字符集是由数据库的字符集来决定的，要修改数据表的字段集，可以使用 ALTER TABLE 语句中 CHARACTER SET 子句来实现。修改字符集的语法格式如下：

```
ALTER TABLE 〈数据表名〉 CHARACTER SET〈字符集名〉;
```

功能：修改指定的数据表的字符集。

8) 修改校对规则

一个数据表创建之后，可以使用 ALTER TABLE 语句中 COLLATE 子句来修改校对规则。修改校对规则的语法格式如下：

```
ALTER TABLE 〈数据表名〉 COLLATE〈校对规则名〉;
```

功能：修改指定的数据表的校对规则。

9) 修改存储引擎

在数据表创建之后，还可以使用 ALTER TABLE 语句中 ENGINE 子句来修改存储引擎。修改存储引擎的语法格式如下：

```
ALTER TABLE 表名 ENGINE=〈新的存储引擎类型〉;
```

功能：为数据表的指定字段添加约束。

10) 添加约束

在数据表创建之后，还可以使用 ALTER TABLE 语句中 PRIMARY 子句来添加约束。添加约束的语法格式如下：

```
ALTER TABLE 表名 ADD CONSTRAINT 约束名 约束类型〈字段名〉;
```

功能：为数据表的指定字段添加约束。

8. 复制数据表

数据表建立成功之后，也可以使用 SELECT 命令复制生成副本。

在 SQL 中，使用 SELECT 语句复制表的语法格式如下：

```
CREATE TABLE 〈新数据表〉 SELECT * FROM 〈原数据表〉;
```

功能：复制生成新的数据表。

说明：

① SELECT 子句是一个标准的查询语句，具体语法格式在后面的数据查询项目中详细讲解。

② 原表与新表不能重名。

③ 如果 SELECT 子句中没有条件项，则生成的新表与原表的结构、数据完全相同。

9. 删除数据表

一个数据库运行一段时间之后，用户的需求可能会有变化，数据库中的某些数据表也可能变成冗余，这样就需要删除数据表。

删除数据表的语法格式如下：

```
DROP TABLE [IF EXISTS] 〈数据表名〉;
```

功能：删除已经存在的数据表。

说明：

① DROP：删除数据表的命令动词。

②〈数据表名〉：已经存在的数据表名称。

③ [IF EXISTS]是可选项，它的作用是执行删除命令时先检测指定的表是否存在，防止删除操作出错。

3.2.3 任务实现

【例3-2-1】 按照前面的设计来创建学生表。

数据表属于数据库，在创建数据表之前使用 USE 命令选择要操作的数据库为当前数据库，如果没有选择数据库则会出现 No database selected 的错误提示。

```
mysql> CREATE TABLE students(
    -> sid INT(10) UNSIGNED PRIMARY KEY AUTO_INCREMENT NOT NULL,
    -> sname VARCHAR(20),
    -> sclass INT(10),
    -> sgender VARCHAR(10),
    -> smajor VARCHAR(20),
    -> sbirthday DATE,
    -> credit_points INT(5));
Query OK, 0 rows affected (0.20 sec)
```

语句执行后，便创建了一个名为 students 的数据表。

PRIMARY KEY 是指定该字段为主键。主键主要是为了保证表中数据的完整性，是在定义表时使用的完整性约束条件。定义了完整性约束条件后数据库管理系统会在使用数据库时自动用设置好的完整性约束条件检查输入的数据是否符合要求。

AUTO_INCREMENT 是指设置字段的值自动增加。

查看学生表结构。

```
mysql> DESCRIBE students;
+---------------+--------------+------+-----+---------+----------------+
| Field         | Type         | Null | Key | Default | Extra          |
+---------------+--------------+------+-----+---------+----------------+
| sid           | int(10) unsigned | NO | PRI | NULL    | auto_increment |
| sname         | varchar(20)  | YES  |     | NULL    |                |
| sclass        | int(10)      | YES  |     | NULL    |                |
| sgender       | varchar(10)  | YES  |     | NULL    |                |
| smajor        | varchar(20)  | YES  |     | NULL    |                |
| sbirthday     | date         | YES  |     | NULL    |                |
| credit_points | int(5)       | YES  |     | NULL    |                |
+---------------+--------------+------+-----+---------+----------------+
7 rows in set (0.09 sec)
```

【例3-2-2】 按照前面的设计来创建教师表。

```
mysql> CREATE TABLE teachers(
    -> tid INT(10) UNSIGNED PRIMARY KEY AUTO_INCREMENT NOT NULL,
    -> tname VARCHAR(20),
    -> tschool VARCHAR(20));
Query OK, 0 rows affected (0.20 sec)
```

查看教师表结构。

```
mysql> desc teachers;
+---------+--------------+------+-----+---------+----------------+
| Field   | Type         | Null | Key | Default | Extra          |
+---------+--------------+------+-----+---------+----------------+
| tid     | int(10) unsigned | NO  | PRI | NULL    | auto_increment |
| tname   | varchar(20)  | YES  |     | NULL    |                |
| tschool | varchar(20)  | YES  |     | NULL    |                |
+---------+--------------+------+-----+---------+----------------+
3 rows in set (0.00 sec)
```

【例3-2-3】 按照前面的设计来创建课程表。

```
mysql> CREATE TABLE courses(
    -> cid INT(10) UNSIGNED PRIMARY KEY AUTO_INCREMENT NOT NULL,
    -> cname VARCHAR(20),
    -> credit_point INT(5),
    -> tid INT(10) UNSIGNED NOT NULL,
    -> FOREIGN KEY(tid) REFERENCES teachers(tid));
Query OK, 0 rows affected (0.12 sec)
```

查看课程表结构。

```
mysql> DESC courses;
+--------------+------------------+------+-----+---------+----------------+
| Field        | Type             | Null | Key | Default | Extra          |
+--------------+------------------+------+-----+---------+----------------+
| cid          | int(10) unsigned | NO   | PRI | NULL    | auto_increment |
| cname        | varchar(20)      | YES  |     | NULL    |                |
| credit_point | int(5)           | YES  |     | NULL    |                |
| tid          | int(10) unsigned | NO   | MUL | NULL    |                |
+--------------+------------------+------+-----+---------+----------------+
4 rows in set (0.00 sec)
```

在创建课程表时,在课程表中引用了教师表的教师tid字段。课程和教师实体是存在联系的,教师与课程是一对多的关系,一门课程与教师是多对一的关系。一个教师可以讲授多门课程,一个课程可能同时由多位教师讲授。所以在课程表中的tid字段的内容都应该来自教师表中的tid字段。这种情况tid字段定义为课程表的外键。外键是为了保证参照完整性。

【例3-2-4】 按照前面的设计来创建成绩表。

```
mysql> CREATE TABLE scores(
    -> sid INT(10) UNSIGNED NOT NULL,
    -> cid INT(10) UNSIGNED NOT NULL,
    -> score DECIMAL(5,2),
    -> PRIMARY KEY(sid,cid),
    -> FOREIGN KEY(sid) REFERENCES students(sid),
    -> FOREIGN KEY(cid) REFERENCES course(cid));
Query OK, 0 rows affected (0.13 sec)
```

查看成绩表结构。

```
mysql> DESC scores;
+-------+---------------+------+-----+---------+-------+
| Field | Type          | Null | Key | Default | Extra |
+-------+---------------+------+-----+---------+-------+
| sid   | int(10) unsigned | NO   | PRI | NULL    |       |
| cid   | int(10) unsigned | NO   | PRI | NULL    |       |
| score | decimal(5,2)  | YES  |     | NULL    |       |
+-------+---------------+------+-----+---------+-------+
3 rows in set (0.00 sec)
```

创建成绩表时使用了两个字段共同组成了主键。每个表都必须有主键,并且只能有一个主键。一个主键可以由一个或者多个字段组成。

所有数据表创建成功后,可以在当前数据库下看到创建的表。

```
mysql> SHOW TABLES;
+----------------------+
| Tables_in_studentscore |
+----------------------+
| courses              |
| scores               |
| students             |
| teachers             |
+----------------------+
4 rows in set (0.00 sec)
```

我们已经成功地创建了学生成绩管理系统数据库所需要的学生表(students)、教师表(teachers)、课程表(courses)、成绩表(scores)。

【例3-2-5】 在students表中添加一个电话字段phone,数据类型为varchar(20),新字段放在"sbirthday"字段后面。

```
mysql> ALTER TABLE students ADD phone varchar(20) AFTER sbirthday;
Query OK, 0 rows affected (0.23 sec)
Records: 0  Duplicates: 0  Warnings: 0
```

查看结果。

```
mysql> DESC students;
+---------------+------------------+------+-----+---------+----------------+
| Field         | Type             | Null | Key | Default | Extra          |
+---------------+------------------+------+-----+---------+----------------+
| sid           | int(10) unsigned | NO   | PRI | NULL    | auto_increment |
| sname         | varchar(20)      | YES  |     | NULL    |                |
| sclass        | int(10)          | YES  |     | NULL    |                |
| sgender       | varchar(10)      | YES  |     | NULL    |                |
| smajor        | varchar(20)      | YES  |     | NULL    |                |
| sbirthday     | date             | YES  |     | NULL    |                |
| phone         | varchar(20)      | YES  |     | NULL    |                |
| credit_points | int(5)           | YES  |     | NULL    |                |
+---------------+------------------+------+-----+---------+----------------+
8 rows in set (0.00 sec)
```

从显示结果可以发现,新字段 phone 已经添加完成。

【例 3-2-6】 将 students 表中的 credit_points 字段的数据类型修改为 smallint。

```
mysql> ALTER TABLE students MODIFY credit_points smallint(5);
Query OK, 0 rows affected (0.17 sec)
Records: 0  Duplicates: 0  Warnings: 0
```

查看结果。

```
mysql> DESC students;
+---------------+------------------+------+-----+---------+----------------+
| Field         | Type             | Null | Key | Default | Extra          |
+---------------+------------------+------+-----+---------+----------------+
| sid           | int(10) unsigned | NO   | PRI | NULL    | auto_increment |
| sname         | varchar(20)      | YES  |     | NULL    |                |
| sclass        | int(10)          | YES  |     | NULL    |                |
| sgender       | varchar(10)      | YES  |     | NULL    |                |
| smajor        | varchar(20)      | YES  |     | NULL    |                |
| sbirthday     | date             | YES  |     | NULL    |                |
| phone         | varchar(20)      | YES  |     | NULL    |                |
| credit_points | smallint(5)      | YES  |     | NULL    |                |
+---------------+------------------+------+-----+---------+----------------+
8 rows in set (0.00 sec)
```

【例 3-2-7】 将 students 表中的 phone 字段删除。

```
mysql> ALTER TABLE students DROP phone;
Query OK, 0 rows affected (0.06 sec)
Records: 0  Duplicates: 0  Warnings: 0
```

查看结果。

```
mysql> DESC students;
+---------------+------------------+------+-----+---------+----------------+
| Field         | Type             | Null | Key | Default | Extra          |
+---------------+------------------+------+-----+---------+----------------+
| sid           | int(10) unsigned | NO   | PRI | NULL    | auto_increment |
| sname         | varchar(20)      | YES  |     | NULL    |                |
| sclass        | int(10)          | YES  |     | NULL    |                |
| sgender       | varchar(10)      | YES  |     | NULL    |                |
| smajor        | varchar(20)      | YES  |     | NULL    |                |
| sbirthday     | date             | YES  |     | NULL    |                |
| credit_points | smallint(5)      | YES  |     | NULL    |                |
+---------------+------------------+------+-----+---------+----------------+
7 rows in set (0.00 sec)
```

【例3-2-8】 将students表中的sbirthday字段名更改为sbirth。

```
mysql> ALTER TABLE students CHANGE sbirthday sbirth date;
Query OK, 0 rows affected (0.14 sec)
Records: 0  Duplicates: 0  Warnings: 0
```

查看结果。

```
mysql> DESC students;
```

Field	Type	Null	Key	Default	Extra
sid	int(10) unsigned	NO	PRI	NULL	auto_increment
sname	varchar(20)	YES		NULL	
sclass	int(10)	YES		NULL	
sgender	varchar(10)	YES		NULL	
smajor	varchar(20)	YES		NULL	
sbirth	date	YES		NULL	
credit_points	smallint(5)	YES		NULL	

7 rows in set (0.00 sec)

【例3-2-9】 将scores表的表名改为sc。

```
mysql> ALTER TABLE scores RENAME sc;
Query OK, 0 rows affected (0.00 sec)
```

查看结果。

```
mysql> SHOW TABLES;
```

Tables_in_studentscore
course
sc
students
teachers

4 rows in set (0.00 sec)

【例3-2-10】 以students表为数据源,复制生成副本st表和sstt表。

```
mysql> CREATE TABLE st select * from students;
Query OK, 0 rows affected (0.09 sec)
Records: 0  Duplicates: 0  Warnings: 0

mysql> CREATE TABLE sstt select * from students;
Query OK, 0 rows affected (0.04 sec)
Records: 0  Duplicates: 0  Warnings: 0
```

查看结果。

```
mysql> SHOW TABLES;
+----------------------+
| Tables_in_studentscore |
+----------------------+
| course               |
| scores               |
| sstt                 |
| st                   |
| students             |
| teachers             |
+----------------------+
6 rows in set (0.00 sec)
```

【例 3-2-11】 删除 sstt 表。

```
mysql> drop table sstt;
Query OK, 0 rows affected (0.02 sec)
```

查看结果。

```
mysql> SHOW TABLES;
+----------------------+
| Tables_in_studentscore |
+----------------------+
| course               |
| scores               |
| st                   |
| students             |
| teachers             |
+----------------------+
5 rows in set (0.00 sec)
```

3.2.4 训练任务

因为我们数据库中是用数据表来管理数据的,所以一个数据库必然要创建数据表来存储数据。请读者按照学习的方法在图书管理数据库中建立管理员表、读者表、图书表、借阅表,如表 3-2-5~表 3-2-8 所示。

表 3-2-5 管理员表 (Manager)

字段名	数据类型	长度	键
name	varchar	5	
id	varchar	20	主键
sex	varchar	5	
age	int	11	
telephone	varchar	22	
password	varchar	16	

表 3-2-6 读者表(Reader)

字段名	数据类型	长度	键
name	varchar	10	
id	varchar	20	主键
sex	varchar	10	
age	int	11	
telephone	varchar	15	
email	varchar	20	

表 3-2-7 图书表(Book)

字段名	数据类型	长度	键
id	varchar	20	主键
Book_name	varchar	20	
Publish_house	varchar	20	
Publish_time	datetime		
Book_price	float	3	
Book_sum	int	3	
Book_left	int	3	

表 3-2-8 借阅表(Borrow)

字段名	数据类型	长度	键
Rnum	varchar	20	主键
Bnum	varchar	20	主键
BorrowTime	datetime		主键
ReturnTime	datetime		

请读者按照学习的方法在图书管理数据库中将读者表改为读者信息表、图书表中 id 字段改为 book_id、在图书表中增加一个 book_author 字段。

3.2.5 任务总结

在本任务中,我们学习了使用 MySQL 语句创建、修改数据库中数据表结构的方法,了解了修改表的语法规则。创建数据表的命令是 CREATE TABLE,使用的方法是 CREATE TABLE〈数据表〉( ),括号里面是要添加的表项(字段)。在创建数据表时,也可以设置此数据表的字符集、校验规则和存储引擎(它就是管理数据表中数据的方法),设置方法就是在设置完表项后,在它后面输入 CHARACTER SET 命令,然后输入要设置的字符集名称,如果还要设置此表的校验规则,那么就输入 collate 命令,然后输入校验规则的名称,最后如果还要设置存储引擎,那么就输入 engine,然后输入存储引擎的名称。在创建好一个数据表后,我们可以使用 SHOW 命令查看此数据表的结构,设置不同的参数。查询数据表结构的命令是 desc,使用方法是:desc〈表名〉。我们在创建一个表后,可能才发现有些地方设置得有问题,或者多加了一个字段(表项)又或者少加了一个字段,或者是有些类型设置得有问题,这时我们就需要使用 alter tabel 命令来修改这个数据表。

任务 3
数 据 操 作

3.3.1 任务描述

数据库是用来存放数据的,数据库中存放的数据主要是存放在表中的。数据库与表创建成功以后,就可以向数据库的表中插入数据了。学生成绩管理系统中新入学的学生信息、课程信息、成绩信息都需要存放到表中。如何向表中插入数据、修改数据就十分关键了。

在本任务中,需要在学生表、教师表、成绩表、课程表中进行数据的增、删、改、查操作。

3.3.2 知识准备

数据表建立成功之后,只是一个空的结构,就像是一个空白的表格,需要向数据表中对应的单元格中插入相应的数据,如果数据发生变化,又要对其进行修改,也要经常地删除垃圾数据和冗余数据,这都是关于数据的操作。数据操作必须通过 MySQL 提供的数据库操作语言来实现,具体可分为插入数据、修改数据、删除数据三类操作。

1. 插入数据

在 MySQL 环境中,向数据表中插入数据,使用 INSERT INTO 语句来完成,可以向数据库中已有的数据表插入一行或多行元组数据。

1) 为数据表中所有字段添加数据

为 MySQL 数据表中所有字段插入一条数据的语法格式如下:

```
INSERT INTO〈数据表名〉 (字段1,字段2,…,字段N) VALUES(值1,值2,…,值N);
```

功能:向指定的数据表中插入一条记录。

说明:

① 为数据表中所有字段插入数据。有(字段1,字段2,…,字段N)短语时,可以调整字段的顺序,字段名列表可以省略。

② 字段名序列与插入值序列的数目、类型、顺序必须一一对应,才能正确插入数据,否则就会插入记录失败。

③ 表中定义的字段数据类型如果是字符型,数据必须使用单引号或者双引号。

④ 插入数据完成后,可以使用"SELECT * FROM〈数据表名〉"来查看数据表中的数据。

2) 为指定字段添加数据

为表的指定字段添加数据,就是只向数据表中的部分字段添加数据,没有被指定的字段的值则为建立数据表时定义的默认值或空值,为指定字段添加数据的语法格式如下:

```
INSERT INTO〈数据表名〉(字段1,字段2,……,字段N) VALUES(值1,值2,……,值N);
```

功能:为数据表中指定的字段添加数据。

说明:每个值的顺序、类型必须与对应的字段相匹配。

3) 同时添加多条数据

要同时向数据表中添加多条数据,可在命令中重复使用值列表,语法格式如下:

```
INSERT INTO〈数据表名〉(字段1,字段2,……,字段N) VALUES(值1,值2,……,值N)
[,(值1,值2,……,值N)];
```

功能:同时为数据表添加多条数据。

说明:每组数据用括号括起来,各组之间用逗号分隔。

2. 修改数据

建立好的数据库系统在使用过程中,数据表中的信息经常会发生变化,这就需要管理员及时地更新数据表中的数据。可以使用 UPDATE 命令来完成。UPDATE 语句可以修改、更新一个或多个表的数据。

UPDATE 命令修改 MySQL 数据表中数据的语法格式如下:

```
UPDATE table_name SET field1=new-value1, field2=new-value2[WHERE Clause];
```

功能:更新数据表中的一列或多列数据。

说明:

① table_name 用于指定要更新的表名称

② SET 子句用于指定表中要修改的列名及其列值。其中,每个指定的列值可以是表达式,也可以是该列对应的默认值。如果指定的是默认值,可用关键字 DEFAULT 表示列值。

③ 修改数据表的多个列值时,SET 子句的每个值用逗号分开即可。

④ 在 UPDATE 命令中涉及 WHERE 子句,WHERE 子句是条件语句表示要修改符合条件记录的某个字段的内容。UPDATE 命令可以同时更新一个或多个字段内容,也可以在 WHERE 子句中指定任何条件。

3. 删除数据

在数据库管理系统中,有时会产生些垃圾数据,这就需要将这些数据删除。例如学生成绩管理系统中如果有老师调离、学生退学、课程停止授课,就需要把相关记录从教师表、学生表、课程表中删除。

在 MySQL 中删除数据使用 DELETE、TRUNCATE 两个命令来实现。

1) DELETE 命令

DELETE 语句删除表中记录的语法格式如下:

```
DELETE  FROM table_name [WHERE Clause];
```

功能:删除数据表中的全部或部分数据记录。

说明:

① table_name 指定要删除数据的表名。

② WHERE 子句表示为删除操作限定删除条件,是可选项。省略该子句则删除全部记录。

这条语句的语义是从 FROM 子句中指定的数据源中删除符合 WHERE 子句条件的记录。注意:如果没有 WHERE 子句,那么就会将表中所有数据都删除。

2) TRUNCATE 命令

TRUNCATE 语句删除表中记录的语法格式如下:

```
TRUNCATE [table] table_name;
```

功能:删除数据表中的全部数据记录。

说明:

① [table]可以省略。

② 使用 TRUNCATE 语句,就会删除指定表中的所有数据,只保留表的空结构,而且无法恢复。

3) DELETE、TRUNCATE 命令的区别

① DELETE 是数据操作语言 DML 语句,TRUNCATE 是数据库模式定义语言 DDL 语句。

② DELETE 后面可以加 WHERE 子句,删除部分记录,TRUNCATE 只能删除所有记录。

③ 用 DELETE 语句删除表中所有记录,再向表中添加记录时,自动增加字段的值为删除时该字段的最大值加 1。

TRUNCATE 语句删除表中的数据,再向表中添加记录时,自动增加字段的默认初始值重新由 1 开始。

④ DELETE 语句每删除一条记录都会在日志中记录。TRUNCATE 语句不会在日志

中记录删除的内容,因此语句的执行效率高。

3.3.3 任务实现

【例3-3-1】 向students表中插入"刘洋"的记录。

```
mysql> INSERT INTO
    -> students(sid,sname,sclass,sgender,smajor,sbirth,credit_points)
    -> VALUES(1933062301,"刘洋",1901,"女","软件技术","2000-09-01",5);
Query OK, 1 row affected (0.01 sec)
```

使用查询语句查看表中的数据内容。

```
mysql> SELECT * FROM students;
+------------+-------+--------+---------+-----------+------------+---------------+
| sid        | sname | sclass | sgender | smajor    | sbirth     | credit_points |
+------------+-------+--------+---------+-----------+------------+---------------+
| 1933062301 | 刘洋  | 1901   | 女      | 软件技术  | 2000-09-01 |             5 |
+------------+-------+--------+---------+-----------+------------+---------------+
1 row in set (0.00 sec)
```

有关查询语句的内容我们将在下一个任务详细讲解。

【例3-3-2】 向students表中插入"邓嘉荟"的记录。

```
mysql> INSERT INTO students VALUES(1933062302,"邓嘉荟",1901,"女","软件技术","2000-03-24",3);
Query OK, 1 row affected (0.00 sec)
```

查看结果。

```
mysql> SELECT * FROM students;
+------------+--------+--------+---------+-----------+------------+---------------+
| sid        | sname  | sclass | sgender | smajor    | sbirth     | credit_points |
+------------+--------+--------+---------+-----------+------------+---------------+
| 1933062301 | 刘洋   | 1901   | 女      | 软件技术  | 2000-09-01 |             5 |
| 1933062302 | 邓嘉荟 | 1901   | 女      | 软件技术  | 2000-03-24 |             3 |
+------------+--------+--------+---------+-----------+------------+---------------+
2 rows in set (0.00 sec)
```

通过观察上面的两个插入记录操作,读者可以体会一下有什么不同。

向数据表中插入数据时,如果插入的数值序列与定义表时的字段序列一致,也可以省略列名序列。

【例3-3-3】 在插入语句中同时书写多组插入值序列,同时插入多条记录。

```
mysql> INSERT INTO students
    -> VALUES(1933062303,"葛佳音",1901,"女","软件技术","2000-08-24",5),
    -> (1933062304,"马国飞",1901,"男","软件技术","2000-09-23",3);
Query OK, 2 rows affected (0.00 sec)
Records: 2  Duplicates: 0  Warnings: 0
```

查看结果。

```
mysql> SELECT * FROM students;
+------------+--------+--------+---------+----------+------------+---------------+
|    sid     | sname  | sclass | sgender |  smajor  |   sbirth   | credit_points |
+------------+--------+--------+---------+----------+------------+---------------+
| 1933062301 | 刘洋   |  1901  |   女    | 软件技术 | 2000-09-01 |             5 |
| 1933062302 | 邓嘉蓉 |  1901  |   女    | 软件技术 | 2000-03-24 |             3 |
| 1933062303 | 葛佳音 |  1901  |   女    | 软件技术 | 2000-08-24 |             5 |
| 1933062304 | 马国飞 |  1901  |   男    | 软件技术 | 2000-09-23 |             3 |
+------------+--------+--------+---------+----------+------------+---------------+
4 rows in set (0.00 sec)
```

【例 3-3-4】 以 students 表为源，复制生成 stud 表。

```
mysql> CREATE TABLE stud SELECT * FROM students;
Query OK, 4 rows affected (0.05 sec)
Records: 4  Duplicates: 0  Warnings: 0
```

查看结果。

```
mysql> SELECT * FROM stud;
+------------+--------+--------+---------+----------+------------+---------------+
|    sid     | sname  | sclass | sgender |  smajor  |   sbirth   | credit_points |
+------------+--------+--------+---------+----------+------------+---------------+
| 1933062301 | 刘洋   |  1901  |   女    | 软件技术 | 2000-09-01 |             5 |
| 1933062302 | 邓嘉蓉 |  1901  |   女    | 软件技术 | 2000-03-24 |             3 |
| 1933062303 | 葛佳音 |  1901  |   女    | 软件技术 | 2000-08-24 |             5 |
| 1933062304 | 马国飞 |  1901  |   男    | 软件技术 | 2000-09-23 |             3 |
+------------+--------+--------+---------+----------+------------+---------------+
4 rows in set (0.00 sec)
```

【例 3-3-5】 将 students 表中葛佳音同学的性别改为男。

```
mysql> UPDATE students SET sgender='男'where sname="葛佳音";
Query OK, 1 row affected (0.00 sec)
Rows matched: 1  Changed: 1  Warnings: 0
```

查看结果。

```
mysql> SELECT * FROM students;
+------------+--------+--------+---------+----------+------------+---------------+
|    sid     | sname  | sclass | sgender |  smajor  |   sbirth   | credit_points |
+------------+--------+--------+---------+----------+------------+---------------+
| 1933062301 | 刘洋   |  1901  |   女    | 软件技术 | 2000-09-01 |             5 |
| 1933062302 | 邓嘉蓉 |  1901  |   女    | 软件技术 | 2000-03-24 |             3 |
| 1933062303 | 葛佳音 |  1901  |   男    | 软件技术 | 2000-08-24 |             5 |
| 1933062304 | 马国飞 |  1901  |   男    | 软件技术 | 2000-09-23 |             3 |
+------------+--------+--------+---------+----------+------------+---------------+
4 rows in set (0.00 sec)
```

【例 3-3-6】 将 students 表中学分不足 5 分的改为 5 分。

```
mysql> UPDATE students SET credit_points=5 WHERE credit_points<5;
Query OK, 2 rows affected (0.00 sec)
Rows matched: 2  Changed: 2  Warnings: 0
```

查看结果。

```
mysql> SELECT * FROM students;
+------------+--------+--------+---------+---------+------------+---------------+
|    sid     | sname  | sclass | sgender | smajor  |   sbirth   | credit_points |
+------------+--------+--------+---------+---------+------------+---------------+
| 1933062301 | 刘洋   |  1901  |   女    | 软件技术 | 2000-09-01 |       5       |
| 1933062302 | 邓嘉蓉 |  1901  |   女    | 软件技术 | 2000-03-24 |       5       |
| 1933062303 | 葛佳音 |  1901  |   男    | 软件技术 | 2000-08-24 |       5       |
| 1933062304 | 马国飞 |  1901  |   男    | 软件技术 | 2000-09-23 |       5       |
+------------+--------+--------+---------+---------+------------+---------------+
4 rows in set (0.00 sec)
```

【例 3-3-7】 使用 DELETE 命令，删除学号为 1933062304 的学生信息。

```
mysql> DELETE FROM students WHERE sid=1933062304;
Query OK, 1 row affected (0.00 sec)
```

查看结果。

```
mysql> SELECT * FROM students;
+------------+--------+--------+---------+---------+------------+---------------+
|    sid     | sname  | sclass | sgender | smajor  |   sbirth   | credit_points |
+------------+--------+--------+---------+---------+------------+---------------+
| 1933062301 | 刘洋   |  1901  |   女    | 软件技术 | 2000-09-01 |       5       |
| 1933062302 | 邓嘉蓉 |  1901  |   女    | 软件技术 | 2000-03-24 |       5       |
| 1933062303 | 葛佳音 |  1901  |   男    | 软件技术 | 2000-08-24 |       5       |
+------------+--------+--------+---------+---------+------------+---------------+
3 rows in set (0.00 sec)
```

从运行结果可以看出学号为 1933062304 的记录已经被删除。

【例 3-3-8】 使用 DELETE 命令，删除所有 students 表中所有学生的信息。

```
mysql> DELETE FROM students;
Query OK, 3 rows affected (0.00 sec)
```

查看结果。

```
mysql> SELECT * FROM students;
Empty set (0.00 sec)
```

【例 3-3-9】 在 students 表中录入一条记录（只录入姓名）。

```
mysql> insert into students(sname) values('李丽');
Query OK, 1 row affected (0.00 sec)
```

查看结果。

```
mysql> SELECT * FROM students;
+------------+-------+--------+---------+--------+--------+---------------+
| sid        | sname | sclass | sgender | smajor | sbirth | credit_points |
+------------+-------+--------+---------+--------+--------+---------------+
| 1933062305 | 李丽  | NULL   | NULL    | NULL   | NULL   | NULL          |
+------------+-------+--------+---------+--------+--------+---------------+
1 row in set (0.00 sec)
```

从上面的显示结果可以看到,虽然删除了所有的记录,但当再插入记录时,sid 的值是从原记录的最大值+1 开始排序的。

【例 3-3-10】 使用 TRUNCADE 命令删除 students 表中所有学生的信息。

```
mysql> TRUNCATE students;
Query OK, 0 rows affected (0.03 sec)
```

查看结果。

```
mysql> SELECT * FROM stud;
Empty set (0.00 sec)
```

【例 3-3-11】 在 students 表中录入一条记录(只录入姓名)。

```
mysql> insert into students(sname) values('李丽');
Query OK, 1 row affected (0.00 sec)
```

查看结果。

```
mysql> SELECT * FROM students;
+-----+-------+--------+---------+--------+--------+---------------+
| sid | sname | sclass | sgender | smajor | sbirth | credit_points |
+-----+-------+--------+---------+--------+--------+---------------+
| 1   | 李丽  | NULL   | NULL    | NULL   | NULL   | NULL          |
+-----+-------+--------+---------+--------+--------+---------------+
1 row in set (0.00 sec)
```

从上面的显示结果可以看到,使用 TRUNCADE 命令删除所有的记录后,再插入记录时,sid 的值是从 1 开始排序的。

3.3.4 训练任务

图书管理系统的数据库用来收集、存储书籍信息、读者信息、图书借阅信息,所以需要向表中添加数据。如果记录存储各个环节信息需要变更,就需要修改或者删除表中信息。

① 按照下面各个数据表中的信息,向管理员表、图书表、读者表、借阅表表中插入记录。如表 3-3-1~表 3-3-4 所示。

表 3-3-1 管理员表(Manager)

name	id	sex	age	telephone	password
程霞	99666091	男	19	13052849321	3195405567
王远平	76837009	男	18	13063778327	7916195841
张鑫彤	66427190	女	18	15373388179	2521575037

表 3-3-2 读者表(Reader)

name	id	sex	age	telephone	email
王东华	4979328308	男	19	18764759893	f3wlafqf3lkr@wbgj.com
周远琳	1501866240	男	20	17883417294	yrkop2sn4@xkkk.com
王晟	9369403270	男	18	17208847981	vske1toneox6@marku.com
张欣然	8166973209	女	18	15527499156	ydxn@ogrlh.com
王江	3580041887	男	19	15060838530	tdulodjufc3jb@ogroii.com
董雨晨	1445640383	女	19	13698333794	rnwy4@fheht.com
赵勇健	4332784312	男	18	13366442401	0dd2@ifhrt.com
赵鑫	9265664183	女	18	18987518873	ewgnmvhl5@ddku.com
吴冰	5672452845	女	19	18748001652	3ojt4n1mdq@ntdk.com
孙宇	8104614274	男	19	18375282145	elwjasqz3e2mfrh@qrba.com

表 3-3-3 图书表(Book)

id	Book_name	Publish_house	Publish_time	Book_price	Book_sum	Book_left
915761343879427	《时间简史》	宏图出版社	2019-10-24	¥45	130	96
703322762702280	《1984》	气韵出版社	2020-09-12	¥50	120	103
850133914915126	《了不起的盖茨比》	独角仙出版社	2020-03-10	¥45	90	52
600162098505673	《人性的枷锁》	鼎盛出版社	2021-05-19	¥54	100	81
365968519114859	《总统班底》	甲骨文出版社	2021-10-07	¥25	150	65
845419205164378	《天生就会跑》	华辰出版社	2021-03-30	¥34	150	84
268065711653304	《安吉拉的灰烬》	琳出版社	2020-12-12	¥37	95	15
413356145679682	《远大前程》	游历出版社	2021-03-25	¥47	90	92
124864894908802	《华氏451度》	九八出版社	2020-08-18	¥42	130	72
996558053112379	《傲慢与偏见》	见久出版社	2021-08-25	¥39	100	52

表 3-3-4 借阅表(Borrow)

Rnum	Bnum	BorrowTime	ReturnTime
9369403270	703322762702280	2021-04-12	2021-06-05
4979328308	365968519114859	2021-05-02	2021-06-31
8104614274	268065711653304	2021-04-10	2021-07-12
3580041887	850133914915126	2021-03-21	2021-06-30
4332784312	996558053112379	2021-03-20	2021-05-14
9265664183	850133914915126	2020-12-01	2020-12-31
4979328308	850133914915126	2020-11-02	2020-12-18
8104614274	845419205164378	2020-11-02	2020-12-09
1501866240	124864894908802	2020-10-23	2020-11-30
1501866240	124864894908802	2020-10-10	2020-12-25

② 修改管理员表、图书表、读者表、借阅表表中的记录。

a. 管理员程霞同学的电话号码更换为 18140113470。

b. 又新购了图书《了不起的盖茨比》5 本供同学们借阅。

c. 读者王东华同学的电子邮箱更换为 wangdonghua@163.com。

d. 王晟同学还书日期延缓一周。

e. 周远琳同学 2010-10-10 所借书籍改为《远大前程》。

③ 删除信息。

a. 张鑫彤同学毕业不再是管理员。

b. 吴冰同学毕业收回借书证。

c. 书库撤回所有琳出版社出版的书籍。

d. 《远大前程》出新版书籍,书库撤回现有书籍。

e. 删除王江同学所有借书记录。

④ 以图书表为源,复制生成副本图书表 1。

⑤ 删除图书表 1 中的所有信息。

3.3.5 任务总结

在本任务中,我们学习了使用 MySQL 语句添加、修改、删除表中数据的方法,了解了各种数据操作的语法规则。我们使用 insert 命令向数据表中插入数据。在往数据表中插入数据时,不小心写错了数据,或者需要更新数据,此时需要修改对应字段的数据,可以使用 update 命令来修改数据。在插入数据时,可能多插入了一个数据,或者是之前的某个数据,要将其从数据库中清理出去,这时就可以使用 delete 命令。在以后的章节中会逐步使用 MySQL 语句完成相应的任务。

项目 4　数据类型与约束

　　MySQL 数据库是用来管理数据的,所有的数据都存储在数据表中,每个数据表都可以存储姓名、性别、年龄、出生日期、照片等不同的数据。那么如何正确、高效地存储这些数据呢? 这是我们本项目讨论的主题。在设计数据表结构时,需要根据实际情况,选择合适的数据类型,设置不同的约束条件,以便更好地维护和管理数据。通过本项目的学习,读者能够理解和掌握数据类型,正确地建立表的约束。

> **学习目标**
>
> - 了解数值类型、字符类型、时间类型等数据类型的种类;
> - 掌握各种常用的数据类型的使用方法和规则;
> - 能够正确选择使用数据类型;
> - 掌握主键约束、唯一约束、默认约束、非空约束的概念、作用;
> - 能够根据实际情况正确设置表的各种约束。

任务 1

数 据 类 型

4.1.1 任务描述

数据类型是指列、存储过程参数、表达式和局部变量的数据特征,它决定了数据的存储格式,代表了不同的信息类型。

在建立数据表的过程中,需要存储不同的数据,有一些数据是要存储为数字的,比如学生的成绩、班级的人数等。也有些数据是要存储为字符的,比如姓名、地址等。我们需要根据具体情况的不同选择不同的数据类型。不同的数据类型决定了 MySQL 存储数据方式的不同,数据库提供了多种数据类型,其中包括数值类型、字符类型、时间日期类型等,下面针对这些数据类型进行学习。

4.1.2 知识准备

1. 数值类型

在 MySQL 数据库中,经常需要存储一些数字有很多但又各有不同的取值范围、精度,对应的字节大小也不相同,可分为下面几种。

1) 整数类型

表 4-1-1 MySQL 整数类型

数据类型	字节数	无符号数据的取值范围	有符号数的取值范围
TINYINT	1	0～255	－128～127
SMALLINT	2	0～65535	－32768～32767
MEDIUMINT	3	0～16777215	－8388608～8388607
INT	4	0～4294967295	－2147483648～2147483647
BIGING	8	0～18446744073709551615	－9223372036854775808～9223372036854775807

从表 4-1-1 中可以发现,不同整数类型所占的字节数和取值范围都不相同,TINYINT 类型字节数最小,BIGING 类型字节数最大,不同整数类型的取值范围可以根据字节数进行计算得出,例如,TINYINT 整数类型的数据占用 1 字节,而 1 字节是 8 个二进制位,用 8 个二进制位表达的最大无符号数就是 2^8-1,即 255,最大有符号数就是 2^7-1,即 127。同理可以计算出其他不同整数类型的取值范围。

在建立数据表的操作中,若使用无符号数据类型,需要在数据类型右边加上UNSIGNED关键字加以说明,即用INT UNSIGNED表示无符号的INT类型。

【例4-1-1】 创建my_table表,分别设置有符号和无符号的INT数据类型进行测试。操作方法及结果如下。

① 打开ZLM数据库。

```
mysql> use zlm;
Database changed
```

② 建立表。

```
mysql> create table my_table(id1 int,id2 int unsigned);
Query OK, 0 rows affected (0.06 sec)
```

③ 插入记录测试。

```
mysql> insert into my_table value(100,200);
Query OK, 1 row affected (0.03 sec)

mysql> insert into my_table value(-100,-200);
ERROR 1264 (22003): Out of range value for column 'id2' at row 1
```

第一条语句插入记录成功,因为两个数据都是无符号的;第二条语句插入失败,因为在建立my_table表时id2字段使用了UNSIGNED短语,所以不能接受"-200"这个数据,导致插入数据失败。

```
mysql> create table my_table(id1 int,id2 int unsigned);
Query OK, 0 rows affected (0.06 sec)
```

④ 查看my_table表结构。

```
mysql> show create table my_table;
+----------+------------------------------------------------+
| Table    | Create Table                                   |
+----------+------------------------------------------------+
| my_table | CREATE TABLE 'my_table' (
  'id1' int(11) DEFAULT NULL,
  'id2' int(10) unsigned DEFAULT NULL
) ENGINE=MyISAM DEFAULT CHARSET=utf8 COLLATE=utf8_bin |
+----------+------------------------------------------------+
1 row in set (0.00 sec)
```

在显示结果中,数据类型的右侧括号内标了显示宽度,符号也占一位,"id1"的显示宽度为11;"id2"的显示宽度为10,多出的1位就是符号位。需要说明的是:显示宽度与取值范围无关,若数值的位数小于显示宽度,会填充空格,反之,则不影响显示结果。

2) 浮点数类型(表 4-1-2)

表 4-1-2　MySQL 浮点数类型

数据类型	字节数	负数的取值范围	非负数的取值范围
FLOAT	4	$-3.402823466E+38 \sim$ $-1.175494351E-38$	0 和 $1.175494351E-38 \sim 3.402823466E+38$
DOUBLE	8	$-1.7976931348623157E+308 \sim$ $-2.2250738585072014E-308$	0 和 $2.2250738585072014E-308 \sim$ $1.7976931348623157E+308$

表 4-1-2 中列举的取值范围是理论上的极值,根据不同软硬件环境,实际范围可能会小。也可以使用 UNSIGNED 短语限定为无符号,此时,取值范围不包括负数。

浮点数类型虽然取值范围很大,但是精度并不高,FLOAT 的精度为 6~7 位,DOUBLE 的精度大约是 15 位,如果超出精度,可能会导致给定的数值与实际保存的数值不一致,产生精度损失。

3) 定点数类型(表 4-1-3)

表 4-1-3　MySQL 定点数类型

数据类型	参数说明		
DECIMAL(M,D)	M-D>=2	M 默认值为 10,最大值为 65;D 默认值为 0,最大值为 30	可加 UNSIGNED 短语限定非负数

使用 DECIMAL 数据类型时,若小数部分超出范围,会自动进行四舍五入,并给出警告,若整数部分超出范围,则会插入失败,并报错。

4) BIT 类型

BIT 类型用于存储二进制数据,语法格式为:BIT(M),M 表示位数,取值范围是 1~64。BIT 类型字段在数字插入时转换为二进制数据保存,进行 SELECT 查询时,会自动转换为对应的字符显示。

2. 字符串类型

MySQL 中的字符串类型有 CHAR、VARCHAR 等多种类型,不同数据类型各有不同的特点,具体情况如表 4-1-4 所示。

表 4-1-4　MySQL 字符串类型

数据类型	格式	类型说明
CHAR	CHAR(M)	固定长度的字符串
VARCHAR	VARCHAR(M)	可变长度的字符串
TEXT	TEXT	长文本数据
ENUM	ENUM('字符串 1','字符串 2',…,'值 N')	枚举类型

（续表）

数据类型	格式	类型说明
SET	SET('字符串1','字符串2',…,'值N')	字符串对象
BINARY	BINARY(M)	固定长度的二进制数据
VARBINARY	VARBINARY(M)	可变长度的二进制数据
BLOB	BLOB	二进制大对象

① CHAR 和 VARCHAR 都用来表示字符串数据，CHAR 表示固定长度的字符串，VARCHAR 表示可变长度的字符串。

② BINARY 和 VARBINARY 与 CHAR 和 VARCHAR 类似，不同之处在于，BINARY 和 VARBINARY 所表示的是二进制数据。

③ TEXT 类型用于表示大文本数据。

④ BLOB 用于表示数据量很大的二进制类型。

⑤ ENUM 类型也称为枚举类型，ENUM 类型的数据只能从枚举列表中取值，并且只能取一个，枚举列表中的每个值都有一个顺序编号，MySQL 中存储的是这个编号，而不是列表中的值。

⑥ SET 类型用于表示字符串对象，它的值可以有零个或多个。

⑦ BIT 类型用于表示二进制数据，格式 BIT(M)，M 表示每个值的位数，范围是 1～64，如果给定的数据长度小于 M，将在数据的左边用 0 补齐。

3. 日期时间类型

为了存储日期和时间，MySQL 提供了用于表示日期和时间的数据类型，表 4-1-5 列举了这些数据类型的相关信息。

表 4-1-5　MySQL 日期时间类型

数据类型	字节数	取值范围	日期格式	零值
YEAR	1	1901～2155	YYYY	0000
DATE	4	1000-01-01～9999-12-3	YYYY-MM-DD	0000-00-00
TIME	3	-838:59:59～838:59:59	HH:MM:SS	00:00:00
DATETIME	8	1000-01-01 00:00:00～ 9999-12-31 23:59:59	YYYY-MM-DD HH:MM:SS	0000-00-00 00:00:00
TIMESTAMP	4	1970-01-01 00:00:01～ 2038-01-19 03:14:07	YYYY-MM-DD HH:MM:SS	0000-00-00 00:00:00

表 4-1-5 中各种日期时间类型的取值范围都是不同的，如果在操作过程中插入的数据不合法，系统将自动插入对应的零值。

① YEAR 类型用于表示年份，用 4 位字符串或数字都可以表示年份；用 2 位字符串表

示年份时,"00"~"69"表示2000~2069,"70"~"99"表示1970~1999;用2位数值表示年份时,1~69表示2001~2069,70~99表示1970~1999;"0"表示2000年,而0表示的是0000。

② DATE类型有"YYYY-MM-DD""YYYYMMDD"两种格式,其中表示年份的YYYY的用法与YEAR相同;可以使用CURRENT_DATE或者NOW()输入当前系统日期。

③ TIME类型有"HH:MM:SS""HHMMSS""D HH:MM:SS"三种格式,其中D表示日,取值范围为0~34,插入数据时,小时的值等于(D×24+HH);可以使用CURRENT_TIME或者NOW()输入当前系统时间。

④ DATETIME类型用于表示日期和时间,与前面的日期、时间的表示方法相同,可以使用NOW()输入当前系统日期和时间。

⑤ TIMESTAMP也叫时间戳,用于表示日期和时间,显示形式与DATETIME相同,但取值范围小,可以使用CURRENT_TIMESTAMP来输入系统当前日期和时间。无输入或输入NULL时,保存的是系统当前日期和时间。

4.1.3 任务总结

数据库管理工作目的在于如何方便、快捷、准确、高效地管理和使用数据,而做好这项工作的前提是掌握数据的类型、取值范围、使用方法等知识。在本任务中,我们学习了数值、字符、时间三种数据类型,也就相当于知道了要在什么样的容器中存放什么样的物体,这在数据表设计过程中有着非常重要的意义。在设置表结构的时候,读者需要根据实际情况正确选择合适的数据类型,从而保证数据库中数据的正确性和有效性。

任务2 表 的 约 束

4.2.1 任务描述

前面学习的数据类型是限制我们可以在表里存储什么数据的一种方法。但是,在很多情况下,还存在些数据类型无法解决的问题。比如,一个包含产品价格的字段应该只接受正数。但是没有哪种标准数据类型只接受正数;或者,需要根据其他字段或者其他行的数据来约束字段数据,比如,在一个包含产品信息的表中,每个产品的名称都不能是空的。对于这些问题,可以通过定义约束来解决。约束允许对数据施加任意控制。如果用户企图在字段里存储违反约束的数据,系统就会报错。这种情况也适用于数值来自缺省值的情况。

4.2.2 知识准备

表约束(Table Constraint)是 2018 年全国科学技术名词审定委员会公布的计算机科学技术名词。它是对表的属性值及其不同属性之间的依赖关系的限制规定,可以认为就是参数限制,是在表中定义的用于维护数据库完整性的一些规则。

约束(Constraint)是 MySQL 提供的自动保持数据库完整性的一种方法,定义可输入表或表的单个列中的数据的限制条件 MySQL 中有 5 种约束:主关键字约束(Primary Key Constraint)、外关键字约束(Foreign Key Constraint,涉及多表操作,在后面章节中讲述)、唯一性约束(Unique Constraint)、非空约束(Not Null Constraint)和缺省约束(Default Constraint)。

1. 主键约束(Primary Key Constraint)

主键约束指定表的一列或几列的组合的值在表中具有唯一性、非空性,即能唯一地确定一条记录。每个表中只能有一列被指定为主关键字,不允许主关键字列有 NULL 属性,且 IMAGE 和 TEXT 类型的列不能被指定为主关键字。主键约束的创建可以是列级约束,也可以是表级约束。

定义主关键字约束的语法如下:

格式 1:

```
CONSTRAINT constraint_name
PRIMARY KEY
(column_name1[,column_name2,…,column_name16])
```

格式 2:

```
column_name 数据类型 PRIMARY KEY
```

功能:用于定义表的主关键字约束。

说明:

① constraint_name:约束的名称。在数据库中应是唯一的。如果不指定,则系统会自动生成一个约束名。

② column_name:指定组成主关键字的列名,也可以由多列组成,最多由 16 个列组成。

③ 使用格式 1 建立的约束是表级约束,是独立于无的定义,可以应用在一个表的多个列上;使用格式 2 建立的约束属于列级约束,只对该列起作用,由此我们可以体会列级约束与表级约束的不同之处。

④ 添加主键约束后,该字段插入重复值或 NULL 值就会失败。

2. 唯一性约束(Unique Constraint)

唯一性约束指定一个或多个列的组合的值具有唯一性,以防止在列中输入重复的值。

唯一性约束指定的列可以有 NULL 属性，但只能有一个空值。由于主关键字值是具有唯一性的，因此主关键字列不能再设定唯一性约束。唯一性约束最多由 16 个列组成。

定义唯一性约束的语法如下：

格式 1：

```
CONSTRAINT constraint_name
    UNIQUE [CLUSTERED | NONCLUSTERED]
    (column_name1[, column_name2,…,column_name16])
```

格式 2：

```
column_name 数据类型 UNIQUE
```

功能：用于定义表的唯一约束。

说明：

① constraint_name：约束的名称。在数据库中应是唯一的。如果不指定，系统也会自动生成一个约束名。

② column_name：指定组成唯一约束的列名，也可以由多列组成。

③ 格式 1 与格式 2 的区别同主关键字约束。

④ [CLUSTERED | NONCLUSTERED]：是可选项，CLUSTERED 可以定义聚集索引，NONCLUSTERED 定义非聚集索引，聚集索引存储记录是物理上连续存在，而非聚集索引是逻辑上的连续，物理存储并不连续。唯一性约束默认的是非聚集索引；主键约束默认的是聚集索引。

⑤ 一个表中唯一性约束可以有多个，但主键约束只能有一个。

3. 非空约束(Not Null Constraint)

非空约束只是简单地声明一个字段的值不能为 NULL，可以通过 NOT NULL 来定义。

定义非空约束的语法如下：

```
column_name 数据类型 NOT NULL
```

功能：用于定义表中某个列的非空约束。

说明：

① column_name：字段名，用于指定设置非空约束的列。

② 一个非空约束总是写成一个字段约束即列的约束，非空约束在功能上等效于在其他数据库环境中创建一个检查约束，创建一个明确的非空约束效率非常高，缺点是不能给它一个明确的约束名。

4. 缺省约束(Default Constraint)

缺省约束通过定义列的缺省值或使用数据库的缺省值对象绑定表的列，来指定列的缺省值。如果一个字段已经指定了默认值，当向表中插入一条新记录时，如果没有给这个字段赋值,数据库系统就会自动地为这个字段插入事先设置好的默认值,默认值通过 DEFAULT 关键字来定义。

定义缺省约束的语法如下:

格式 1：

```
CONSTRAINT constraint_name
DEFAULT constant_expression [FOR column_name]
```

格式 2：

```
column_name 数据类型 DEFAULT 默认值
```

功能:用于定义表中某个列的默认值。

说明:

① column_name:字段名,用于指定设置非空约束的列。

② BLOB、TEXT 数据类型不支持默认约束。

前面讲述的主键约束、唯一性约束、非空约束、默认约束既可以在建立表的时候创建,也可以通过使用 ALTER TABLE 命令修改表的方法来创建。

5. 列值自动增长(AUTO_INCREMENT)

为数据表设置主键约束后,每次插入记录时,都检查主键的值,防止插入一个重复值而导致插入数据操作失败。这就会给操作者带来很大的麻烦。MySQL 系统提供了列值的自动增长功能,可以解决这个问题。

定义列自动增长的语法格式如下:

```
column_name 数据类型 AUTO_INCREMENT
```

功能:插入一条新记录时,该字段的值会自动增加。

说明:

① 一个表中只能有一个自动增长的列,该列的数据类型必须是整数类型,且已经被定义为键(UNIQUE KEY)或主键(PRIMARY KEY)。

② 如果插入记录操作时该字段插入的是一个具体值,就不会使用自动增长值,只有当插入 0、NULL、DEFAULT 或省略该字段时,才会使用自动增长值。

③ 对于一个空表,自动增长值默认从 1 开始自增,每次加 1,对于已经存在记录的表,如果插入的值大于自动增长值,系统会接受用户插入的值;反之,如果插入的值小于自动增长值,而正好该值没有出现的话,系统会接受用户插入的值,用户可以插入任意一个大于自动增长值的合法的值。

④ 如果已经存在的记录的值不是连续排列的,中间有空缺时,自动增长值不会减小或补空缺。

建立 LKN 表,设置 id 字段值自动增长。

```
mysql> create table lkn(id int(2) primary key auto_increment,
    -> name varchar(4));
Query OK, 0 rows affected (0.04 sec)
```

输入记录，ID字段值分别为0、null、default。

```
mysql> insert into lkn value(0,'a1');
Query OK, 1 row affected (0.00 sec)

mysql> insert into lkn value(null,'a2');
Query OK, 1 row affected (0.00 sec)

mysql> insert into lkn value(default,'a3');
Query OK, 1 row affected (0.00 sec)
```

查看结果。

```
mysql> select * from lkn;
+----+------+
| id | name |
+----+------+
|  1 | a1   |
|  2 | a2   |
|  3 | a3   |
+----+------+
3 rows in set (0.00 sec)
```

录入记录，ID字段值为6。

```
mysql> insert into lkn value(6,'a6');
Query OK, 1 row affected (0.00 sec)
```

查看结果。

```
mysql> select * from lkn;
+----+------+
| id | name |
+----+------+
|  1 | a1   |
|  2 | a2   |
|  3 | a3   |
|  6 | a6   |
+----+------+
4 rows in set (0.00 sec)
```

录入记录，ID字段值为空。

```
mysql> insert into lkn(name) value('a7');
Query OK, 1 row affected (0.02 sec)
```

查看结果。

```
mysql> select * from lkn;
+----+------+
| id | name |
+----+------+
|  1 | a1   |
|  2 | a2   |
|  3 | a3   |
|  6 | a6   |
|  7 | a7   |
+----+------+
5 rows in set (0.00 sec)
```

录入记录,ID 字段值为 5。

```
mysql> insert into lkn value(5,'a5');
Query OK, 1 row affected (0.02 sec)
```

查看结果。

```
mysql> select * from lkn;
+----+------+
| id | name |
+----+------+
|  1 | a1   |
|  2 | a2   |
|  3 | a3   |
|  6 | a6   |
|  7 | a7   |
|  5 | a5   |
+----+------+
6 rows in set (0.00 sec)
```

4.2.3 任务实现

【例 4-2-1】 使用 CREATE TABLE 方式建立员工信息表结构,同时建立约束。

创建一个员工信息表,包括员工编号、姓名、性别、出生日期、职称、身份证号等字段,定义员工编号为主键且自动增长,姓名字段不能为空,性别字段默认值为"男",身份证号字段值不能重复。

建立表结构。

```
mysql> create table emp
    -> (empid int(6) primary key  auto_increment,
    -> name varchar(8) not null,
    -> ssex enum('男','女') not null default '男',
    -> sbirthday date,
    -> prof varchar(10),
    -> idcard char(18) unique);
Query OK, 0 rows affected (0.05 sec)
```

查看员工信息表结构。

```
mysql> DESC emp;
+-----------+----------------+------+-----+---------+----------------+
|Field      |Type            |Null  |Key  |Default  |Extra           |
+-----------+----------------+------+-----+---------+----------------+
|empid      |int(6)          |NO    |PRI  |NULL     |auto_increment  |
|name       |varchar(8)      |NO    |     |NULL     |                |
|ssex       |enum('男','女') |NO    |     |男       |                |
|sbirthday  |date            |YES   |     |NULL     |                |
|prof       |varchar(10)     |YES   |     |NULL     |                |
|idcard     |char(18)        |YES   |UNI  |NULL     |                |
+-----------+----------------+------+-----+---------+----------------+
6 rows in set (0.00 sec)
```

【例4-2-2】 使用CREATE TABLE方式建立学生成绩表结构,同时建立约束。

建立学生成绩表,包括学号、课程号、成绩、学分、开课学期等字段,定义学号、课程号为组合主键,开课学期不能为空。

建立表命令。

```
mysql> create table scoure
    -> (sno char(10),
    -> cno char(4),
    -> degree float,
    -> xf float,
    -> term varchar(10) not null,
    -> primary key (sno,cno) );
Query OK, 0 rows affected (0.09 sec)
```

查看学生成绩表结构。

```
mysql> DESC scoure;
+--------+-------------+------+-----+---------+-------+
|Field   |Type         |Null  |Key  |Default  |Extra  |
+--------+-------------+------+-----+---------+-------+
|sno     |char(10)     |NO    |PRI  |NULL     |       |
|cno     |char(4)      |NO    |PRI  |NULL     |       |
|degree  |float        |YES   |     |NULL     |       |
|xf      |float        |YES   |     |NULL     |       |
|term    |varchar(10)  |NO    |     |NULL     |       |
+--------+-------------+------+-----+---------+-------+
5 rows in set (0.00 sec)
```

4.2.4 任务总结

在数据库管理工作中,数据的正确性是至关重要的,要保证数据表中插入数据的正确性,可以使用数据库完整性的规则来控制这些数据,也就是主键约束、唯一约束、默认约束、非空约束,利用这些约束对表中的字段进行限制,从而保证数据表中数据的有效性、正确性和唯一性。通过本任务的学习,读者应掌握约束的作用和使用方法,能够正确设置表的约束,从而防止错误数据的产生,保证数据的完整性。

项目5 数据查询

在数据库使用过程中,查询工作是数据库操作的核心任务之一。数据库的查询是数据库检索、数据统计输出、向表中添加数据、对现有数据进行修改、无用数据删除的基础性工作。数据查询操作占数据库实际操作额50%以上。具体地讲,查询是对已有表中的数据按照某种条件进行筛选,将满足条件的数据筛选出来形成一个新的记录集进行显示,也称为查询结果记录集。这个记录集的结构与表的结构类似,由行和列组成,但它并不是真正存放在数据库中的表,而是一种存放在计算机内存中的虚拟表。查询是查找和筛选的扩充,它不但能实现数据检索,而且可以在查询过程中进行计算、合并不同数据源,甚至可以添加、更改或删除基本表中的数据。查询的数据源是存放在数据库中的基本表或已经创建好的视图,可以是一个或多个数据源。若是多个数据源,就需要指定这些数据源之间的关联关系,以保证查询结果的正确性。查询结果只有在运行查询语句后才会产生,因此也称为动态记录结果集。

在学生成绩管理系统中经常要查询各科成绩最高分、最低分、平均分。查询各科成绩,按各科进行排序。查询某位老师所讲授课程的学生中,该门课程成绩最高分学生信息。

> **学习目标**
>
> - 了解查询语句的作用及类型;
> - 掌握各类查询的特点和使用方法;
> - 掌握查询语句的语法规则;
> - 掌握聚合函数的使用方法;
> - 掌握分组和排序的方法;
> - 理解多表连接的使用规则;
> - 掌握嵌套查询的使用规则;
> - 理解集合查询的含义;
> - 能够利用查询语句解决具体问题。

任务 1
单表无条件查询

5.1.1 任务描述

在学生成绩管理系统中,如果想从学生表中查询出学生的基本信息,从教师表中查询出教师的姓名和所在院系,从课程表中查询出课程名称和课程的学分,该如何操作呢？在MySQL 中,可以使用 SELECT 语句进行查询数据。查询数据是指根据用户需求,使用查询语句从数库的数据源中查询出用户需要的数据。查询操作时数据库使用过程中使用频率最高的操作是数据库的主要使用形式。

5.1.2 知识准备

在 MySQL 环境中,从数据表中查询数据的基本语句是 SELECT,可以根据自己的需求,设置不同的查询条件。使用 SELECT 语句进行查询的语法格式如下：

```
SELECT
{*|〈字段列名〉}
FROM〈表1〉,〈表2〉…
[WHERE〈表达式〉]
[GROUP BY〈group by definition〉]
[HAVING〈expression〉[{〈operator〉〈expression〉}…]]
[ORDER BY〈order by definition〉]
[LIMIT[〈offset〉,]〈row count〉]
```

功能:在指定的数据表中根据不同的条件筛选出满足条件的记录数据。

说明：

① {*|〈字段列名〉}包含星号通配符的字段列表,表示所要查询字段的名称。

② 〈表1〉,〈表2〉…中表 1 和表 2 表示查询数据的来源,可以是单个或多个。

③ WHERE〈表达式〉是可选项,如果选择该项,将限定查询数据必须满足该查询条件。

④ GROUP BY〈字段〉子句告诉 MySQL 如何显示查询出来的数据,并按照指定的字段分组。

⑤ [ORDER BY〈字段〉]子句告诉 MySQL 按什么样的顺序显示查询出来的数据,可以进行的排序有升序(ASC)和降序(DESC),默认情况下是升序。

⑥ [LIMIT[〈offset〉,]〈row count〉]子句告诉 MySQL 每次显示查询出来的数据条数。

本任务以学生成绩管理系统数据库 studentscore 为例,数据库包括学生表、教师表、课程表、成绩表。查询中所用基本表中数据如表 5-1-1～表 5-1-4 所示。

表 5-1-1 学生表（students）

sid	sname	sclass	sgender	smajor	sbirthday	credit_points
1933062301	刘洋	1901	女	软件技术	2000-09-01	12
1933062302	邓嘉荟	1901	女	软件技术	2000-04-02	12
1933062303	葛佳音	1901	女	软件技术	2000-07-09	12
1933062304	索凤洋	1901	女	软件技术	2000-08-04	12
1933062305	曹宏美	1901	女	软件技术	2000-11-07	12
1933062306	郭思宇	1901	女	软件技术	2000-09-23	12
1933062307	周慧敏	1901	女	软件技术	2000-03-21	12

表 5-1-2 教师表（teachers）

tid	tname	tschool
20100004	张三	信息工程学院
19990001	李四	信息工程学院
19980002	赵六	信息工程学院
20050011	王五	信息工程学院
20030008	刘一	信息工程学院
20080009	陈二	信息工程学院
20070003	孙七	信息工程学院

表 5-1-3 课程表（courses）

cid	tid	cname	credit_points
0101	20070003	数据库基础与应用	5
0102	20080009	PHP 程序设计	3
0103	20030008	Java 程序设计	5
0104	20050011	Python 程序设计	3
0105	19980002	计算机基础	3
0106	19990001	android 程序设计	4
0107	20100004	网页制作	4

表 5-1-4 成绩表(scores)

sid	cid	score
1933062301	0101	80
1933062301	0102	90
1933062301	0103	99
1933062302	0101	70
1933062302	0102	60
1933062302	0103	80
1933062303	0101	80
1933062303	0102	80
1933062303	0103	80
1933062304	0101	50
1933062304	0102	50
1933062304	0103	50
1933062305	0101	76
1933062305	0102	87
1933062305	0103	31
1933062306	0101	34
1933062306	0102	88
1933062306	0103	82
1933062307	0101	90
1933062307	0102	77
1933062307	0103	85

使用 INSERT 语句将下面的几条学生记录数据插入学生表中。

```
mysql> INSERT INTO students VALUES(1933062301,"刘洋",1901,"女","软件技术","2000-09-01",12),
    -> (1933062302,"邓嘉蓉",1901,"女","软件技术","2000-04-02",12),
    -> (1933062303,"葛佳音",1901,"女","软件技术","2000-07-09",12),
    -> (1933062304,"索凤洋",1901,"女","软件技术","2000-08-04",12),
    -> (1933062305,"曹宏美",1901,"女","软件技术","2000-11-07",12),
    -> (1933062306,"郭思宇",1901,"女","软件技术","2000-03-21",12),
    -> (1933062307,"周慧敏",1901,"女","软件技术","2000-03-23",12);
Query OK, 7 rows affected (0.09 sec)
Records: 7  Duplicates: 0  Warnings: 0
```

使用 INSERT 语句将下面的几条教师记录数据插入教师表中。

```
mysql> INSERT INTO teachers VALUES(20100004,"张三","信息工程学院"),
    -> (19900001,"李四","信息工程学院"),
    -> (19980002,"赵六","信息工程学院"),
    -> (20050011,"王五","信息工程学院"),
    -> (20030008,"刘一","信息工程学院"),
    -> (20080009,"陈二","信息工程学院"),
    -> (20070003,"孙七","信息工程学院");
Query OK, 7 rows affected (0.00 sec)
Records: 7  Duplicates: 0  Warnings: 0
```

使用 INSERT 语句将下面的几条课程记录数据插入课程表中。

```
mysql> INSERT INTO course VALUES(0103,"java 程序设计",5,20030008),
    -> (0104,"python 程序设计",3,20050011),
    -> (0105,"计算机基础",3,19980002),
    -> (0106,"android 程序设计",4,19900001),
    -> (0107,"网页制作",4,20100004);
Query OK, 5 rows affected (0.00 sec)
Records: 5  Duplicates: 0  Warnings: 0
```

使用 INSERT 语句将下面的几条成绩记录数据插入成绩表中。

```
mysql> INSERT INTO scores VALUES(1933062301,0101,80),
    -> (1933062301,0102,90),
    -> (1933062301,0103,99),
    -> (1933062302,0101,70),
    -> (1933062302,0102,60),
    -> (1933062302,0103,80),
    -> (1933062303,0101,80),
    -> (1933062303,0102,80),
    -> (1933062303,0103,80),
    -> (1933062304,0101,50),
    -> (1933062304,0102,50),
    -> (1933062304,0103,50),
    -> (1933062305,0101,76),
    -> (1933062305,0102,87),
    -> (1933062305,0103,31),
    -> (1933062306,0101,34),
    -> (1933062306,0102,88),
    -> (1933062306,0103,82),
    -> (1933062307,0101,90),
    -> (1933062307,0102,77),
    -> (1933062307,0103,85);
Query OK, 21 rows affected (0.00 sec)
Records: 21  Duplicates: 0  Warnings: 0
```

5.1.3 任务实现

【例 5-1-1】 查询 students 表中全体学生的学号和姓名。

```
mysql> SELECT sid,sname FROM students;
+------------+--------+
| sid        | sname  |
+------------+--------+
| 1933062301 | 刘洋   |
| 1933062302 | 邓嘉蓉 |
| 1933062303 | 葛佳音 |
| 1933062304 | 索凤洋 |
| 1933062305 | 曹宏美 |
| 1933062306 | 郭思宇 |
| 1933062307 | 周慧敏 |
+------------+--------+
7 rows in set (0.00 sec)
```

输出结果显示了 students 表中学号和姓名字段下的所有数据。

【例 5-1-2】 查询全体学生的学号、姓名、班级、学分。

```
mysql> SELECT sid,sname,sclass,credit_points FROM students;
+------------+--------+--------+---------------+
| sid        | sname  | sclass | credit_points |
+------------+--------+--------+---------------+
| 1933062301 | 刘洋   | 1901   | 12            |
| 1933062302 | 邓嘉蓉 | 1901   | 12            |
| 1933062303 | 葛佳音 | 1901   | 12            |
| 1933062304 | 索凤洋 | 1901   | 12            |
| 1933062305 | 曹宏美 | 1901   | 12            |
| 1933062306 | 郭思宇 | 1901   | 12            |
| 1933062307 | 周慧敏 | 1901   | 12            |
+------------+--------+--------+---------------+
7 rows in set (0.00 sec)
```

输出结果显示了 students 表中学号、姓名、班级、学分字段的所有数据。使用 select 子句可以获取多个字段下的数据,只需要在 select 子句后面指定要查找的字段名称,不同字段名称之间用逗号分隔开,最后一个字段后面不需要加逗号。

【例 5-1-3】 查询出选修了课程的学生的学号。

要查询学生的选课情况,可以从学生成绩表中查找学生学号的出现情况。但是一名学生可能选修了多门课程,也就是说同样的学号在表中可能出现多次。在 MySQL 中使用 SELECT 语句执行简单的数据查询语句时,返回所有的匹配结果集。如果表中的某些字段没有唯一性约束,那么在这些字段中就可能出现重复值。这里我们可以使用 DISTINCT 去掉重复值。

DISTINCT 关键字的主要作用是对表中一个或多个字段重复的数据进行过滤,只返回其中的一条数据给用户。

使用 DISTINCT 关键字时需要注意以下几点：

① DISTINCT 关键字只能在 SELECT 语句中使用。

② 在对一个或多个字段去重时，DISTINCT 关键字必须在所有字段的最前面。

③ 如果 DISTINCT 关键字后有多个字段，则会对多个字段进行组合去重，也就是说，只有多个字段组合起来完全是一样的情况下才会被去重。

```
mysql> SELECT DISTINCT sid FROM scores;
+------------+
| sid        |
+------------+
| 1933062301 |
| 1933062302 |
| 1933062303 |
| 1933062304 |
| 1933062305 |
| 1933062306 |
| 1933062307 |
+------------+
7 rows in set (0.12 sec)
```

【例 5-1-4】 查询全体学生的详细信息。

要查询全体学生的详细信息，就要在 SELECT 子句中写出全部的字段名，由于字段名比较多，书写会比较烦琐，这时使用"*"来代替所有字段。

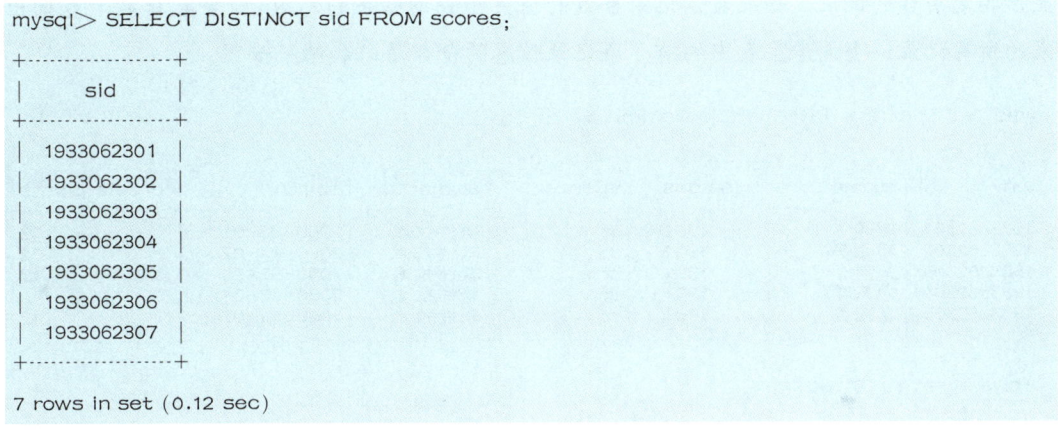

使用"*"查询时，只能按照数据表中字段的顺序进行排列，不能改变字段的排列顺序。我们从查询结果中可以看出，使用"*"通配符后，查询结果返回了所有的列，数据列按照创建表时的顺序显示。使用"*"通配符可以在不知道所需列名称时直接获取它们。使用"*"通配符可以节约查询语句的输入时间，但是也会因返回全表数据而降低查询和使用的应用程序的效率。所以除非需要使用表中所有字段数据，否则一般情况下最好不要使用"*"通配符。

【例 5-1-5】 查询出学生表前 5 条学生记录。

数据库的表中会保存大量的记录，如果一次性地查询出这些数据，会耗费大量的时间，同时也会给数据库服务器造成很大的压力,而用户可能仅仅是需要其中的一部分数据。SQL 语言中使用 LIMIT 关键字指定输出记录范围。LIMIT 是 MySQL 中的一个特殊关键字,用于指定查询结果从哪条记录开始显示,一共显示多少条记录。LIMIT 关键字可以指定初始位置或不指定初始位置。

LIMIT 关键字不指定初始位置是,记录从第一条记录开始显示,显示记录的条数由 LIMIT 关键字指定。如果显示记录的条数小于查询结果的总数,则会从第一条记录开始,显示指定条数的记录。如果显示记录条数的值大于查询结果的总数,则会直接显示查询出来的所有记录。查询学生表中的前 5 条记录信息的命令语句和结果如下:

```
mysql> SELECT * FROM students limit 5;
+------------+--------+--------+---------+----------+------------+---------------+
| sid        | sname  | sclass | sgender | smajor   | sbirth     | credit_points |
+------------+--------+--------+---------+----------+------------+---------------+
| 1933062301 | 刘洋   | 1901   | 女      | 软件技术 | 2000-09-01 |            12 |
| 1933062302 | 邓嘉蓉 | 1901   | 女      | 软件技术 | 2000-04-02 |            12 |
| 1933062303 | 葛佳音 | 1901   | 女      | 软件技术 | 2000-07-09 |            12 |
| 1933062304 | 索凤洋 | 1901   | 女      | 软件技术 | 2000-08-04 |            12 |
| 1933062305 | 曹宏美 | 1901   | 女      | 软件技术 | 2000-11-07 |            12 |
+------------+--------+--------+---------+----------+------------+---------------+
5 rows in set (0.00 sec)
```

结果中只显示了 5 条记录,说明"LIMIT 5"限制了显示条数为 5。

LIMIT 关键字可以指定查询结果从哪条记录开始,显示多少条记录。如查询输出第 3~4 条记录的命令语句和操作结果如下:

```
mysql> SELECT * FROM students limit 2,4;
+------------+--------+--------+---------+----------+------------+---------------+
| sid        | sname  | sclass | sgender | smajor   | sbirth     | credit_points |
+------------+--------+--------+---------+----------+------------+---------------+
| 1933062303 | 葛佳音 | 1901   | 女      | 软件技术 | 2000-07-09 |            12 |
| 1933062304 | 索凤洋 | 1901   | 女      | 软件技术 | 2000-08-04 |            12 |
| 1933062305 | 曹宏美 | 1901   | 女      | 软件技术 | 2000-11-07 |            12 |
| 1933062306 | 郭思宇 | 1901   | 女      | 软件技术 | 2000-03-21 |            12 |
+------------+--------+--------+---------+----------+------------+---------------+
4 rows in set (0.00 sec)
```

由结果可以看出,该语句返回从第 3 条记录开始的 4 条记录。LIMIT 后面第一个数字"2"表示从第 3 行开始,注意记录的位置是从 0 开始,第 3 行的位置为 2,第二个数字表示返回的行数。

【例 5-1-6】 查询出所有同学的姓名和年龄。

在查询语句中不仅可以查询出数据库表中已有的信息,还可以通过对表中数据进行处理查询出表中数据隐含的信息。在本例中学生表中并不存在学生的年龄信息,但是表中的出生日期字段存储了学生的出生日期。我们可以通过当前年份减去出生年份得出学生的当前年龄。其中我们可以使用 CURDATE() 函数返回系统当前日期,使用 YEAR() 函数返回指定日期的年份。命令语句和操作结果如下:

```
mysql> SELECT sname,YEAR(CURDATE())-YEAR(sbirthday) FROM students;
+--------+---------------------------------+
| sname  | YEAR(CURDATE())-YEAR(sbirthday) |
+--------+---------------------------------+
| 刘洋    |                              21 |
| 邓嘉蓉  |                              21 |
| 葛佳音  |                              21 |
| 索凤洋  |                              21 |
| 曹宏美  |                              21 |
| 郭思宇  |                              21 |
| 周慧敏  |                              21 |
+--------+---------------------------------+
7 rows in set (0.00 sec)
```

注意:SQL只能进行一些简单的基础处理,而且SQL的处理时间会明显降低,大家在进行实际项目时要注意不要在查询语句中进行过于复杂的运算。

在上面的查询结果集中,第二个字段的字段名并不是表中现有的字段名,也不能清晰地表达查询结果的含义,这时我们可以给这个字段起一个别名。

【例5-1-7】 查询全体学生姓名、年龄,并为"sname"字段指定别名"姓名",年龄字段指定别名"年龄"。

在使用 SELECT 语句查询数据时,MySQL 会显示每个 SELECT 后面指定输出的字段。为了能更加直观地显示结果,我们可以为字段指定一个别名。

为字段指定别名的语法规则为:

〈字段名〉[AS]〈字数别名〉

其中,各子句的语法含义如下:

① 〈字段名〉为数据表中字段定义时的名称。

② 〈字段别名〉为字段新的名称。

③ AS 关键字可以省略,省略后需要将字段名和别名用空格隔开。

```
mysql> SELECT sname 姓名,YEAR(CURDATE())-YEAR(sbirthday) 年龄 FROM students;
+--------+------+
| 姓名   | 年龄 |
+--------+------+
| 刘洋    |   21 |
| 邓嘉蓉  |   21 |
| 葛佳音  |   21 |
| 索凤洋  |   21 |
| 曹宏美  |   21 |
| 郭思宇  |   21 |
| 周慧敏  |   21 |
+--------+------+
7 rows in set (0.00 sec)
```

5.1.4 训练任务

查询是图书管理系统的基础性操作,图书信息查询、图书借阅查询、图书归还查询、读者信息查询等都是图书管理系统运行的主要工作。请完成以下查询操作。

① 查询图书表中图书编号和作者姓名。
② 查询借阅表所有信息。
③ 查询借阅了图书的同学的姓名。
④ 查询图书表前五条记录信息。
⑤ 查询借阅表第 2~10 条借阅信息。
⑥ 查询图书表图书编号和价格,并为价格字段指定别名价格。
⑦ 查询借阅表每条借阅记录的借阅天数。
⑧ 查询图书表中图书编号和图书名称。
⑨ 查询读者表的读者姓名和电话。
⑩ 查询借阅表中读者编号和借阅时间。

5.1.5 任务总结

在本任务中,我们学习了单表查询的语法格式,了解了 SELECT 语句的基本概念和特点。查询是 SQL 语言的核心,用于表达 SQL 查询的 SELECT 查询命令是功能最强也是最为复杂的 SQL 语句,它的作用就是从数据库中检索数据,并将查询结果返回给用户。SELECT 语句由 SELECT 子句(查询内容)、FROM 子句(查询对象)、WHERE 子句(查询条件)、ORDER BY 子句(排序方式)、GROUP BY 子句(分组方式)等组成。我们会在以后的章节中使用查询语句的其他功能完成相应的任务。

任务 2
单表有条件查询

5.2.1 任务描述

在 MySQL 环境中,使用 SELECT 语句来查询数据表中的数据,通过设置查询的字段名表对数据表中的数据进行列向筛选,这属于投影操作。一般情况下,数据库中会包含大量的数据,需要根据一定的条件来获取相应的数据,或者对查询数据重新进行组合排列,这就需要按照指定的条件来查询数据,这属于选择操作,只要将 WHERE 子句添加到查询语句中就可以完成。

5.2.2 知识准备

条件查询的基本语法规则为：

SELECT〈列名称〉FROM〈数据表〉WHERE〈列〉〈运算符〉〈值〉；

功能：在指定的数据表中筛选出满足条件的记录，显示列名称指定的列的值。

说明：条件查询中使用 WHERE 语句来设定查询条件。在 WHERE 子句中可以指定任何条件，或者可以使用 AND 或者 OR 指定一个或多个条件。WHERE 子句也可以使用在 SQL 其他语句中，比如 DELETE 或者 UPDATE 语句中。WHERE 子句类似于程序语言中的 IF 条件语句。从数据库中查询指定条件的数据，WHERE 子句是非常有用的。如果数据库中没有符合 WHERE 子句指定条件的记录，那么查询不会返回任何数据。

WHERE 子句指定的条件查询的方法可以分为下面几种情况。

1. 带关系运算符的查询

SELECT 字段名1,字段名2,…
FROM〈数据表〉
WHERE〈条件表达式〉

关系运算
知识拓展

功能：在指定表中查找满足条件的记录。

WHERE 子句的条件表达式一般是由比较运算符组成的，常见的比较运算符如表 5-2-1 所示。

表 5-2-1　MySQL 中常见比较运算符

操作符	描述
=	等于
<>	不等于
>	大于
<	小于
<=	大于等于
<=	小于等于
BETWEEN	在某个范围内
LIKE	搜索某种模式

2. 带逻辑运算符的查询

在实际项目查询中很多时候查询需求是由多个查询条件组成的，这时就需要用逻辑运算符进行连接。WHERE 子句中可以使用 AND、OR、NOT 这三个逻辑运算符对条件表达式进行连接。这样能够使查询结果更加精确。语法格式如下：

```
SELECT 字段名 1,字段名 2,…
FROM〈数据表〉
WHERE[NOT]〈条件表达式 1〉AND(OR、XOR)〈条件表达式 2〉
```

说明：

① AND 记录满足所有查询条件时，才会被查询出来。

② OR 记录满足任意一个查询条件时，才会被查询出来。

③ XOR 记录满足其中一个条件，并且不满足另一个条件时，才会被查询出来。

查询条件越多，符合查询条件的记录就会越少。这是由于设置的条件越多，查询语句的限制就越多，能够满足所有条件的记录就更少，因此如何用 WHERE 子句描述能够查询出符合要求的记录就更为重要。

3. 带 BETWEEN AND 的查询

如果查询的条件是在一个范围内，也可以使用 BETWEEN 关键字。在 WHERE 子句中使用 BETWEEN 关键字可以查找指定范围内的数据。BETWEEN 需要两个参数，即范围的起始点和终止值。如果字段值在指定范围内，则返回这些记录。如果字段值不在指定范围内，则不会返回这些记录。

使用 BETWEEN 的语法格式如下：

```
SELECT 字段名 1,字段名 2,…
FROM〈数据表〉
WHERE [NOT] BETWEEN 值 1 AND 值 2;
```

说明：

① NOT 为可选参数，表示指定范围之外的值。如果字段值不满足指定范围内的值，则这些记录被返回。

② 值 1 表示范围的起始值。

③ 值 2 表示范围的终止值。

4. 带 LIKE 的查询

一些情况下的查询条件是模糊的，并不能使用条件表达式明确地表示出来。这时我们用 LIKE 关键字进行模糊查询。使用 LIKE 关键字进行模糊匹配主要应用在字符串的模糊查找上。在模糊匹配时需要一些占位符来作为通配符，这样可以组合成模糊查询需要的条件表达式。在 MySQL 中主要有"％"和"_"两个通配符。通配符是一种特殊的语句，主要用来模糊查询。当需要表达满足某种格式的字符串时，可以使用通配符来代替一个或多个真正的字符。其中"％"代表任意多个字符。"_"代表单个字符。

LIKE 关键字主要用于搜索匹配字段中的指定内容。语法格式如下：

```
SELECT 字段名 1,字段名 2,…
FROM〈数据表〉
WHERE〈字段名〉[NOT] LIKE '字符串'
```

说明：

① NOT 为可选参数，字段中的内容与指定的字符串不匹配时满足条件。

② 字符串是指定用来匹配的字符串。"字符串"可以是一个很完整的字符串，也可以是包含通配符的字符串。"％"和"_"是 MySQL 中最常用的通配符，"％"代表任何长度的字符串，字符串的长度可以为 0；"_"代表一个字符。

5. 带 REGEXP 的查询

正则表达式通常用来检索或替换符合某个模式的文本内容，根据指定的匹配模式匹配文本中符合要求的特殊字符串。例如从一个文本文件中提取电话号码，查找一篇文章中重复的单词或者替换用户输入的某些词语等。正则表达式强大而且灵活，可以应用于非常复杂的查询。

MySQL 中使用 REGEXP 关键字指定正则表达式的字符匹配模式。语法格式如下：

```
WHERE 字段名 REGEXP '操作符'
```

说明：

① 这里只是正则匹配所有包含指定字符的数据。

② 如果是英文字符串，要注意大小写的正则匹配，like 是不区分大小写的，但是 REGEXP 区分，需要指定大小写。

表 5-2-2 是 REGEXP 操作符中常用的字符匹配选项。

表 5-2-2 REGEXP 操作符中常用的字符匹配选项

选项	说明	示例
^	匹配文本的开始字符	^b：匹配以字母 b 为开头的字符串，如 book、big、banana
$	匹配文本的结束字符	st$：匹配以 st 结尾的字符串，如 test、resist、persist
.	匹配任何单个字符	b.t：匹配任何 b 和 t 之间有一个字符，如 bit、bat、but、bite
*	匹配零个或多个在它前面的字符	*n：匹配字符 n 前面有任意个字符，如 fn、ann、faan、abcdn
+	匹配前面的字符 1 次或多次	ba+：匹配以 b 开头后面紧跟至少有一个 a，如 ba、bay、bare、battle
〈字符串〉	匹配包含指定的字符串的文本	fa：字符串至少要包含 fa，如 fan、afa、faad
[字符集合]	匹配字符集合中的任何一个字符	[xz]：匹配 x 或 z，如 dizzy、zebra、x-ray、extra
[^]	匹配不在括号中的任何字符	[^abc]：匹配任何不包含 a、b 或 c 的字符串
字符串{n,}	匹配单面的字符串至少 n 次	b{2,}：匹配 2 个或更多的 b，如 bb、bbbbb、bbbbbbb
字符串{m,n}	匹配前面的字符串至少 m 次，至多 n 次。如果 n 为 0，m 为可选参数	b{2,4}：匹配至少 2 个 b，最多 4 个 b，如 bb、bbbb、bbb

6. 带 IN 的查询

如果查询的条件是判断是否在一个列表内，可以使用 IN 关键字组成条件表达式。语法格式如下：

```
SELECT 字段名 1,字段名 2,…
FROM〈数据表〉
WHERE〈字段名〉[NOT]IN  (值 1,值 2,…)
```

说明：

① IN 关键字用于判断某个字段的值是否在指定的集合中，如果在集合中则满足条件，把该字段查询出来。

② 值 1,值 2,…表示集合中的元素，即指定的条件范围 NOT,可选参数,表示查询不在 IN 关键字指定集合,范围中的记录。

7. 空值查询

如果查询条件是判断记录的特定字段内容为空或不为空,可以使用 IS NULL 和 IS NOT NULL 来判断。空值是没有确定的类型和值,它不同于 0,也不等同于空字符串。如果字段的值是空值,则满足查询条件,如果字段的值不是空值,则不满足查询条件。语法格式如下：

```
SELECT 字段名 1,字段名 2,…
FROM〈数据表〉
WHERE〈字段名〉IS [NOT] NULL；
```

说明：

① NOT 为可选参数,用于判断字段不是空值。

② IS NULL 来判断字段的值是否为空值。

5.2.3 任务实现

【例 5-2-1】 查询学生表中所有女生的信息。

```
mysql> SELECT * FROM students WHERE sgender='女';
+------------+--------+--------+---------+----------+------------+---------------+
|    sid     | sname  | sclass | sgender |  smajor  |   sbirth   | credit_points |
+------------+--------+--------+---------+----------+------------+---------------+
| 1933062301 | 刘洋   |  1901  |   女    | 软件技术 | 2000-09-01 |      12       |
| 1933062302 | 邓嘉蓉 |  1901  |   女    | 软件技术 | 2000-04-02 |      12       |
| 1933062303 | 葛佳音 |  1901  |   女    | 软件技术 | 2000-07-09 |      12       |
| 1933062304 | 索凤洋 |  1901  |   女    | 软件技术 | 2000-08-04 |      12       |
| 1933062305 | 曹宏美 |  1901  |   女    | 软件技术 | 2000-11-07 |      12       |
| 1933062306 | 郭思宇 |  1901  |   女    | 软件技术 | 2000-03-21 |      12       |
| 1933062307 | 周慧敏 |  1901  |   女    | 软件技术 | 2000-03-23 |      12       |
+------------+--------+--------+---------+----------+------------+---------------+
7 rows in set (0.00 sec)
```

可以看到查询结果中包含了所有女生的信息。如果根据指定的条件进行查询，数据表中没有符合查询条件的记录，系统会提示"Empty set(0.00sec)"。

【例5-2-2】 查询成绩表中所有成绩大于85分的学生的学号和成绩。

```
mysql> SELECT sid AS '学号',score as '成绩' FROM scores WHERE score>85;
+------------+--------+
| 学号       | 成绩   |
+------------+--------+
| 1933062301 | 90.00  |
| 1933062301 | 99.00  |
| 1933062305 | 87.00  |
| 1933062306 | 88.00  |
| 1933062307 | 90.00  |
+------------+--------+
5 rows in set (0.00 sec)
```

可以看到，查询结果中所有记录的成绩都大于85分，而小于或等于85分的记录没有被返回。

【例5-2-3】 查询学生有中所有女生的学号、姓名、专业以及出生日期。

```
mysql> SELECT sid ,sname,smajor,sbirthday FROM students WHERE sgender='女';
+------------+--------+----------+------------+
| sid        | sname  | smajor   | sbirthday  |
+------------+--------+----------+------------+
| 1933062301 | 刘洋   | 软件技术 | 2000-09-01 |
| 1933062302 | 邓嘉荟 | 软件技术 | 2000-04-02 |
| 1933062303 | 葛佳音 | 软件技术 | 2000-07-09 |
| 1933062304 | 索凤洋 | 软件技术 | 2000-08-04 |
| 1933062305 | 曹宏美 | 软件技术 | 2000-11-07 |
| 1933062306 | 郭思宇 | 软件技术 | 2000-03-21 |
| 1933062307 | 周慧敏 | 软件技术 | 2000-03-23 |
+------------+--------+----------+------------+
7 rows in set (0.00 sec)
```

【例5-2-4】 查询信息工程学院所有教师名单。

```
mysql> SELECT tname FROM teachers WHERE tschool="信息工程学院";
+--------+
| tname  |
+--------+
| 张胜利 |
| 李云芳 |
| 赵大志 |
| 王亮   |
| 刘铁男 |
| 陈成功 |
| 孙月悦 |
+--------+
7 rows in set (0.00 sec)
```

【例 5-2-5】 查询考试成绩不及格学生的学号。

```
mysql> SELECT sid FROM scores WHERE score<60;
+------------+
| sid        |
+------------+
| 1933062304 |
| 1933062304 |
| 1933062304 |
| 1933062305 |
| 1933062306 |
+------------+
5 rows in set (0.00 sec)
```

在查询结果中我们发现出现重复的学生学学号,这说明该学生有多门成绩不及格。我们可以使用 DISTINCT 关键字去掉重复值,这时查询的结果就是有不及格成绩的学生学号,操作命令作运行结果如下:

```
mysql> SELECT DISTINCT sid FROM scores WHERE score<60;
+------------+
| sid        |
+------------+
| 1933062304 |
| 1933062305 |
| 1933062306 |
+------------+
3 rows in set (0.00 sec)
```

注意:使用非数字类型的数据时需要用单引号括起来。

【例 5-2-6】 查询出软件技术专业女生的信息。

```
mysql> SELECT * FROM students WHERE smajor='软件技术' and sgender='女';
+------------+-------+--------+---------+----------+------------+---------------+
| sid        | sname | sclass | sgender | smajor   | sbirth     | credit_points |
+------------+-------+--------+---------+----------+------------+---------------+
| 1933062301 | 刘洋  | 1901   | 女      | 软件技术 | 2000-09-01 |            12 |
| 1933062302 | 邓嘉蓉| 1901   | 女      | 软件技术 | 2000-04-02 |            12 |
| 1933062303 | 葛佳音| 1901   | 女      | 软件技术 | 2000-07-09 |            12 |
| 1933062304 | 索凤洋| 1901   | 女      | 软件技术 | 2000-08-04 |            12 |
| 1933062305 | 曹宏美| 1901   | 女      | 软件技术 | 2000-11-07 |            12 |
| 1933062306 | 郭思宇| 1901   | 女      | 软件技术 | 2000-03-21 |            12 |
| 1933062307 | 周慧敏| 1901   | 女      | 软件技术 | 2000-03-23 |            12 |
+------------+-------+--------+---------+----------+------------+---------------+
7 rows in set (0.01 sec)
```

从查询结果中可以看出,查询结果集中所有记录都是软件技术专业且性别都为女。

【例 5-2-7】 查询出成绩在 90 分以上或不及格的学生的学号、姓名、课程号、成绩。

```
mysql> SELECT sid,cid,score FROM scores WHERE score>90 OR score<60;
+------------+-----+-------+
| sid        | cid | score |
+------------+-----+-------+
| 1933062301 | 103 | 99.00 |
| 1933062304 | 101 | 50.00 |
| 1933062304 | 102 | 50.00 |
| 1933062304 | 103 | 50.00 |
| 1933062305 | 103 | 31.00 |
| 1933062306 | 101 | 34.00 |
+------------+-----+-------+
6 rows in set (0.00 sec)
```

从查询出的结果集中可以看出学生成绩或者大于 90 分或者小于 60 分。

【例 5-2-8】 查询非 1901 班的学生信息。

```
mysql> SELECT * FROM students WHERE sclass<>1901;
Empty set (0.00 sec)
```

【例 5-2-9】 查询成绩在 60~80 分的学生学号和成绩。

```
mysql> SELECT sid,score FROM scores WHERE score BETWEEN 60 AND 80;
+------------+-------+
| sid        | score |
+------------+-------+
| 1933062301 | 80.00 |
| 1933062302 | 70.00 |
| 1933062302 | 60.00 |
| 1933062302 | 80.00 |
| 1933062303 | 80.00 |
| 1933062303 | 80.00 |
| 1933062303 | 80.00 |
| 1933062305 | 76.00 |
| 1933062307 | 77.00 |
+------------+-------+
9 rows in set (0.00 sec)
```

从查询结果集中可以看出成绩为 60 分和 80 分的记录,这说明 BETWEEN 能够匹配指定范围内的所有值,包括起始值和终止值。

【例 5-2-10】 查询所有刘姓学生的个人信息。

```
mysql> SELECT * FROM students WHERE sname LIKE '刘%';
+------------+--------+--------+---------+----------+------------+---------------+
| sid        | sname  | sclass | sgender | smajor   | sbirth     | credit_points |
+------------+--------+--------+---------+----------+------------+---------------+
| 1933062301 | 刘洋   | 1901   | 女      | 软件技术 | 2000-09-01 |            12 |
+------------+--------+--------+---------+----------+------------+---------------+
1 row in set (0.01 sec)
```

从查询结果集中可以看到查询结果中返回了所有以汉字"刘"开头的学生姓名。

【例 5-2-11】 查询名字中第二个字是"佳"的学生的信息。

```
mysql> SELECT * FROM students WHERE sname LIKE '_佳%';
+------------+--------+--------+---------+----------+------------+---------------+
| sid        | sname  | sclass | sgender | smajor   | sbirth     | credit_points |
+------------+--------+--------+---------+----------+------------+---------------+
| 1933062303 | 葛佳音 | 1901   | 女      | 软件技术 | 2000-07-09 |            12 |
+------------+--------+--------+---------+----------+------------+---------------+
1 row in set (0.17 sec)
```

【例 5-2-12】 查询 stud 表中出现"嘉"字的学生信息。

```
mysql> select * from stud1 where sname regexp '嘉';
+------------+--------+--------+---------+----------+------------+---------------+
| sid        | sname  | sclass | sgender | smajor   | sbirth     | credit_points |
+------------+--------+--------+---------+----------+------------+---------------+
| 1933062302 | 邓嘉蓉 | 1901   | 女      | 软件技术 | 2000-03-24 |             3 |
+------------+--------+--------+---------+----------+------------+---------------+
1 row in set (0.17 sec)
```

【例 5-2-13】 查询所有专业是计算机应用、软件技术、动漫制作的学生的姓名和性别。

```
mysql> SELECT sname,sgender FROM students WHERE smajor IN('计算机应用','软件技术',
'动漫制作');
+--------+---------+
| sname  | sgender |
+--------+---------+
| 刘洋   | 女      |
| 邓嘉蓉 | 女      |
| 葛佳音 | 女      |
| 索凤洋 | 女      |
| 曹宏美 | 女      |
| 郭思宇 | 女      |
| 周慧敏 | 女      |
+--------+---------+
7 rows in set (0.00 sec)
```

【例 5-2-14】查询在成绩表中没有成绩的学生的学号和相应的课程号。

```
mysql> SELECT sid,cid FROM scores WHERE score IS NULL;
Empty set (0.00 sec)
```

【例 5-2-15】查询所有有成绩的学生的学号和课程号。

```
mysql> SELECT sid,cid FROM scores WHERE score IS NOT NULL;
+------------+-----+
| sid        | cid |
+------------+-----+
| 1933062301 | 101 |
| 1933062301 | 102 |
| 1933062301 | 103 |
| 1933062302 | 101 |
| 1933062302 | 102 |
| 1933062302 | 103 |
| 1933062303 | 101 |
| 1933062303 | 102 |
| 1933062303 | 103 |
| 1933062304 | 101 |
| 1933062304 | 102 |
| 1933062304 | 103 |
| 1933062305 | 101 |
| 1933062305 | 102 |
| 1933062305 | 103 |
| 1933062306 | 101 |
| 1933062306 | 102 |
| 1933062306 | 103 |
| 1933062307 | 101 |
| 1933062307 | 102 |
| 1933062307 | 103 |
+------------+-----+
21 rows in set (0.00 sec)
```

注意：IS NULL 是一个整体，不能将 IS 换成"＝"，否则将不能查询出任何结果，数据库系统会出现"Empty set(0.00 sec)"这样的提示。

5.2.4 训练任务

在图书管理系统的很多功能需求是根据条件查询出特定信息。图书管理系统数据库中一般包含大量信息，实际使用中很少会一次全部查询，都会使用 WHERE 子句加过滤条件查询我们需要的数据。请完成下列查询。

① 查询出书名为数据结构的书籍信息。
② 查询出单价大于 45 元的书籍编号和书名。
③ 查询出出版社为人民邮电出版社的图书。
④ 查询出清华大学出版社出版的作者为谭浩强的图书。
⑤ 查询出书名以 JAVA 开始的图书信息。
⑥ 查询出没有单价的图书信息。
⑦ 查询出出版社为人民邮电出版社、清华大学出版社的图书。
⑧ 查询出已被借出还未被归还的图书信息。
⑨ 查询可以借出的书籍的信息。
⑩ 查询 2020-10-10 前归还的书籍编号。

5.2.5　任务总结

在本任务中我们主要学习了 WHERE 子句的作用和使用方法，了解了条件表达式的使用，强化了我们对条件查询用于过滤数据的认识，查询结果中只包含满足条件的记录。在实际操作中，WHERE 子句多用关系运算符和逻辑运算符构造查询条件，以满足查询的需要。

任务 3
聚合函数的使用

5.3.1　任务描述

我们已经知道可以通过在查询中执行存储数据的计算来获得想要的查询结果。这是由于在数据库存储的数据中隐藏一些有价值的信息，需要对综合信息进行统计后才能得出。一般需要对一组值进行计算并返回单个值，要在查询中执行这样的计算，就要使用聚合函数了。聚合函数是分组查询语法中重要的一部分。实际工作中出现需要汇总数据而不用把它们实际查询出来的情况，由此 MySQL 提供了专门的函数。聚合函数又称为组函数，作用是将当前所在数据表当作一个组进行统计。

5.3.2　知识准备

在实际项目开发过程中，我们经常需要对某些数据进行统计计算，例如要统计出某门课程平均分、某单位员工的最高工资、最低工资等数据。MySQL 提供了一些函数来帮助我们实现这些功能。如表 5-3-1 所示。

表 5-3-1　Mysql 中聚合函数表

序号	函数名称	作用
1	COUNT()	返回数据表中某列的行数,省略参数时,返回表的总行数
2	SUM()	返回数据表中数值型列值的和
3	AVG()	返回数据表中数值型列值的平均值
4	MAX()	返回数据表中数值型列值的最大值
5	MIN()	返回数据表中数值型列值的最小值

聚合函数的特点是：

① 每个聚合函数接收一个参数（字段名或者表达式），在统计结果中默认忽略字段为 NULL 的记录。

② 如果想要把列值为 NULL 的行也参与到聚合函数的计算中，需要使用 IFNULL 函数对 NULL 值进行转换。

③ 聚合函数不能够嵌套使用。

MySQL 提供了许多聚合函数，包括 AVG、COUNT、SUM、MIN、MAX 等。除 COUNT 函数外，其他聚合函数在执行计算时会忽略 NULL 值。

1. COUNT( )函数

COUNT()函数在查询语句 SELECT 中起到计数的作用。语法格式如下：

```
SELECT COUNT(*) FROM〈数据表名〉;
```

功能：返回数据表中记录的个数。

说明：

① COUNT(列名)，返回 SELECT 语句检索的行中指定字段的非 NULL 值的总数。如果没有匹配的行，COUNT(列名)返回 0，即空值的行不计入统计结果。

② COUNT(*)，返回数据表中的总行数，无论它们是否包含 NULL 值，无论某列有数值或者为空值。

2. AVG( )函数

AVG()函数通过计算返回的行数和每一行数据的和来计算指定列数据的平均值。语法格式如下：

```
SELECT DISTINCT AVG(*) FROM〈数据表名〉;
```

功能：返回数据表中数值型参数列的平均值。

说明：

① 命令中的列名必须是数据型的字段名。

② DISTINCT 用于返回字段的不同值的平均值。如果没有匹配的行，AVG()返回

NULL。

③ AVG()函数忽略列值为 NULL 的行。

3. SUM()函数

SUM()函数计算参数列数据的总和。语法格式如下：

```
SELECT DISTINCT SUM(*) FROM 〈数据表名〉;
```

功能：返回数据表中数值型参数列的平均值。

说明：

① 命令中的列名必须是数据型的字段名。

② DISTINCT 用于返回字段的不同值的总和。如果没有匹配的行，SUM()返回 NULL，而不是 0。

③ SUM()函数忽略列值为 NULL 的行。

4. MAX()函数

MAX()函数可以对字母大小进行判断，并返回最大的字符或者字符串值。语法格式如下：

查找极值

```
SELECT MAX(*) FROM 〈数据表名〉;
```

功能：返回数据表中参数列的最大值。

说明：

① MAX()函数在对字符类型的数据进行比较时，按照字符的 ASCII 码值大小进行比较。在比较时，先比较第一个字符，如果相等，继续比较下一个字符，一直到两个字符不相等或者字符结束为止。

② MAX()函数忽略列值为 NULL 的行。

③ MAX()函数不仅适用于查询数值类型，也可以应用于字符类型。

5. MIN()函数

MIN()函数可以对字母大小进行判断，并返回最小的字符或者字符串值。语法格式如下：

```
SELECT MIN(*) FROM 〈数据表名〉;
```

功能：返回数据表中参数列的最小值。

说明：

① MIN()函数在对字符类型的数据进行比较时，按照字符的 ASCII 码值大小进行比较。在比较时，先比较第一个字符，如果相等，继续比较下一个字符，一直到两个字符不相等或者字符结束为止。

② MIN()函数忽略列值为 NULL 的行。

③ MIN()函数不仅适用于查询数值类型,也可以应用于字符类型。

5.3.3 任务实现

【例 5-3-1】 查询学生表中的学生总数。

```
mysql> SELECT COUNT(*) FROM students;
+----------+
| COUNT(*) |
+----------+
|        7 |
+----------+
1 row in set (0.00 sec)
```

如果是下面的数据表:

```
mysql> select * from students;
+-----+-------+--------+---------+--------+--------+---------------+
| sid | sname | sclass | sgender | smajor | sbirth | credit_points |
+-----+-------+--------+---------+--------+--------+---------------+
|   1 | 李丽  | NULL   | NULL    | NULL   | NULL   |          NULL |
+-----+-------+--------+---------+--------+--------+---------------+
1 row in set (0.00 sec)
```

那么在使用 COUNT(*)和 COUNT(class)、COUNT(sname)命令时,结果会有不同。

```
mysql> select count(*) from students;
+----------+
| count(*) |
+----------+
|        1 |
+----------+
1 row in set (0.00 sec)

mysql> select count(sclass) from students;
+---------------+
| count(sclass) |
+---------------+
|             0 |
+---------------+
1 row in set (0.00 sec)

mysql> select count(sname) from students;
+--------------+
| count(sname) |
+--------------+
|            1 |
+--------------+
1 row in set (0.00 sec)
```

通过查询结果可以看出,无论字段的值是什么,COUNT(*)都能返回students表中记录的总行数。

在统计计算时可以使用DISTINCT剔除字段值重复的条数。函数返回不相同且非NULL的字段值的行数。如果没有匹配的行,则COUNT(DISTINCT)返回0。

【例5-3-2】 查询选修了课程的学生人数。

```
mysql> SELECT COUNT(DISTINCT sid) FROM scores;
+---------------------+
| COUNT(DISTINCT sid) |
+---------------------+
|                   7 |
+---------------------+
1 row in set (0.00 sec)
```

【例5-3-3】 计算出0101号课程学生的平均成绩。

```
mysql> SELECT AVG(score) FROM scores WHERE cid=0101;
+------------+
| AVG(score) |
+------------+
|  68.571429 |
+------------+
1 row in set (0.00 sec)
```

【例5-3-4】 计算出0101号课程学生的总成绩。

```
mysql> SELECT SUM(score) FROM scores WHERE cid=0101;
+------------+
| sum(score) |
+------------+
|     480.00 |
+------------+
1 row in set (0.05 sec)
```

【例5-3-5】 查询出学生表中的sname字段的最大值。

```
mysql> SELECT MAX(sname) FROM students;
+------------+
| MAX(sname) |
+------------+
| 郭思宇     |
+------------+
1 row in set (0.00 sec)
```

【例5-3-6】 查询选修了0101号课程的学生的最高分和最低分。

```
mysql> SELECT MAX(score) AS 最高分,MIN(score) AS 最低分 FROM scores WHERE cid=0101;
+--------+--------+
| 最高分 | 最低分 |
+--------+--------+
|  90.00 |  34.00 |
+--------+--------+
1 row in set (0.00 sec)
```

从运行的结果可以看出 MAX( )函数查询出了某门课的最高分,MIN( )函数查询出了某门课程的最低分。

【例 5-3-7】 查询学号为 1933062301 的学生的总成绩和平均成绩。

```
mysql> SELECT SUM(score) AS 总成绩,AVG(score) AS 平均成绩 FROM scores WHERE sid=1933062301;
+--------+-----------+
| 总成绩 | 平均成绩  |
+--------+-----------+
| 269.00 | 89.666667 |
+--------+-----------+
1 row in set (0.00 sec)
```

由查询结果可以看出 SUM( )函数返回学生的所有成绩之和。

5.3.4 训练任务

图书管理系统应该具有完善的查询统计功能,从而提高图书管理人员工作效率,减轻图书管理员以往繁忙的工作。请查询统计出以下信息。

① 查询图书表中的图书总数。
② 查询图书最高单价图书信息。
③ 查询图书最低单价图书信息。
④ 查询图书平均单价。
⑤ 查询图书馆图书价格总额。
⑥ 查询图书被借阅次数。
⑦ 查询办理图书证人数。
⑧ 查询图书馆读者平均年龄。
⑨ 查询图书平均被借阅时间。
⑩ 统计图书出版社数量。

5.3.5 任务总结

通过本任务我们了解了常见聚合函数的使用方法和作用。SELECT 语句及其表达式都可以使用这些聚合函数。聚合函数可以帮助用户更加方便地处理表中的数据,使 MySQL 数据库的功能更加强大。

任务 4

分 组 与 排 序

5.4.1 任务描述

如果要对查询的结果集按数据表中的某个字段内容进行分组归类统计,就需要用到 GROUP BY 子句。GROUP BY 子句根据一个或多个字段对查询结果集进行分组。注意 GROUP BY 分组查询的只能是分组字段或者是聚合函数。分组的目的是将结果集中的数据行根据选择列的值进行逻辑分组,以便能汇总表内容的子集,实现对每个组而不是对整个结果集进行整合。如果希望查询结果能够按照某列的顺序输出,就要使用 ORDER BY 子句来完成这个任务。

5.4.2 知识准备

1. 分组

分组查询的语法格式如下:

```
GROUP BY {〈列名〉|〈表达式〉|〈位置〉} [ASC | DESC]
```

功能:将查询结果按指定列或表达式的值进行升序(降序)排序输出。

说明:

①〈列名〉:指定用于分组的列。可以指定多个列,彼此间用逗号分隔。

②〈表达式〉:指定用于分组的表达式。通常与聚合函数一块使用,例如可将表达式 COUNT(*)AS'人数'作为 SELECT 选择列表清单的一项。

③〈位置〉:指定用于分组的选择列在 SELECT 语句结果集中的位置,通常是一个正整数。例如,GROUP BY 2 表示根据 SELECT 语句列清单上的第 2 列的值进行逻辑分组。

④[ASC|DESC]:关键字 ASC 表示按升序分组,关键字 DESC 表示按降序分组,其中 ASC 为默认值,注意这两个关键字必须位于对应的列名、表达式、列的位置之后。

注意:GROUP BY 子句中的各选择列必须也是 SELECT 语句的选择列清单中的一项。

对于 GROUP BY 子句的使用,需要注意以下几点。

① GROUP BY 子句可以包含任意数目的列,使其可以对分组进行嵌套,为数据分组提供更加细致的控制。

② GROUP BY 子句列出的每个列都必须是检索列或有效的表达式,但不能是聚合函数。若在 SELECT 语句中使用表达式,则必须在 GROUP BY 子句中指定相同的表达式。

③ 除聚合函数之外,SELECT 语句中的每个列都必须在 GROUP BY 子句中给出。

④ 若用于分组的列中包含有 NULL 值,则 NULL 将作为一个单独的分组返回;若该列中存在多个 NULL 值,则将这些 NULL 值所在的行分为一组。

2. 排序

数据库查询操作中,通过条件查询语句可以查询到符合用户需求的数据,但是查询到的数据一般都是按照数据最初被添加到表中的顺序来显示。如果使查询结果的顺序满足使用的要求,查询出的结果集进行排序可以使用 ORDER BY 子句。ORDER BY 子句可以把查询结果按照指定字段进行排序。指定排序的字段可以是一个或多个,排序规则可以是升序或者降序。

在 SQL SELECT 语句中使用 ORDER BY 子句查询的语法格式如下:

```
SELECT field1,field2,…,fieldN FROM table_name1,table_name2,…
ORDER BY field1 [ASC [DESC][默认 ASC]], [field2…] [ASC|DESC]
```

功能:将查询结果按照指定列的列值有序输出。

说明:

① 字段名指定用于排序的列。可以指定多个列,列名之间用逗号分隔。在字段名后可以指定用于排序的表达式。可以指定用于排序的列在 SELECT 语句结果集中的位置,通常是一个正整数。

② 关键字 ASC 表示按升序分组,关键字 DESC 表示按降序分组,其中 ASC 为默认值。这两个关键字必须位于对应的列名、表达式、列的位置之后。

对于 ORDER BY 子句的使用,需要注意以下几点。

① ORDER BY 子句中可以包含子查询。

② 当排序的值中存在空值时,ORDER BY 子句会将该空值作为最小值来对待。

③ 可以使用任何字段来作为排序的条件,从而返回排序后的查询结果。可以设定多个字段来排序。当在 ORDER BY 子句中指定多个列进行排序时,MySQL 会按照列的顺序从左到右依次进行排序。

④ 查询的数据并没有以一种特定的顺序显示,如果没有对它们进行排序,则将根据插入到数据表中的顺序显示。使用 ORDER BY 子句对指定的列数据进行排序。

5.4.3 任务实现

【例 5-4-1】 统计各班学生人数。

```
mysql> SELECT sclass ,COUNT(*) AS 各班人数 FROM students GROUP BY sclass;
+--------+----------+
| sclass | 各班人数 |
+--------+----------+
|  1901  |    7     |
+--------+----------+
1 row in set (0.00 sec)
```

【例 5-4-2】 统计学校男生、女生人数。

```
mysql> SELECT sgender ,COUNT( * ) AS 人数 FROM students GROUP BY sgender;
+---------+------+
| sgender | 人数 |
+---------+------+
| 女      |    7 |
+---------+------+
1 row in set (0.00 sec)
```

【例 5-4-3】 统计各班男生、女生人数。

```
mysql> SELECT sclass,sgender,COUNT( * ) FROM students GROUP BY sclass,sgender;
+--------+---------+----------+
| sclass | sgender | COUNT(*) |
+--------+---------+----------+
| 1901   | 女      |        7 |
+--------+---------+----------+
1 row in set (0.00 sec)
```

【例 5-4-4】 统计各专业女生人数。

```
mysql> SELECT smajor,COUNT( * ) AS 各专业女生人数
    -> FROM students WHERE sgender='女'GROUP BY sgender='女';
+----------+----------------+
| smajor   | 各专业女生人数 |
+----------+----------------+
| 软件技术 |              7 |
+----------+----------------+
1 row in set (0.00 sec)
```

【例 5-4-5】 查询选修了 2 门以上课程的学生学号。

```
mysql> SELECT sid FROM scores GROUP BY sid HAVING COUNT( * )>2;
+------------+
| sid        |
+------------+
| 1933062301 |
| 1933062302 |
| 1933062303 |
| 1933062304 |
| 1933062305 |
| 1933062306 |
| 1933062307 |
+------------+
7 rows in set (0.00 sec)
```

注意：HAVING 关键字是过滤条件，而 WHERE 子句是查询条件。也就是说两个子句的作用对象不同。工作顺序是按照 WHERE 子句的条件从数据源中查询出结果集后再使用 HAVING 子句指定的过滤条件过滤出符合条件的结果集。

【例 5-4-6】查询 0101 号课程的学生学号和成绩，查询结果按分数的降序排列。

```
mysql> SELECT sid,score FROM scores WHERE cid=0101 ORDER BY score DESC;
+------------+-------+
| sid        | score |
+------------+-------+
| 1933062307 | 90.00 |
| 1933062301 | 80.00 |
| 1933062303 | 80.00 |
| 1933062305 | 76.00 |
| 1933062302 | 70.00 |
| 1933062304 | 50.00 |
| 1933062306 | 34.00 |
+------------+-------+
7 rows in set (0.00 sec)
```

由查询结果可以看出，MySQL 对查询的结果按 score 字段的数据按数值大小进行了降序排序。

升序用（ASC）关键字，降序用（DESC）关键字。默认值为升序，即若列名后有 DESC，则以该字段值降序排序，否则以该字段值升序排序。

【例 5-4-7】查询全体学生情况，查询结果首先按班级升序排列，同班学生按出生日期降序排列。

```
mysql> SELECT * FROM students ORDER BY sclass ASC,sbirth desc;
+------------+--------+--------+---------+----------+------------+---------------+
| sid        | sname  | sclass | sgender | smajor   | sbirth     | credit_points |
+------------+--------+--------+---------+----------+------------+---------------+
| 1933062305 | 曹宏美 | 1901   | 女      | 软件技术 | 2000-11-07 | 12            |
| 1933062301 | 刘洋   | 1901   | 女      | 软件技术 | 2000-09-01 | 12            |
| 1933062304 | 索凤洋 | 1901   | 女      | 软件技术 | 2000-08-04 | 12            |
| 1933062303 | 葛佳音 | 1901   | 女      | 软件技术 | 2000-07-09 | 12            |
| 1933062302 | 邓嘉蓉 | 1901   | 女      | 软件技术 | 2000-04-02 | 12            |
| 1933062307 | 周慧敏 | 1901   | 女      | 软件技术 | 2000-03-23 | 12            |
| 1933062306 | 郭思宇 | 1901   | 女      | 软件技术 | 2000-03-21 | 12            |
+------------+--------+--------+---------+----------+------------+---------------+
7 rows in set (0.02 sec)
```

这里涉及日期型数值排序，日期型数值按年、月、日的数值大小排序。其他的非数值型数据排序规则为：字符型按照 ASCII 编码大小进行排序，逻辑型数据 false 小于 true，记录中的 NULL，升序时排在最前，降序时排在最后。

5.4.4 训练任务

在图书管理系统使用过程中有时需要分组查询或者对查询结果进行排序输出，方便管理人员了解图书馆的运行状况。请完成下列查询。

① 查询图书表中各个出版社的图书数目。

② 查询办理借书证的男生女生人数。

③ 查询图书信息按单价降序排序。

④ 查询作者为谭浩强的书籍,按书名升序排序。

⑤ 查询未还书籍按借出时间降序排序。

⑥ 图书表按出版时间升序排序。

⑦ 读者表按年龄升序排序。

⑧ 查询借阅表中每个人借书的数目。

⑨ 查询图书表中每个作者出版图书数目。

⑩ 按编号排序输出管理员信息。

5.4.5 任务总结

通过本任务的学习,我们了解了分组和排序的作用,分组其实就是按照某一列进行分类分组。GROUP BY 关键词可以根据一个或多个字段对查询结果进行分组。单独使用 GROUP BY 关键字时,查询结果会只显示每个分组的第一条记录。排序是对查询结果集按某一列进行排序,以便查询结果有顺序。

任务 5
多表连接查询

5.5.1 任务描述

在设计数据库表的时候,使用实体关系模式,确定了一个表中只包含一个实体的属性。在关系型数据库中,表与表之间是有联系的,在实际项目开发过程中,查询的数据可能涉及多个实体,也就是查询的数据可能会分布在多个表中,经常使用多表查询。这样查询数据的数据源可能由多个表中的数据构成。多表查询就是同时查询两个或两个以上的表。例如在学生成绩管理系统中学生的信息和成绩信息分别存放在学生表和成绩表中,而课程的信息存放在课程表中。这时就需要使用多表连接查询来完成查询任务。

5.5.2 知识准备

多表连接查询是数据查询操作中经常使用的查询方式。在 MySQL 中多表连接包括交叉连接(Cross Join)、内连接(Inner Join)、自连接(Self Join)、外连接(Outer Join)。其中外连接还分为左外连接(Left Join)、右外连接(Right Join)和全外连接(Full Join)。

1. 交叉连接

交叉连接(CROSS JOIN)一般用来返回连接表的笛卡尔积。是把第一个表的每行依次与第二个表的每一行进行连接。得到的查询结果集的行数是两个表行数的乘积。

交叉连接的语法格式如下：

```
SELECT〈字段名〉FROM〈表1〉CROSS JOIN〈表2〉[WHERE 子句]
```

功能：将两个表进行笛卡儿积连接后，再查询数据表中的数据。

说明：

① 字段名为需要查询的字段名称。

② 〈表1〉〈表2〉为需要交叉连接的表名。

③ WHERE 子句为用来设置交叉连接的查询条件。

④ 交叉连接可以查询两个或两个以上的表。多个表交叉连接时，在 FROM 后连续使用 CROSS JOIN 或","即可。以上两种语法的返回结果是相同的。

2. 内连接

内连接(INNER JOIN)主要通过设置连接条件的方式，来移除查询结果集中某些数据行的交叉连接。内连接使用 INNER JOIN 关键字连接两张表，并使用 ON 子句来设置连接条件。如果没有连接条件，INNER JOIN 和 CROSS JOIN 在语法上是等同的，内连接是按照一定的条件把一个表中的行与另一个表中的行进行匹配。

内连接的语法格式如下：

```
SELECT〈字段名〉 FROM〈表1〉INNER JOIN〈表2〉ON join_condition1 INNER JOIN〈表3〉ON join_condition2 … WHERE where_conditions;
```

功能：将两个表按一定的条件进行连接后，再查询数据表中的满足条件的数据。

说明：

① 字段名为需要查询的字段名称。

② 〈表1〉〈表2〉为需要内连接的表名。

③ ON 子句用来设置内连接的连接条件。

④ 如果要连接多个表，可以通连两两连接的方式来完成。

对于表 1 中的每一行，INNER JOIN 子句将它与表 2 的每一行进行比较，以检查它们是否都满足连接条件。当满足连接条件时，INNER JOIN 将返回由表 1 和表 2 中的列组成的新行。注意，表 1 和表 2 中的行必须根据连接条件进行匹配。如果找不到匹配项，查询将返回一个空结果集。当连接超过 2 个表时，也应用此逻辑。INNER JOIN 也可以使用 WHERE 子句指定连接条件，但是 INNER JOIN ... ON 语法是官方标准语法，而且 WHERE 子句在某些时候会影响查询性能。

多表查询之自连接

3. 自连接

多表连接查询是不同表之间行与行之间进行连接。当需要同一个表的不同行之间进行连接时,也可以同一个表进行自身的连接操作。将表中行与同一表中的其他行组合时,叫作自连接。要执行自连接操作,必须使用表别名来帮助 MySQL 在单个查询中区分连接的是逻辑上的两个表。自连接是以多表连接的方式获得同一张表中数据的复杂关系表示或者关系处理。自连接的语法格式如下:

```
SELECT〈字段名〉 FROM〈表1〉INNER JOIN〈表1〉ON join_condition1 …
WHERE where_conditions;
```

功能:将同一个表按一定的条件进行连接后,再查询数据表中的满足条件的数据。

说明:

① 字段名为需要查询的字段名称。

②〈表1〉为需要自连接的表名。

③ ON 子句用来设置内连接的连接条件。

4. 外连接

内连接的查询结果都是符合连接条件的记录,而外连接会先将连接的表分为基表和参考表,再以基表为依据返回满足和不满足条件的记录。外连接可以是左向外连接、右向外连接或完整外部连接。

在 FROM 子句中指定外连接时,可以由下列几组关键字中的一组指定。

1) 左连接(LEFT JOIN 或 LEFT OUTER JOIN)

左外连接使用 LEFT OUTER JOIN 关键字连接两个表,并使用 ON 子句来设置连接条件。左向外连接的结果集包括 LEFT OUTER 子句中指定的左表的所有行,而不仅仅是连接列所匹配的行。如果左表的某行在右表中没有匹配行,则在相关联的结果集行中右表的所有选择列表列均为空值。

左连接的语法格式如下:

```
SELECT〈字段名〉FROM〈表1〉LEFT OUTER JOIN〈表2〉〈ON 子句〉
```

说明:

① 字段名为需要查询的字段名称。

②〈表1〉〈表2〉为需要左连接的表名。

③ LEFT OUTER JOIN 为左连接关键字,在左连接中可以省略 OUTER,直接使用 LEFT JOIN。ON 子句为用来设置左连接的连接条件,不能省略。

2) 右连接(RIGHT JOIN 或 RIGHT OUTER JOIN)

右向外连接是左向外连接的反向连接,将返回右表的所有行。如果右表的某行在左表中没有匹配行,则将为左表返回空值。

右连接的语法格式如下:

```
SELECT〈字段名〉FROM〈表1〉RIGHT OUTER JOIN〈表2〉〈ON 子句〉
```

说明：

① 字段名为需要查询的字段名称。

② 〈表1〉〈表2〉需要右连接的表名。

③ RIGHT OUTER JOIN 为右连接关键字，其中 OUTER 可以省略，只使用关键字 RIGHT JOIN。

④ ON 子句为用来设置右连接的连接条件，不能省略。

3) 全连接(FULL JOIN 或 FULL OUTER JOIN)

完整外部连接返回左表和右表中的所有行。当某行在另一个表中没有匹配行时，则另一个表的选择列表列包含空值。如果表之间有匹配行，则整个结果集行包含基表的数据值。

全连接的语法格式如下：

```
SELECT 〈字段名〉FROM〈表1〉FULL JOIN〈表2〉〈ON子句〉
```

说明：

① 字段名为需要查询的字段名称。

② 〈表1〉〈表2〉为需要全连接的表名。

③ FULL OUTER JOIN 为右连接关键字，其中 OUTER 可以省略，只使用关键字 FULL JOIN。

④ ON 子句为用来设置右连接的连接条件，不能省略。

5.5.3 任务实现

【例 5-5-1】 教师表和课程表进行交叉连接。

为了方便观察成绩表和课程表交叉连接后的运行结果，我们先分别查询出这两个表的数据，再进行交叉连接查询。

① 查询教师表中的数据。

```
mysql> SELECT * FROM teachers;
+----------+-------+--------------+
| tid      | tname | tschool      |
+----------+-------+--------------+
| 20100004 | 张三  | 信息工程学院 |
| 19900001 | 李四  | 信息工程学院 |
| 19980002 | 赵六  | 信息工程学院 |
| 20050011 | 王五  | 信息工程学院 |
| 20030008 | 刘一  | 信息工程学院 |
| 20080009 | 陈二  | 信息工程学院 |
| 20070003 | 孙七  | 信息工程学院 |
+----------+-------+--------------+
7 rows in set (0.00 sec)
```

② 查询课程表中的数据。

```
mysql> SELECT * FROM course;
+-----+------------------+--------------+----------+
| cid | cname            | credit_point | tid      |
+-----+------------------+--------------+----------+
| 103 | java 程序设计    |            5 | 20030008 |
| 104 | python 程序设计  |            3 | 20050011 |
| 105 | 计算机基础       |            3 | 19980002 |
| 106 | android 程序设计 |            4 | 19900001 |
| 107 | 网页制作         |            4 | 20100004 |
+-----+------------------+--------------+----------+
5 rows in set (0.10 sec)
```

③ 使用 CROSS JOIN 查询出两张表中的笛卡尔积。

```
mysql> SELECT * FROM teachers CROSS JOIN course;
+----------+-------+--------------+-----+------------------+--------------+----------+
| tid      | tname | tschool      | cid | cname            | credit_point | tid      |
+----------+-------+--------------+-----+------------------+--------------+----------+
| 20100004 | 张三  | 信息工程学院 | 103 | java 程序设计    |            5 | 20030008 |
| 20100004 | 张三  | 信息工程学院 | 104 | python 程序设计  |            3 | 20050011 |
| 20100004 | 张三  | 信息工程学院 | 105 | 计算机基础       |            3 | 19980002 |
| 20100004 | 张三  | 信息工程学院 | 106 | android 程序设计 |            4 | 19900001 |
| 20100004 | 张三  | 信息工程学院 | 107 | 网页制作         |            4 | 20100004 |
| 19900001 | 李四  | 信息工程学院 | 103 | java 程序设计    |            5 | 20030008 |
| 19900001 | 李四  | 信息工程学院 | 104 | python 程序设计  |            3 | 20050011 |
| 19900001 | 李四  | 信息工程学院 | 105 | 计算机基础       |            3 | 19980002 |
| 19900001 | 李四  | 信息工程学院 | 106 | android 程序设计 |            4 | 19900001 |
| 19900001 | 李四  | 信息工程学院 | 107 | 网页制作         |            4 | 20100004 |
| 19980002 | 赵六  | 信息工程学院 | 103 | java 程序设计    |            5 | 20030008 |
| 19980002 | 赵六  | 信息工程学院 | 104 | python 程序设计  |            3 | 20050011 |
| 19980002 | 赵六  | 信息工程学院 | 105 | 计算机基础       |            3 | 19980002 |
| 19980002 | 赵六  | 信息工程学院 | 106 | android 程序设计 |            4 | 19900001 |
| 19980002 | 赵六  | 信息工程学院 | 107 | 网页制作         |            4 | 20100004 |
| 20050011 | 王五  | 信息工程学院 | 103 | java 程序设计    |            5 | 20030008 |
| 20050011 | 王五  | 信息工程学院 | 104 | python 程序设计  |            3 | 20050011 |
| 20050011 | 王五  | 信息工程学院 | 105 | 计算机基础       |            3 | 19980002 |
| 20050011 | 王五  | 信息工程学院 | 106 | android 程序设计 |            4 | 19900001 |
| 20050011 | 王五  | 信息工程学院 | 107 | 网页制作         |            4 | 20100004 |
| 20030008 | 刘一  | 信息工程学院 | 103 | java 程序设计    |            5 | 20030008 |
| 20030008 | 刘一  | 信息工程学院 | 104 | python 程序设计  |            3 | 20050011 |
| 20030008 | 刘一  | 信息工程学院 | 105 | 计算机基础       |            3 | 19980002 |
| 20030008 | 刘一  | 信息工程学院 | 106 | android 程序设计 |            4 | 19900001 |
| 20030008 | 刘一  | 信息工程学院 | 107 | 网页制作         |            4 | 20100004 |
| 20080009 | 陈二  | 信息工程学院 | 103 | java 程序设计    |            5 | 20030008 |
| 20080009 | 陈二  | 信息工程学院 | 104 | python 程序设计  |            3 | 20050011 |
| 20080009 | 陈二  | 信息工程学院 | 105 | 计算机基础       |            3 | 19980002 |
| 20080009 | 陈二  | 信息工程学院 | 106 | android 程序设计 |            4 | 19900001 |
| 20080009 | 陈二  | 信息工程学院 | 107 | 网页制作         |            4 | 20100004 |
| 20070003 | 孙七  | 信息工程学院 | 103 | java 程序设计    |            5 | 20030008 |
| 20070003 | 孙七  | 信息工程学院 | 104 | python 程序设计  |            3 | 20050011 |
| 20070003 | 孙七  | 信息工程学院 | 105 | 计算机基础       |            3 | 19980002 |
| 20070003 | 孙七  | 信息工程学院 | 106 | android 程序设计 |            4 | 19900001 |
| 20070003 | 孙七  | 信息工程学院 | 107 | 网页制作         |            4 | 20100004 |
+----------+-------+--------------+-----+------------------+--------------+----------+
35 rows in set (0.01 sec)
```

由运行结果可以看出，teachers 表和 courses 表交叉连接查询后，返回了 49 条记录。如果表中的数据较多，得到的运行结果会非常多，交叉连接会非常慢，而且得到的运行结果也没有多大意义。所以，通过交叉连接的方式进行多表查询一般并不常用，在实际项目中应该尽量避免交叉查询的使用。一般情况下不建议使用交叉连接。

【例 5-5-2】 查询每个学生及其选课的情况。

学生的基本信息存放在学生表中,学生的选课信息存放在成绩表中,观察两个表的结构,学生表和成绩表同时拥有 sid 字段。我们可以通过 sid 字段内容是否相等对两个表中的记录进行匹配。

```
mysql> SELECT * FROM students,scores WHERE students.sid=scores.sid;
+------------+--------+--------+---------+-----------+------------+---------------+------------+-----+-------+
| sid        | sname  | sclass | sgender | smajor    | sbirth     | credit_points | sid        | cid | score |
+------------+--------+--------+---------+-----------+------------+---------------+------------+-----+-------+
| 1933062301 | 刘洋   | 1901   | 女      | 软件技术  | 2000-09-01 | 12            | 1933062301 | 101 | 80.00 |
| 1933062301 | 刘洋   | 1901   | 女      | 软件技术  | 2000-09-01 | 12            | 1933062301 | 102 | 90.00 |
| 1933062301 | 刘洋   | 1901   | 女      | 软件技术  | 2000-09-01 | 12            | 1933062301 | 103 | 99.00 |
| 1933062302 | 邓嘉蓉 | 1901   | 女      | 软件技术  | 2000-04-02 | 12            | 1933062302 | 101 | 70.00 |
| 1933062302 | 邓嘉蓉 | 1901   | 女      | 软件技术  | 2000-04-02 | 12            | 1933062302 | 102 | 60.00 |
| 1933062302 | 邓嘉蓉 | 1901   | 女      | 软件技术  | 2000-04-02 | 12            | 1933062302 | 103 | 80.00 |
| 1933062303 | 葛佳音 | 1901   | 女      | 软件技术  | 2000-07-09 | 12            | 1933062303 | 101 | 80.00 |
| 1933062303 | 葛佳音 | 1901   | 女      | 软件技术  | 2000-07-09 | 12            | 1933062303 | 102 | 80.00 |
| 1933062303 | 葛佳音 | 1901   | 女      | 软件技术  | 2000-07-09 | 12            | 1933062303 | 103 | 80.00 |
| 1933062304 | 索凤洋 | 1901   | 女      | 软件技术  | 2000-08-04 | 12            | 1933062304 | 101 | 50.00 |
| 1933062304 | 索凤洋 | 1901   | 女      | 软件技术  | 2000-08-04 | 12            | 1933062304 | 102 | 50.00 |
| 1933062304 | 索凤洋 | 1901   | 女      | 软件技术  | 2000-08-04 | 12            | 1933062304 | 103 | 50.00 |
| 1933062305 | 曹宏美 | 1901   | 女      | 软件技术  | 2000-11-07 | 12            | 1933062305 | 101 | 76.00 |
| 1933062305 | 曹宏美 | 1901   | 女      | 软件技术  | 2000-11-07 | 12            | 1933062305 | 102 | 87.00 |
| 1933062305 | 曹宏美 | 1901   | 女      | 软件技术  | 2000-11-07 | 12            | 1933062305 | 103 | 31.00 |
| 1933062306 | 郭思宇 | 1901   | 女      | 软件技术  | 2000-03-21 | 12            | 1933062306 | 101 | 34.00 |
| 1933062306 | 郭思宇 | 1901   | 女      | 软件技术  | 2000-03-21 | 12            | 1933062306 | 102 | 88.00 |
| 1933062306 | 郭思宇 | 1901   | 女      | 软件技术  | 2000-03-21 | 12            | 1933062306 | 103 | 82.00 |
| 1933062307 | 周慧敏 | 1901   | 女      | 软件技术  | 2000-03-23 | 12            | 1933062307 | 101 | 90.00 |
| 1933062307 | 周慧敏 | 1901   | 女      | 软件技术  | 2000-03-23 | 12            | 1933062307 | 102 | 77.00 |
| 1933062307 | 周慧敏 | 1901   | 女      | 软件技术  | 2000-03-23 | 12            | 1933062307 | 103 | 85.00 |
+------------+--------+--------+---------+-----------+------------+---------------+------------+-----+-------+
21 rows in set (0.01 sec)
```

【例 5-5-3】 查询所有女生的学号、姓名、课程号及成绩。

```
mysql> SELECT students.sid,sname,cid,score FROM students,scores WHERE
students.sid=scores.sid AND sgender='女';
+------------+--------+-----+-------+
| sid        | sname  | cid | score |
+------------+--------+-----+-------+
| 1933062301 | 刘洋   | 101 | 80.00 |
| 1933062301 | 刘洋   | 102 | 90.00 |
| 1933062301 | 刘洋   | 103 | 99.00 |
| 1933062302 | 邓嘉蓉 | 101 | 70.00 |
| 1933062302 | 邓嘉蓉 | 102 | 60.00 |
| 1933062302 | 邓嘉蓉 | 103 | 80.00 |
| 1933062303 | 葛佳音 | 101 | 80.00 |
| 1933062303 | 葛佳音 | 102 | 80.00 |
| 1933062303 | 葛佳音 | 103 | 80.00 |
| 1933062304 | 索凤洋 | 101 | 50.00 |
| 1933062304 | 索凤洋 | 102 | 50.00 |
| 1933062304 | 索凤洋 | 103 | 50.00 |
| 1933062305 | 曹宏美 | 101 | 76.00 |
| 1933062305 | 曹宏美 | 102 | 87.00 |
| 1933062305 | 曹宏美 | 103 | 31.00 |
| 1933062306 | 郭思宇 | 101 | 34.00 |
| 1933062306 | 郭思宇 | 102 | 88.00 |
| 1933062306 | 郭思宇 | 103 | 82.00 |
| 1933062307 | 周慧敏 | 101 | 90.00 |
| 1933062307 | 周慧敏 | 102 | 77.00 |
| 1933062307 | 周慧敏 | 103 | 85.00 |
+------------+--------+-----+-------+
21 rows in set (0.11 sec)
```

【例 5-5-4】 输出软件技术专业同学的学号、姓名、课程名和成绩。

```
mysql> SELECT students.sid,sname,cname,score
    -> FROM students,scores,course
    -> WHERE students.sid=scores.sid AND scores.cid=course.cid
    -> AND smajor="软件技术";
+------------+--------+----------------+-------+
| sid        | sname  | cname          | score |
+------------+--------+----------------+-------+
| 1933062301 | 刘洋   | java 程序设计  | 99.00 |
| 1933062302 | 邓嘉蓉 | java 程序设计  | 80.00 |
| 1933062303 | 葛佳音 | java 程序设计  | 80.00 |
| 1933062304 | 索凤洋 | java 程序设计  | 50.00 |
| 1933062305 | 曹宏美 | java 程序设计  | 31.00 |
| 1933062306 | 郭思宇 | java 程序设计  | 82.00 |
| 1933062307 | 周慧敏 | java 程序设计  | 85.00 |
+------------+--------+----------------+-------+
7 rows in set (0.00 sec)
```

【例 5-5-5】 查询同时选修了课程 101 和 103 的学生的学号。

```
mysql> select distinct s1.sid from students s1 join
    ->     scores s2 on s1.sid=s2.sid join scores s3 on
    ->     s2.sid=s3.sid and s2.cid=101 and s3.cid=103;
+------------+
| sid        |
+------------+
| 1933062301 |
| 1933062302 |
| 1933062303 |
| 1933062304 |
| 1933062305 |
| 1933062306 |
| 1933062307 |
+------------+
7 rows in set (0.00 sec)
```

【例 5-5-6】 查询与刘洋在同一班的学生的学号、姓名和专业。

```
mysql> SELECT B.sid,B.sname,B.smajor
    -> FROM students A,students B
    -> WHERE A.sclass=B.sclass AND A.sname='刘洋'AND B.sname! ='刘洋';
```

```
+------------+--------+-----------+
| sid        | sname  | smajor    |
+------------+--------+-----------+
| 1933062302 | 邓嘉蓉 | 软件技术  |
| 1933062303 | 葛佳音 | 软件技术  |
| 1933062304 | 索凤洋 | 软件技术  |
| 1933062305 | 曹宏美 | 软件技术  |
| 1933062306 | 郭思宇 | 软件技术  |
| 1933062307 | 周慧敏 | 软件技术  |
+------------+--------+-----------+
6 rows in set (0.00 sec)
```

【例 5-5-7】 利用左外连接查询学生选修课程情况。

在左连接之前我们先查看学生表和成绩表中的数据。

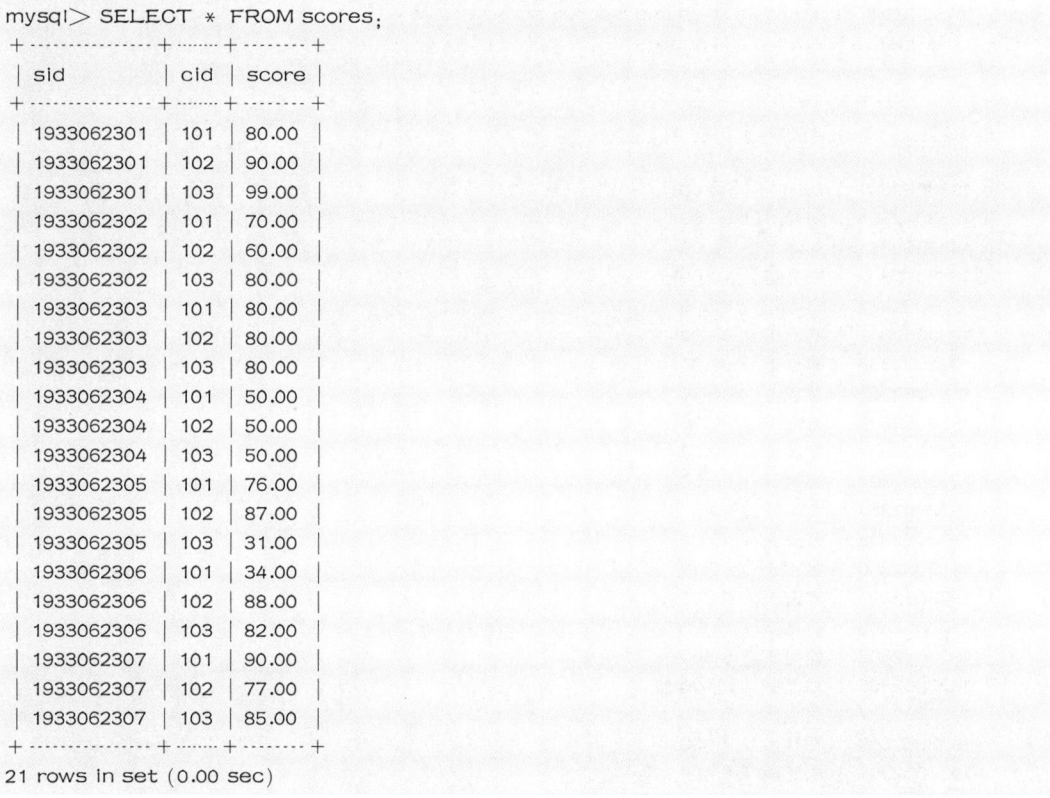

【例 5-5-8】 在学生表和成绩表中查询所有学生信息和对应的课程编号和成绩。

```
mysql> SELECT students.sid,sname,sgender,sbirth,sclass,cid,score
    -> FROM students LEFT JOIN scores
    -> ON students.sid=scores.sid;
+------------+--------+---------+------------+--------+-----+-------+
| sid        | sname  | sgender | sbirth     | sclass | cid | score |
+------------+--------+---------+------------+--------+-----+-------+
| 1933062301 | 刘洋   | 女      | 2000-09-01 | 1901   | 101 | 80.00 |
| 1933062301 | 刘洋   | 女      | 2000-09-01 | 1901   | 102 | 90.00 |
| 1933062301 | 刘洋   | 女      | 2000-09-01 | 1901   | 103 | 99.00 |
| 1933062302 | 邓嘉蓉 | 女      | 2000-04-02 | 1901   | 101 | 70.00 |
| 1933062302 | 邓嘉蓉 | 女      | 2000-04-02 | 1901   | 102 | 60.00 |
| 1933062302 | 邓嘉蓉 | 女      | 2000-04-02 | 1901   | 103 | 80.00 |
| 1933062303 | 葛佳音 | 女      | 2000-07-09 | 1901   | 101 | 80.00 |
| 1933062303 | 葛佳音 | 女      | 2000-07-09 | 1901   | 102 | 80.00 |
| 1933062303 | 葛佳音 | 女      | 2000-07-09 | 1901   | 103 | 80.00 |
| 1933062304 | 索凤洋 | 女      | 2000-08-04 | 1901   | 101 | 50.00 |
| 1933062304 | 索凤洋 | 女      | 2000-08-04 | 1901   | 102 | 50.00 |
| 1933062304 | 索凤洋 | 女      | 2000-08-04 | 1901   | 103 | 50.00 |
| 1933062305 | 曹宏美 | 女      | 2000-11-07 | 1901   | 101 | 76.00 |
| 1933062305 | 曹宏美 | 女      | 2000-11-07 | 1901   | 102 | 87.00 |
| 1933062305 | 曹宏美 | 女      | 2000-11-07 | 1901   | 103 | 31.00 |
| 1933062306 | 郭思宇 | 女      | 2000-03-21 | 1901   | 101 | 34.00 |
| 1933062306 | 郭思宇 | 女      | 2000-03-21 | 1901   | 102 | 88.00 |
| 1933062306 | 郭思宇 | 女      | 2000-03-21 | 1901   | 103 | 82.00 |
| 1933062307 | 周慧敏 | 女      | 2000-03-23 | 1901   | 101 | 90.00 |
| 1933062307 | 周慧敏 | 女      | 2000-03-23 | 1901   | 102 | 77.00 |
| 1933062307 | 周慧敏 | 女      | 2000-03-23 | 1901   | 103 | 85.00 |
+------------+--------+---------+------------+--------+-----+-------+
21 rows in set (0.01 sec)
```

【例5-5-9】 利用右外连接查询学生课程成绩情况。

```
mysql> SELECT students.sid,sname,scores.score FROM
    -> students RIGHT JOIN scores
    -> ON students.sid=scores.sid;
+------------+--------+-------+
| sid        | sname  | score |
+------------+--------+-------+
| 1933062301 | 刘洋   | 80.00 |
| 1933062301 | 刘洋   | 90.00 |
| 1933062301 | 刘洋   | 99.00 |
| 1933062302 | 邓嘉蓉 | 70.00 |
| 1933062302 | 邓嘉蓉 | 60.00 |
| 1933062302 | 邓嘉蓉 | 80.00 |
| 1933062303 | 葛佳音 | 80.00 |
| 1933062303 | 葛佳音 | 80.00 |
| 1933062303 | 葛佳音 | 80.00 |
| 1933062304 | 索凤洋 | 50.00 |
| 1933062304 | 索凤洋 | 50.00 |
| 1933062304 | 索凤洋 | 50.00 |
| 1933062305 | 曹宏美 | 76.00 |
| 1933062305 | 曹宏美 | 87.00 |
| 1933062305 | 曹宏美 | 31.00 |
| 1933062306 | 郭思宇 | 34.00 |
| 1933062306 | 郭思宇 | 88.00 |
| 1933062306 | 郭思宇 | 82.00 |
| 1933062307 | 周慧敏 | 90.00 |
| 1933062307 | 周慧敏 | 77.00 |
| 1933062307 | 周慧敏 | 85.00 |
+------------+--------+-------+
21 rows in set (0.00 sec)
```

5.5.4　训练任务

如果要查询的数据分布在多张表中，每张表取一列或多列数据，就需要多表连接查询。多表连接分为内连接和外连接。下面请使用多表连接查询完成下面的任务。

① 查询读者表和借阅表的所有组合。
② 查询借过书的读者编号 497932808，借书的图书编号、借书日期和还书日期。
③ 查询借过《时间简史》的读者信息。
④ 查询图书名称相同，但作者不同的图书信息。
⑤ 查询所有读者的借阅信息。
⑥ 查询所有图书和读者的借阅信息。
⑦ 查询男生的借阅信息。
⑧ 查询 2021-04-12 借书的学生信息。
⑨ 查询 2021-04-12 借出的图书信息。
⑩ 查询 2020-11-02 借出图书总价值。

5.5.5　任务总结

通过本任务的学习，我们了解了多表查询的作用和使用方法。内连接是用左边表的记录去匹配右边表的记录，如果符合条件则显示。左外连接是查询的是左表的所有数据以及其交集部分，如果有符合条件的则显示具体数据，否则显示 NULL。右连接是查询右表的所有数据以及交集的部分，如果有符合条件的则显示具体数据，否则显示 NULL。在实际应用中大部分情况下，查询语句都会涉及多张表格，所以多表查询任务的学习有很大的实际意义。

任务 6

子　查　询

5.6.1　任务描述

子查询是指一个查询语句嵌套在另一个查询语句内部的查询，也就是查询语句中还包含另一个查询语句。我们把外层查询语句称为父查询或者主查询，内层查询语句称为子查询或者从查询。子查询嵌套在父查询的 WHERE 子句或者 HAVING 子句处。子查询是由标准的 SELECT 子句，包含一个或多个数据源的 FROM 子句以及可选的 WHERE 子句、可选的 GROUP BY 子句、可选的 HAVING 子句组成的。子查询需要用圆括号括起来，并且不能使用 ORDER BY 排序。当子查询的返回值只有一个时，可以使用比较运算符如＝、＜、＞、＞＝、＜＝、！＝等将父查询和子查询连接起来。

5.6.2 知识准备

子查询的作用是可以先对查询的数据源进行过滤和创建计算字段。在 WHERE 子句中对于嵌套子查询的层数是没有限制的,但是嵌套查询层数太多会影响性能。并且列必须匹配,也就是说在 WHERE 子句中使用子查询,应该保证 SELECT 语句具有与 WHERE 子句相同数目的列。一般情况下,子查询返回单个列与单个列匹配,但如果需要也可以使用多个列。涉及外部查询的子查询当列名可能有多义性时必须重新命名。

因为查询运行的方式不同,子查询可分为嵌套子查询和相关子查询两类。

1. 嵌套子查询

嵌套子查询简称嵌套查询,最常见的使用是在 WHERE 子句的 IN 操作符中,以及用来填充计算列,也称为嵌套查询,建立子查询最可靠的方法是逐渐进行,这与 MySQL 处理它们的方法非常相似。首先,建立和测试最内层的查询。然后建立和测试外层查询,并且仅在确认它正常后才嵌入子查询。这时,再次对其进行测试。对于要增加的每个查询,重复这些步骤。

嵌套查询的求解方法是由里向外处理的,即每个子查询在其上一级查询处理之前求解,子查询的结果用于建立其父查询的查找条件。

嵌套查询可以使一系列简单查询构成复杂的查询,从而明显地增强了 SQL 的查询能力。以层层嵌套的方式来构造程序正是 SQL(Structured Query Language)中"结构化"的含义所在。

1) 带 IN 关键字的子查询

结合关键字 IN 所使用的子查询主要用于判断一个给定值是否存在于子查询的结果集中。语法格式如下:

[NOT] IN

说明:

① 用于指定表达式。当表达式与子查询返回的结果集中的某个值相等时,返回 TRUE,否则返回 FALSE;若使用关键字 NOT,则返回的值正好相反。

② 用于指定子查询。这里的子查询只能返回一列数据。对于比较复杂的查询要求,可以使用 SELECT 语句实现子查询的多层嵌套。

2) 带 ANY、ALL 关键字的子查询

带有 ANY(SOME)或 ALL 操作符的子查询。ANY 和 SOME 关键字是同义词。ANY 和 ALL 操作符在使用时必须和比较运算符一起使用。

语法格式:

〈字段〉〈比较符〉[ANY|ALL]〈子查询〉

说明：ANY 表示满足其中任意一个条件，它允许创建一个表达式，对子查询的返回值列表进行比较，只要满足内层子查询中的任意一个比较条件，就返回一个结果作为外层查询条件；ALL 则表示满足其中所有的条件。

ANY 关键词可以理解为"对于子查询返回的列中的任一数值，如果比较结果为 TRUE，则返回 TRUE；ALL 关键词可以理解为"对于子查询返回的列中的所有数值，如果比较结果为 TRUE，则返回 TRUE。

表 5-6-1 ANY 和 ALL 的用法和具体含义

用法	含义
>ANY	大于子查询结果中的某个值
>ALL	大于子查询结果中的所有值
<ANY	小于子查询结果中的某个值
<ALL	小于子查询结果中的所有值
>=ANY	大于等于子查询结果中的某个值
>=ALL	大于等于子查询结果中的所有值
<=ANY	小于等于子查询结果中的某个值
<=ALL	小于等于子查询结果中的所有值
=ANY	等于子查询结果中的某个值
=ALL	等于子查询结果中的所有值（通常无实际意义）
!=ANY 或<>ANY	不等于子查询结果中的某个值
!=ALL 或<>ALL	不等于子查询结果中的任何一个值

2. 相关子查询

在相关子查询中，子查询的执行依赖于外部查询，即子查询的查询条件依赖于外部查询的某个属性值。

相关子查询的执行过程与嵌套子查询完全不同，嵌套子查询中子查询只执行一次，而相关子查询中的子查询需要重复地执行。相关子查询中一般使用 EXISTS 关键字，用于判断查询子句是否有记录，如果有一条或多条记录存在则返回 TRUE，否则返回 FALSE，与子查询后的 SELECT 语句中的字段列表无关。

语法格式如下：

[NOT]EXISTS

相关子查询
操作示例

相关子查询的执行过程如下：

① 子查询为外部查询的每一个元组（行）执行一次，外部查询将子查询引用列的值传给子查询。

② 如果子查询的任何行与其匹配，外部查询则取此行放入结果表。

③ 再回到①，直到处理完外部表的每一行。

在相关子查询中，经常要用到 EXISTS 操作符，EXISTS 代表存在量词，"∃"为存在量词符号。带有 EXISTS 的子查询不需要返回任何实际数据，而只需要返回一个逻辑真值"TRUE"或逻辑假值"FALSE"。也就是说，它的作用是在 WHERE 子句中测试子查询返回的行是否存在。如果存在则返回真值，如果不存在则返回假值。

在相关子查询中，子查询的作用相当于测试，它不产生任何数据，只返回 TRUE 或 FALSE。当返回值为 TRUE 时，外层的查询才会执行 EXISTS 关键字。

5.6.3 任务实现

【例 5-6-1】 查询所有年龄大于平均年龄的学生姓名。

```
mysql> SELECT sname FROM students
    -> WHERE YEAR(CURDATE())-YEAR(sbirthday)
    -> >(SELECT AVG(YEAR(CURDATE())-YEAR(sbirthday)) FROM students);
Empty set (0.12 sec)
```

【例 5-6-2】 查询与刘洋在同一个班的同学。

```
mysql> SELECT sid,sname,sclass
    -> FROM students
    -> WHERE sclass=(SELECT sclass FROM students WHERE sname='刘洋');
+------------+--------+--------+
| sid        | sname  | sclass |
+------------+--------+--------+
| 1933062301 | 刘洋   | 1901   |
| 1933062302 | 邓嘉蓉 | 1901   |
| 1933062303 | 葛佳音 | 1901   |
| 1933062304 | 索凤洋 | 1901   |
| 1933062305 | 曹宏美 | 1901   |
| 1933062306 | 郭思宇 | 1901   |
| 1933062307 | 周慧敏 | 1901   |
+------------+--------+--------+
7 rows in set (0.00 sec)
```

如果子查询的返回值不止一个，而是一个集合，则不能直接使用比较运算符，可以在比较运算符和子查询之间插入 ANY、SOME 或 ALL。其中等值关系可以用 IN 操作符。

【例 5-6-3】 查询与刘洋同班同学的学号、姓名和所在班级。

```
mysql> SELECT sid,sname,sclass
    -> FROM students
    -> WHERE sclass IN(SELECT sclass FROM students WHERE sname='刘洋');
```

```
+------------+--------+--------+
| sid        | sname  | sclass |
+------------+--------+--------+
| 1933062301 | 刘洋   | 1901   |
| 1933062302 | 邓嘉蓉 | 1901   |
| 1933062303 | 葛佳音 | 1901   |
| 1933062304 | 索凤洋 | 1901   |
| 1933062305 | 曹宏美 | 1901   |
| 1933062306 | 郭思宇 | 1901   |
| 1933062307 | 周慧敏 | 1901   |
+------------+--------+--------+
7 rows in set (0.10 sec)
```

【例5-6-4】 查询出没有选修网页制作的学生学号和姓名。

这个查询的执行步骤是首先在课程表中查询出网页制作的课程号,然后根据课程号从成绩表中查询出学生的学号,再根据学号在学生表中查询出学生相应的姓名。

```
mysql> SELECT sid,sname
    -> FROM students
    -> WHERE sid NOT IN(SELECT sid
    -> FROM scores
    -> WHERE cid IN(SELECT cid
    -> FROM course
    -> WHERE cname='网页制作'));
+------------+--------+
| sid        | sname  |
+------------+--------+
| 1933062301 | 刘洋   |
| 1933062302 | 邓嘉蓉 |
| 1933062303 | 葛佳音 |
| 1933062304 | 索凤洋 |
| 1933062305 | 曹宏美 |
| 1933062306 | 郭思宇 |
| 1933062307 | 周慧敏 |
+------------+--------+
7 rows in set (0.00 sec)
```

【例5-6-5】 查询其他专业中比软件技术专业某一学生年龄小的学生的姓名和年龄。

```
mysql> SELECT sname,YEAR(CURDATE())-YEAR(sbirthday)
    ->    FROM students
    ->    WHERE YEAR(CURDATE())-YEAR(sbirthday)<ANY(SELECT YEAR(CURDATE())-
    ->    YEAR(sbirthday) FROM students
    ->    WHERE smajor='软件技术') AND smajor<>'软件技术';
Empty set (0.00 sec)
```

【例 5-6-6】 查询其他班级比 1901 班级学生年龄都小的学生。

```
mysql> SELECT *
    -> FROM students
    -> WHERE YEAR(CURDATE())-YEAR(sbirthday)<ALL
    -> (SELECT YEAR(CURDATE())-YEAR(sbirthday)
    -> FROM students
    -> WHERE sclass=1901)
    -> AND sclass<>1901;
Empty set (0.00 sec)
```

【例 5-6-7】 查询与其他同学班级不同的学生的学号和姓名。

```
select sid,sname from students A where not exists (select * from students B where
    -> A.sclass=B.sclass and A.sid<>B.sid);
+------------+--------+
| sid        | sname  |
+------------+--------+
| 1933062308 | 宏宇   |
+------------+--------+
1 row in set (0.02 sec)    -> );
Query OK, 0 rows affected (0.42 sec)
```

在前面的例子中，注意子查询是独立的。这意味着可以将子查询作为独立查询执行。与独立子查询不同，相关子查询是使用外部查询中的数据的子查询。换句话说，相关的子查询取决于外部查询。对外部查询中的每一行对相关子查询进行一次评估。相关子查询是一个子查询中引用了某张表且这张表也在子查询外部被使用到。相关子查询是使用外部查询中的值的子查询（嵌套在另一个查询中的查询）。因为子查询需为外部查询返回的每一行执行一次，所以它可能会很慢。相关子查询被用来做逐行的处理，子查询会为外部查询出来的每一行执行内部 SQL（外部语句也可为 UPDATE 或 DELETE 语句）。相关子查询是由外部查询驱动内部查询。而正常的嵌套查询中，内部查询首先被立即执行，返回的值被外部查询使用并执行外部查询。相关子查询内部查询依赖于外部查询进行处理，而在嵌套查询中外部查询依赖于内部查询。使用相关子查询会使性能降低，因为它执行的次数远远大于嵌套查询的次数。

【例 5-6-8】 查询所有选修了 0101 课程的学生姓名。

```
mysql> SELECT sname FROM students
    -> WHERE EXISTS(SELECT *
    -> FROM scores
    -> WHERE sid=students.sid AND cid=0101);
```

```
+--------+
| sname  |
+--------+
| 刘洋   |
| 邓嘉荟 |
| 葛佳音 |
| 索凤洋 |
| 曹宏美 |
| 郭思宇 |
| 周慧敏 |
+--------+
7 rows in set (0.00 sec)
```

【例 5-6-9】 查询选修了所有课程的学生姓名。

```
mysql> SELECT sname FROM students
    ->      WHERE NOT EXISTS
    ->      (SELECT * FROM course
    ->      WHERE NOT EXISTS
    ->      (SELECT * FROM scores
    ->      WHERE sid=students.sid AND cid=course.cid));
Empty set (0.00 sec)
```

5.6.4 任务训练

在图书管理系统中一些业务功能需要进行嵌套查询，请按照嵌套查询完成以下查询。

① 查询王东华借书的图书编号。
② 使用子查询查询所有借阅过《数据结构》的学生姓名。
③ 查询书籍单价大于书籍平均单价的书籍信息。
④ 查询所有未还书籍学生姓名。
⑤ 查询借阅过大于 3 本的学生的借阅证号、姓名、年龄。
⑥ 查询被借阅次数最多的书籍信息。
⑦ 查询书名包括"计算机"的图书的名称、单价、平均价格。
⑧ 查询借过图书的读者编号和姓名。
⑨ 查询借过最多书的读者信息。
⑩ 查询没有借过书的读者信息。

5.6.5 任务总结

通过本任务的学习，我们了解了嵌套查询的语法规则和作用，嵌套查询是多条 SQL 语句嵌套使用，一条 SQL 语句的查询结果作为另一条查询语句的条件或查询结果。先处理内查询，由内向外处理，外层查询利用内层查询的结果嵌套查询不仅仅可以用于父查询 SELECT 语句使用。

任务 7

集 合 查 询

5.7.1 任务描述

SELECT 语句的查询结果是元组的集合,所以多个 SELECT 语句的结果可进行集合操作。集合操作主要包括并操作 UNION、交操作 INTERSECT、差操作 EXCEPT。注意,参加集合操作的各查询结果的列数必须相同,对应的数据类型也必须相同。MySQL 数据库并不支持交操作和差操作。

5.7.2 知识准备

如果需要将两个 SELECT 语句的结果作为一个整体显示出来,那么就需要用到 UNION 或者 UNION ALL 关键字。UNION(或称为联合)的作用是将多个结果合并在一起显示出来。

集合查询的语法格式如下:

```
(SELECT 列名 FROM 表名)
UNION[ALL]
(SELECT 列名 FROM 表名);
```

功能:

显示多个 SELECT 语句的合并结果。

说明:

① UNION 因为要进行重复值扫描,所以效率低。如果合并没有刻意要删除重复行,那么就使用 UNION All。具体地讲:UNION 对两个结果集进行并集操作,不包括重复行,同时进行默认规则的排序。UNION All 对两个结果集进行并集操作,包括重复行,不进行排序。

② 两个 SELECT 语句中的列名序列应该是相同的。

5.7.3 任务实现

【例 5-7-1】 查询软件技术专业的学生以及年龄不大于 20 岁的学生。

```
mysql> SELECT * FROM students
    -> WHERE smajor='软件技术'
    -> UNION
    -> SELECT * FROM students
    -> WHERE YEAR(CURDATE())-YEAR(sbirthday)<=19;
```

```
+------------+--------+--------+---------+----------+------------+---------------+
| sid        | sname  | sclass | sgender | smajor   | sbirth     | credit_points |
+------------+--------+--------+---------+----------+------------+---------------+
| 1933062301 | 刘洋   | 1901   | 女      | 软件技术 | 2000-09-01 | 12            |
| 1933062302 | 邓嘉荟 | 1901   | 女      | 软件技术 | 2000-04-02 | 12            |
| 1933062303 | 葛佳音 | 1901   | 女      | 软件技术 | 2000-07-09 | 12            |
| 1933062304 | 索凤洋 | 1901   | 女      | 软件技术 | 2000-08-04 | 12            |
| 1933062305 | 曹宏美 | 1901   | 女      | 软件技术 | 2000-11-07 | 12            |
| 1933062306 | 郭思宇 | 1901   | 女      | 软件技术 | 2000-03-21 | 12            |
| 1933062307 | 周慧敏 | 1901   | 女      | 软件技术 | 2000-03-23 | 12            |
+------------+--------+--------+---------+----------+------------+---------------+
7 rows in set (0.00 sec)
```

5.7.4 任务训练

图书馆里系统使用中,图书馆工作人员在进行图书信息统计时需要完成下面的任务。

① 查询人民邮电出版社出版的以及价格大于 50 元的图书。

② 查询读者表中性别为男生以及年龄不大于 20 岁的读者信息。

③ 查询图书表中今年出版的以及藏书量大于 20 册的图书。

④ 查询机械出版社出版的以及借出过 20 次的图书。

⑤ 查询今年出版的以及单价不大于 50 元的图书。

5.7.5 任务总结

通过本任务的学习,我们了解了集合查询的使用方法和作用,了解了集合查询的语法规则。在数据库中集合的概念与数学中的一样,同样有交集、并集、差集。我们会在以后的章节中使用 SQL 语句完成相应的任务。

聚合函数与主查询、子查询的使用实例

项目 6　数据库的高级应用

通过前面的学习,读者对数据库的基本操作有了一定的了解,在数据库中还有一些高级的数据库对象,如索引、外键、视图、存储过程、触发器等,用来提升数据库性能,维护数据的一致性和完整性。此外,有时还需要进行事务管理,用来进行数据的备份和还原等,本项目将针对这些知识点进行详细的讲解。

学习目标

- 了解索引的作用和使用原则;
- 掌握创建索引、查看索引、删除索引的方法;
- 掌握外键的概念;
- 熟练掌握添加外键约束和删除外键约束的方法;
- 了解使用外键约束的关联表操作;
- 了解视图的概念和作用;
- 掌握视图的创建、查看、修改和删除操作;
- 熟悉存储过程的概念;
- 掌握存储过程的创建和调用、查看、修改和删除方法;
- 了解触发器的概念、触发器与存储过程的区别;
- 掌握触发器的创建与执行方法;
- 了解事务的概念和基本特性;
- 掌握事务的开启、提交和回滚操作以及保存点设置;
- 熟悉事务的四种隔离级别。

任务 1

索　引

6.1.1　任务描述

索引是 MySQL 中十分重要的数据库对象,常用于实现数据的快速检索。在 MySQL 中通常有以下两种方式访问数据库表的行数据。

1. 顺序访问

顺序访问是在表中实行全表扫描,从头到尾逐行遍历,直到在无序的行数据中找到符合条件的目标数据。当表中有大量数据的时候,顺序访问的效率会非常低下。比如要在几千万条数据中查找数据时,使用顺序访问方式将会遍历所有的数据,这需要花费大量的时间,显然会影响数据库的处理性能。

2. 索引访问

索引访问是通过遍历索引来直接访问表中记录行的方式。使用这种方式的前提是对表建立一个索引,创建了索引之后,查找数据时可以直接根据该列上的索引找到对应记录行的位置,从而快捷地查找到数据。

简单来说,若是不使用索引,MySQL 则必须从第一条记录开始读完整个表,直到找出相关的行。表越大,查询数据所花费的时间就越多。如果表中查询的列建立了索引,MySQL 就能快速到达一个位置去搜索数据文件,而不必查看所有数据,这比按顺序读每一行要快得多。

比如,在学生基本信息表 students 中,如果基于学生的学号建立了索引,系统就建立了一张索引列到实际记录的映射表。当用户需要查找学号为 19980002 的数据的时候,系统先在索引上找到该记录,然后通过映射表直接找到数据行,并且返回该行数据。因为扫描索引的速度一般远远大于扫描实际数据行的速度,所以采用索引的方式可以大大提高数据库的工作效率。

本次任务主要在学生成绩管理数据库中创建索引来提升查询效率,在完成任务之前,首先进行索引相关知识点的讲解。

6.1.2　知识准备

1. 索引概述

索引是一种数据结构,用来快速查询数据表中有某一特定值的记录。索引可以看作是根据表中的一列或多列按照一定顺序建立的列值与记录行之间的对应关系表,其实质是一张描述索引列的列值与原表中记录行之间一一对应关系的有序表。

索引的原理很简单,就是把无序的数据变成有序的查询。首先把创建了索引的列的内容进行排序,然后对排序结果生成倒排表,并在倒排表中拼接数据地址链。在查询的时候,先拿到倒排表内容,再取出数据地址链,从而拿到具体数据。我们也可以将索引理解成书中的目录,为了方便查找书中的内容,通过对内容建立索引形成目录。创建索引之后,查询数据时不用读完记录的所有信息,而只需查询索引列。因此,使用索引可以很大程度上提高数据库的查询速度,提高数据库系统的性能。需要注意的是,索引是一个文件,它是要占据物理空间的。

索引可分为多种类型,分类方法也不相同,总的来说,有两种分类方法,一是按照创建索引的字段个数划分,另一种是按照实现索引的语法和功能来划分。

按照索引创建的字段个数,可以将索引分为单列索引和复合索引。

1) 按照字段个数划分

(1) 单列索引

建立在单个列上的索引被称为单列索引。

(2) 复合索引

建立在多个列上的索引被称为复合索引,又叫联合索引、组合索引。在建立复合索引的时候应该注意索引列的顺序,一般情况下,将查询需求频繁或者字段选择性高的列放在前面。复合索引列值的组合必须唯一。

按照索引实现的语法不同,可以将索引分为主键索引、唯一索引、普通索引、全文索引、空间索引。

2) 按语法和功能划分

(1) 主键索引

建立在主键上的索引被称为主键索引,一张数据表只能有一个主键索引,索引列值不允许有空值,MySQL 在创建表时通常会自动创建主键索引。

(2) 唯一索引

建立在 UNIQUE 字段上的索引被称为唯一索引,一张表可以有多个唯一索引,索引列的值必须唯一,但允许有空值。

(3) 普通索引

建立在普通字段上的索引被称为普通索引。普通索引是 MySQL 中的基本索引类型,允许在定义索引的列中插入重复值和空值。

(4) 全文索引

全文索引类型为 FULLTEXT,在定义索引的列上支持值的全文查找,允许在这些索引列中插入重复值和空值。全文索引可以在 CHAR、VARCHAR 或者 TEXT 类型的列上创建。从 MySQL 5.7 开始,MySQL 内置了 ngram 全文检索插件,用来支持中文分词,并且对 MyISAM 和 InnoDB 引擎有效。

(5) 空间索引

空间索引是对空间数据类型的字段建立的索引,MySQL 中的空间数据类型有 4 种,分别是 GEOMETRY、POINT、LINESTRING 和 POLYGON。MySQL 使用 SPATIAL 关键字进行扩展,创建空间索引的列必须声明为 NOT NULL。

2. 索引的基本操作

1) 创建索引

创建索引主要有以下三种方式。

(1) 直接创建索引

使用 CREATE INDEX 命令直接创建索引的语法格式如下:

```
CREATE [UNIQUE | FULLTEXT | SPATIAL] INDEX 〈索引名称〉
ON 〈表名〉(〈列名〉[(〈长度〉)] [ ASC | DESC]);
```

这是创建索引的基本格式,CREATE INDEX 表示创建索引,〈索引名称〉用来指定索引名。一个表可以创建多个索引,但每个索引在该表中的名称是唯一的。

我们可以用关键字 UNIQUE、FULLTEXT、SPATIAL 来分别指明创建唯一索引、全文索引或者空间索引,如果未指定,则表示创建普通索引。

〈表名〉指明了要在哪张表上创建索引,〈列名〉指定要创建索引的列名,〈长度〉是可选项,指定使用列前的多少个字符来创建索引。ASC|DESC 也是可选项,ASC 指定索引按照升序来排列,DESC 指定索引按照降序来排列,默认为 ASC。

如果只在一列上创建的索引创建的是单列索引,若是将多列进行组合创建索引,则创建的是复合索引,多个列名之间用逗号隔开。

(2) 在创建表的同时添加索引

索引也可以在创建表的同时创建。使用 CREATE TABLE 创建表时,除了可以定义列的数据类型,还可以定义主键约束、外键约束或者唯一性约束,而不论创建哪种约束,在定义约束的同时相当于在指定列上创建了一个索引。使用 CREATE TABLE 时创建索引语法格式如下:

```
CREATE TABLE 〈表名〉(
…
[UNIQUE | FULLTEXT | SPATIAL] [ INDEX | KEY ] [〈索引名称〉] (〈列名〉,…)
);
```

在 CREATE TABLE 语句中添加此语句,INDEX 或者 KEY 关键字表示在创建新表的同时创建该表的索引。〈列名〉指定要创建索引的列名,表示在哪一列或哪几列上创建索引。

(3) 通过修改表结构的方式创建索引

如果数据表已经创建完成,还可以通过修改表结构的方式添加索引,使用 ALTER

TABLE 命令创建索引语法格式如下：

```
ALTER TABLE 〈表名〉
ADD [UNIQUE | FULLTEXT | SPATIAL ] INDEX [〈索引名称〉](〈列名〉,…);
```

ALTER TABLE 语句可以在一个已有的表上创建索引，〈表名〉指明了要增加索引的表的名称。ADD INDEX 表示添加索引，〈索引名称〉可自己命名，缺省时，MySQL 将根据第一个索引列赋一个名称。〈列名〉表示在哪一列或哪几列上创建索引。

2）查看索引

创建索引后可以查看索引，查看索引的 SQL 语句格式如下：

```
SHOW INDEX FROM 〈表名〉;
```

〈表名〉指明查看哪张表上的索引。比如，使用 SHOW INDEX 查看 students 表中的索引。

```
mysql> show index from students;
+--------+------------+----------+--------------+-------------+-----------+-------------+----------+--------+------+------------+---------+---------------+
| Table  | Non_unique | Key_name | Seq_in_index | Column_name | Collation | Cardinality | Sub_part | Packed | Null | Index_type | Comment | Index_comment |
+--------+------------+----------+--------------+-------------+-----------+-------------+----------+--------+------+------------+---------+---------------+
|students|          0 | PRIMARY  |            1 | sid         | A         |           7 | NULL     | NULL   |      | BTREE      |         |               |
+--------+------------+----------+--------------+-------------+-----------+-------------+----------+--------+------+------------+---------+---------------+
1 row in set (0.02 sec)
```

可以看到 MySQL 自动为学生表中的主键 sid 添加了主键索引。这里查看到的索引信息可读性比较弱，可以加上 \G 来将查到的结构变成纵向。

```
mysql> show index from students \G
*************************** 1. row ***************************
        Table: students
   Non_unique: 0
     Key_name: PRIMARY
 Seq_in_index: 1
  Column_name: sid
    Collation: A
  Cardinality: 7
     Sub_part: NULL
       Packed: NULL
         Null: 
   Index_type: BTREE
      Comment: 
Index_comment: 
1 row in set (0.00 sec)
```

其中，每个索引描述信息的属性如表 6-1-1 所示。

表 6-1-1　索引描述属性信息

属性名	属性含义
Table	表示索引所在的表的名称
Non_unique	表示索引是否包括重复词,不唯一,值为 1;唯一,值为 0
Key_name	表示索引的名称
Seq_in_index	表示索引的列序号
Column_name	列名称
Collation	表示列以什么方式存储在索引中,在 MySQL 中有值 'A'(升序)或者 NULL(无分类)
Cardinality	索引在唯一值的数据的估值
Sub_part	如果列只是部分编入索引,则为被编入索引的字符的数目,如果整列被编入索引,则为 NULL
Packed	指示关键词如何被压缩,如果没有被压缩,则为 NULL
NULL	如果列含有 NULL,则为 YES,如果没有,则该列为 NO
Index_type	索引类型,包含 BTREE、FULLTEXT、HASH、RTREE 等
Comment	注释
Index_comment	索引注释

3) 删除索引

对于数据表中已经创建但不再使用的索引,应该及时删除,避免占用系统资源,删除索引的 SQL 语法格式如下:

格式 1:

```
DROP INDEX <索引名称> ON <表名>;
```

格式 2:

```
ALTER TABLE <表名> DROP INDEX <索引名称>;
```

3. 索引的使用原则

数据库索引是一种数据结构,它以额外的写入和存储空间来维护索引数据结构为代价,提高了数据库表上数据检索操作的速度。索引并非越多越好,一个表中如果有大量的索引,不仅占用磁盘空间,而且会影响 INSERT、DELETE、UPDATE 等语句的性能,因为在表中的数据更改的同时,索引也会进行调整和更新。使用索引一般有以下原则。

(1) 最适合索引的列是出现在 WHERE 子句中的列或连接子句中指定的列

索引的作用类似于指向表行的指针,可以快速确定哪些行与 WHERE 子句中的条件匹配,并检索这些行的其他列值,较频繁作为查询条件的字段才需要创建索引。我们通常将查询语句中在 JOIN 子句和 WHERE 子句里经常出现的列作为索引列,可以大大加快数据的查询速度,这是使用索引最主要的原因。

(2) 当唯一性是某种数据本身的特征时,适合建立索引

通过创建唯一索引可以保证数据库表中每一行数据的唯一性,使用唯一索引需能确保定义的列的数据完整性,提高查询速度。

(3) 在频繁进行排序或分组的列上适合建立索引

在频繁进行 GROUP BY 或 ORDER BY 操作的列上适合建立索引,在使用分组和排序子句进行数据查询时,可以显著减少查询中分组和排序的时间。如果待排序的列有多个,可以在这些列上建立复合索引。

(4) 尽量使用短索引

索引字段越小越好,较小的索引涉及的磁盘 I/O 较少,较短的值比较起来更快。更为重要的是,对于较短的键值,索引高速缓存中的块能容纳更多的键值,因此,MySQL 也可以在内存中容纳更多的值,这样就增加了找到行而不用读取索引中较多块的可能性。如果对长字符串列进行索引,应该指定一个前缀长度,这样能够节省大量索引空间。

(5) 索引需要额外的磁盘空间,并降低写操作的性能,不要过度索引

创建和维护索引组要耗费时间,并且随着数据量的增加,所耗费的时间也会增加,所以不要过度使用索引。

对于那些查询中很少涉及的列,重复值比较多的列不要建立索引。

更新频繁字段不适合创建索引,因为在修改表内容的时候,索引会进行更新甚至重构,索引列越多,这个时间就会越长。所以只保持需要的索引有利于查询即可。

数据量小的表最好不要使用索引。由于数据较少,查询花费的时间可能比遍历索引的时间还要短,索引可能不会产生优化效果。

不能有效区分数据的列不适合建立索引。比如性别字段,最多也就三种(男、女、未知),区分度太低。这种字段如果建立索引,不但不会提高查询效率,反而会严重降低数据更新速度。

总的来说,索引虽然加快了查询速度,但索引也是有代价的。索引文件本身要消耗存储空间,同时索引会加重插入、删除和修改记录时的负担,另外,MySQL 在运行时也要消耗资源维护索引,因此索引并不是越多越好,要合理使用索引。

6.1.3 任务实现

【例 6-1-1】 在创建表时添加索引。

在学生成绩管理数据库中新创建一个表 resume,用来存放学生的简历信息,使用 CREATE TABLE 创建数据表 resume,表中加入四个字段 id、author_no、title、content,分别表示简历编号、作者编号、简历名称、简历内容,并设置 id 作为主键,通过 INDEX 关键字为字段 title 添加索引,并设置索引名称为 title_index。

```
mysql> CREATE TABLE `resume` (
    -> `id` INT NOT NULL AUTO_INCREMENT,
    -> `author_no`  int(20) unsigned NOT NULL,
    -> `title` varchar(50) NOT NULL,
    -> `content` text DEFAULT NULL,
    -> PRIMARY KEY (`id`),
    -> INDEX title_index (`title`)
    -> );
Query OK, 0 rows affected (0.42 sec)
```

添加索引成功后,通过 SHOW CREATE TABLE 查看建表信息。

```
mysql> SHOW CREATE TABLE resume \G
*************************** 1. row ***************************
       Table: resume
Create Table: CREATE TABLE `resume` (
  `id` int(11) NOT NULL AUTO_INCREMENT,
  `author_no` int(20) unsigned NOT NULL,
  `title` varchar(50) NOT NULL,
  `content` text,
  PRIMARY KEY (`id`),
  KEY `title_index` (`title`)
) ENGINE=InnoDB DEFAULT CHARSET=utf8
1 row in set (0.00 sec)
```

可以看到,创建了普通索引 title_index。

【例 6-1-2】 使用 CREATE INDEX 直接创建索引。

在学生成绩管理数据库的 resume 表中为 title 字段添加一个唯一索引,命名为 title_unique_index。

```
mysql> CREATE UNIQUE INDEX title_unique_index on resume(title);
Query OK, 0 rows affected (0.67 sec)
Records: 0  Duplicates: 0  Warnings: 0
```

其中,CREATE UNIQUE INDEX 表示创建唯一索引,title_unique_index 表示索引名称,resume(title) 指明了为 resume 表中的字段 title 添加索引。添加索引成功后,通过 SHOW CREATE TABLE 查看建表信息。

```
mysql> SHOW CREATE TABLE resume \G
*************************** 1. row ***************************
       Table: resume
Create Table: CREATE TABLE `resume` (
  `id` int(11) NOT NULL AUTO_INCREMENT,
  `author_no` int(20) unsigned NOT NULL,
```

```
  'title' varchar(50) NOT NULL,
  'content' text,
  PRIMARY KEY ('id'),
  UNIQUE KEY 'title_unique_index' ('title'),
  KEY 'title_index' ('title')
) ENGINE=InnoDB DEFAULT CHARSET=utf8
1 row in set (0.00 sec)
```

可以看到,创建了唯一索引 title_unique_index。

【例 6-1-3】 通过修改表的方式创建索引。

在学生成绩管理数据库的 resume 表的 title 和 content 两个列上添加一个全文索引,命名为 resume_fulltext_index。

```
mysql> ALTER TABLE resume
    -> ADD FULLTEXT INDEX resume_fulltext_index (title, content);
Query OK, 0 rows affected, 1 warning (2.79 sec)
Records: 0  Duplicates: 0  Warnings: 1
```

其中,ALTER TABLE 给已经存在的表的指定字段创建全文索引,ADD FULLTEXT INDEX 表示添加全文索引,resume_fulltext_index 表示索引名称,括号内包含两个字段 title 和 content,指明创建的是复合索引。添加索引成功后,通过 SHOW CREATE TABLE 查看建表信息如下。

```
mysql> SHOW CREATE TABLE resume \G
*************************** 1. row ***************************
       Table: resume
Create Table: CREATE TABLE 'resume' (
  'id' int(11) NOT NULL AUTO_INCREMENT,
  'author_no' int(20) unsigned NOT NULL,
  'title' varchar(50) NOT NULL,
  'content' text,
  PRIMARY KEY ('id'),
  UNIQUE KEY 'title_unique_index' ('title'),
  KEY 'title_index' ('title'),
  FULLTEXT KEY 'resume_fulltext_index' ('title','content')
) ENGINE=InnoDB DEFAULT CHARSET=utf8
1 row in set (0.00 sec)
```

【例 6-1-4】 使用 SHOW INDEX 查看索引。

查看学生成绩管理数据库的 resume 表中的索引,使用 SHOW INDEX 查看 resume 表中的索引。

```
mysql> SHOW INDEX from resume;
+--------+------------+--------------------+--------------+-------------+-----------+-------------+----------+--------+------+------------+---------+---------------+
| Table  | Non_unique | Key_name           | Seq_in_index | Column_name | Collation | Cardinality | Sub_part | Packed | Null | Index_type | Comment | Index_comment |
+--------+------------+--------------------+--------------+-------------+-----------+-------------+----------+--------+------+------------+---------+---------------+
| resume |    0       | PRIMARY            |      1       |     id      |     A     |      0      |   NULL   |  NULL  |      |   BTREE    |         |               |
| resume |    0       | title_unique_index |      1       |    title    |     A     |      0      |   NULL   |  NULL  |      |   BTREE    |         |               |
| resume |    1       | title_index        |      1       |    title    |     A     |      0      |   NULL   |  NULL  |      |   BTREE    |         |               |
| resume |    1       | resume_fulltext_index |   1       |    title    |   NULL    |      0      |   NULL   |  NULL  |      |  FULLTEXT  |         |               |
| resume |    1       | resume_fulltext_index |   2       |   content   |   NULL    |      0      |   NULL   |  NULL  | YES  |  FULLTEXT  |         |               |
+--------+------------+--------------------+--------------+-------------+-----------+-------------+----------+--------+------+------------+---------+---------------+
5 rows in set (0.00 sec)
```

可以看到，resume 表中自动为主键 id 创建了主键索引，此外还包含上述案例中创建的普通索引 title_index、唯一索引 title_unique_index、全文索引 resume_fulltext_index。这里加上\G 来将查到的结构变成纵向。

```
mysql> SHOW INDEX from resume \G
*************************** 1. row ***************************
        Table: resume
   Non_unique: 0
     Key_name: PRIMARY
 Seq_in_index: 1
  Column_name: id
    Collation: A
  Cardinality: 0
     Sub_part: NULL
       Packed: NULL
         Null: 
   Index_type: BTREE
      Comment: 
Index_comment: 
*************************** 2. row ***************************
        Table: resume
   Non_unique: 0
     Key_name: title_unique_index
 Seq_in_index: 1
  Column_name: title
    Collation: A
  Cardinality: 0
     Sub_part: NULL
       Packed: NULL
         Null: 
   Index_type: BTREE
      Comment: 
Index_comment: 
*************************** 3. row ***************************
        Table: resume
   Non_unique: 1
     Key_name: title_index
 Seq_in_index: 1
```

```
        Column_name: title
         Collation: A
       Cardinality: 0
          Sub_part: NULL
            Packed: NULL
              Null:
        Index_type: BTREE
           Comment:
     Index_comment:
*************************** 4. row ***************************
             Table: resume
        Non_unique: 1
          Key_name: resume_fulltext_index
      Seq_in_index: 1
       Column_name: title
         Collation: NULL
       Cardinality: 0
          Sub_part: NULL
            Packed: NULL
              Null:
        Index_type: FULLTEXT
           Comment:
     Index_comment:
*************************** 5. row ***************************
             Table: resume
        Non_unique: 1
          Key_name: resume_fulltext_index
      Seq_in_index: 2
       Column_name: content
         Collation: NULL
       Cardinality: 0
          Sub_part: NULL
            Packed: NULL
              Null: YES
        Index_type: FULLTEXT
           Comment:
     Index_comment:
5 rows in set (0.00 sec)
```

【例6-1-5】 使用 DROP INDEX 删除索引。

删除 resume 表中的全文索引 resume_fulltext_index。

```
mysql> DROP INDEX resume_fulltext_index ON resume;
Query OK, 0 rows affected (0.55 sec)
Records: 0  Duplicates: 0  Warnings: 0
```

重新查看 resume 中的索引结果。

```
mysql> SHOW INDEX from resume;
+--------+--------+-------------------+--------+--------+-----------+-------------+----------+--------+------+--------+---------+---------------+
| Table  | Non_   | Key_name          | Seq_in | Column | Collation | Cardinality | Sub_part | Packed | Null | Index  | Comment | Index_        |
|        | unique |                   | _index | _name  |           |             |          |        |      | _type  |         | comment       |
+--------+--------+-------------------+--------+--------+-----------+-------------+----------+--------+------+--------+---------+---------------+
| resume | 0      | PRIMARY           | 1      | id     | A         | 0           | NULL     | NULL   |      | BTREE  |         |               |
| resume | 0      | title_unique_index| 1      | title  | A         | 0           | NULL     | NULL   |      | BTREE  |         |               |
| resume | 1      | title_index       | 1      | title  | A         | 0           | NULL     | NULL   |      | BTREE  |         |               |
+--------+--------+-------------------+--------+--------+-----------+-------------+----------+--------+------+--------+---------+---------------+
3 rows in set (0.01 sec)
```

可以看到，全文索引 resume_fulltext_index 删除成功。

【例6-1-6】 使用 ALTER TABLE 方式删除索引。

删除中 resume 表中的唯一索引 title_unique_index。

```
mysql> ALTER TABLE resume DROP INDEX title_unique_index;
Query OK, 0 rows affected (0.36 sec)
Records: 0  Duplicates: 0  Warnings: 0
```

重新查看 resume 中的索引结果。

```
mysql> show index from resume;
+--------+--------+-------------+--------+--------+-----------+-------------+----------+--------+------+--------+---------+---------------+
| Table  | Non_   | Key_name    | Seq_in | Column | Collation | Cardinality | Sub_part | Packed | Null | Index  | Comment | Index_        |
|        | unique |             | _index | _name  |           |             |          |        |      | _type  |         | comment       |
+--------+--------+-------------+--------+--------+-----------+-------------+----------+--------+------+--------+---------+---------------+
| resume | 0      | PRIMARY     | 1      | id     | A         | 0           | NULL     | NULL   |      | BTREE  |         |               |
| resume | 1      | title_index | 1      | title  | A         | 0           | NULL     | NULL   |      | BTREE  |         |               |
+--------+--------+-------------+--------+--------+-----------+-------------+----------+--------+------+--------+---------+---------------+
2 rows in set (0.00 sec)
```

可以看到，唯一索引 title_unique_index 删除成功。

6.1.4 训练任务

在图书管理数据库中对图书表 Book 中的列字段创建索引，并优化查询。具体要求如下。

① 创建索引名为 index_publish_house 的单列索引。

② 查询图书表 Book 中甲骨文出版社的图书信息。

③ 创建索引名为 index_name_publish 的索引，包含图书名称和出版社名称两列。

④ 查询图书表中鼎盛出版社书名为《人性的枷锁》的图书信息。

6.1.5 任务总结

索引用于快速定位数据,无须在每次访问数据库表时搜索数据库表中的每一行。我们可以使用数据库表的一个或多个列创建索引,为快速随机查找和有效访问有序记录提供基础。

虽然索引大大提高了查询速度,同时却会降低更新表的速度,如对表进行 INSERT、UPDATE 和 DELETE。因为更新表时,MySQL 不仅要保存数据,还要保存一下索引文件,建立索引会占用磁盘空间的索引文件,所以合理地创建索引是必要的。

本次任务主要讲解了索引的概念及分类、索引的基本操作、索引的使用原则等知识点。通过本次任务的学习,读者应掌握索引的创建和使用,能够运用索引提升查询速度。

任务 2

外 键 约 束

6.2.1 任务描述

在数据库设计时,为了保证不同的表中相同含义数据的一致性和完整性,可以为数据库表添加外键约束。比如,数据库中有一张员工表和一张部门表,如果在添加员工信息时不小心为某个员工设置了不存在的部门,此时就会出现数据信息保存不对等的情况。为了将两表之间的数据建立关系,可以用外键约束将这两张表关联起来,关联后的两个表就会对相关的操作产生约束。比如,员工表只能插入部门表中已存在的部门编号,保证数据的一致性。

MySQL 的外键约束是用来在两个表之间建立连接的,其中一个表发生变化,另外一个表也会相应发生变化。外键约束的主要作用就是能够让表与表之间的数据建立关联,使数据更加完整,关联性更强。

本次任务主要在学生成绩管理数据库中设置外键约束来保证数据的完整性,首先进行外键约束相关知识点的讲解。

6.2.2 知识准备

1. 外键约束的概念

MySQL 的外键约束用来在两个表的数据之间建立连接,表示一张表中的一个字段被另一张表中对应的字段约束。外键对相关表中的数据造成了限制,使 MySQL 能够保持数据的参照完整性。

一个外键关系通常包含一个主表和一个从表,主表包含原始的字段数据,从表引用主表中该字段的数据,外键约束定义在从表上。一般来说,主表中的关联字段是主键,从表中的关联字段是外键,外键是通过定义外键约束实现的。也就是说,相关联字段中主键所在的表是主表,外键所在的表是从表。

定义外键的时候需要注意:外键中列的数据类型必须和主表主键中对应列的数据类型相同。一个表可以有一个或多个外键,外键可以为空值。若不为空值,则每一个外键的值必须等于主表中主键的某个值。

比如,在学生成绩管理数据库中,课程表 courses 和教师表 teachers 可以建立关联关系。此门课程的授课教师需由教师表中已存在的教师所讲授,即课程表中授课教师编号 tid 关联到教师表的主键 tid。此时,将课程表中的授课教师编号 tid 定义成外键,关联到教师表中的主键 tid。教师表是主表,课程表是从表。当两个表之间定义了外键约束时,课程表只能插入教师表中已存在的教师编号,保证数据的一致性。表的关联关系如图 6-2-1 所示。

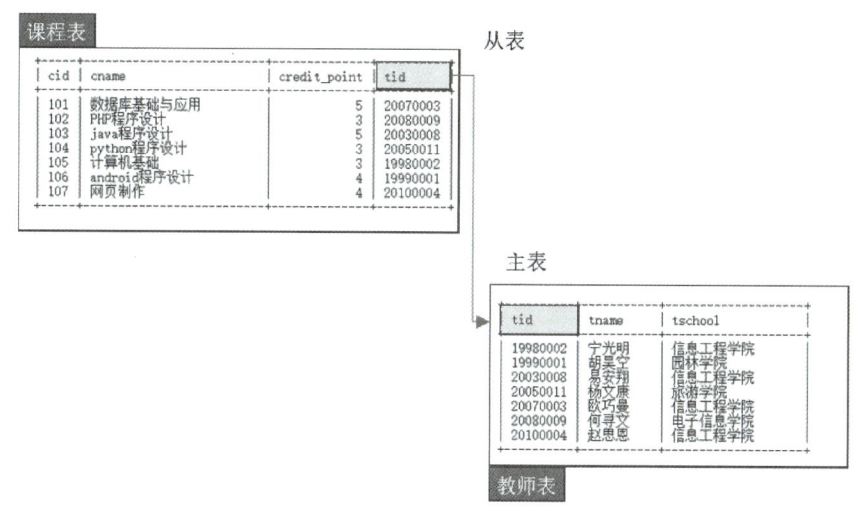

图 6-2-1 课程表和教师表的关联关系

2. 外键操作

1) 添加外键约束

添加外键约束有两种常用方法。我们可以在创建数据表(CREATE TABLE 语句)时添加外键约束,还可以通过修改表结构(ALTER TABLE 语句)在相应的位置添加外键约束。

外键操作示例

① 在创建表时设置外键约束。在 CREATE TABLE 语句中,通过 FOREIGN KEY 关键字来指定外键的语法格式如下:

```
CREATE TABLE 〈表名〉(
...
[CONSTRAINT 〈外键约束名〉] FOREIGN KEY  [〈索引名〉](〈列名〉)
REFERENCES 〈主表名〉(〈列名〉)
);
```

在 CREATE TABLE 语句中，关键字 CONSTRAINT 用于定义外键约束的名称，如果省略，MySQL 将会自动生成一个名字。FOREIGN KEY 关键字用来指明从表中需要引用外键的字段。其中，〈索引名〉也是可选参数，用来表示外键索引名称，如果省略，MySQL 也会在建立外键时自动创建一个外键索引，加快查询速度。REFERENCES 用来指明主表中关联的字段。

② 在修改表时添加外键约束。外键约束也可以在修改表时添加。使用 ALTER TABLE 语句添加外键约束的语法格式如下：

```
ALTER TABLE〈表名〉
ADD [CONSTRAINT〈外键约束名〉] FOREIGN KEY [〈索引名〉](〈列名〉)
REFERENCES〈主表名〉(〈列名〉);
```

在修改表时添加外键约束的前提是，从表中外键列中的数据必须与主表中主键列中的数据一致或者是没有数据。也就是说，在为已经创建好的数据表添加外键约束时，要确保添加外键约束的列的值全部源于主键列，并且外键列不能为空。

2) 查看外键约束

查看外键约束也可以有两种方法。

① 通过 SHOW CREATE TABLE 语句查看。通过 SHOW CREATE TABLE 语句可以查看到这张表上定义的所有外键约束。语法格式如下：

```
SHOW CREATE TABLE〈表名〉;
```

比如，查看学生成绩管理数据库中课程表 courses 上定义的外键约束信息，语句如下：

```
mysql> SHOW CREATE TABLE courses \G
*************************** 1. row ***************************
       Table: courses
Create Table: CREATE TABLE 'courses' (
  'cid' int(10) unsigned NOT NULL AUTO_INCREMENT,
  'cname' varchar(20) DEFAULT NULL,
  'credit_point' int(5) DEFAULT NULL,
  'tid' int(10) unsigned NOT NULL,
  PRIMARY KEY ('cid'),
  KEY 'courses_ibfk_1' ('tid'),
  CONSTRAINT 'courses_ibfk_1' FOREIGN KEY ('tid') REFERENCES 'teachers' ('tid')
) ENGINE = InnoDB AUTO_INCREMENT = 108 DEFAULT CHARSET = utf8
1 row in set (0.00 sec)
```

可以看到，在课程表 courses 上建立了外键约束 courses_ibfk_1，将课程表中授课教师 tid 关联到教师表 teachers 的主键 tid。

② 通过库 information_schema 的 key_column_usage 表来查看。有时候，在测试库中需要对某个数据表做清空数据操作，需要查看这张表的约束关系，这时可以通过库

information_schema 的 key_column_usage 表来查看这个表被那些表所关联。语法格式如下：

```
SELECT * FROM INFORMATION_SCHEMA.KEY_COLUMN_USAGE
WHERE REFERENCED_TABLE_NAME='<表名>';
```

比如，查看学生成绩管理数据库中教师表 teachers 被哪些表引用，语句如下：

```
mysql> select * from INFORMATION_SCHEMA.KEY_COLUMN_USAGE
    -> where REFERENCED_TABLE_NAME='teachers' \G
*************************** 1. row ***************************
           CONSTRAINT_CATALOG: def
            CONSTRAINT_SCHEMA: studentscore
              CONSTRAINT_NAME: courses_ibfk_1
                TABLE_CATALOG: def
                 TABLE_SCHEMA: studentscore
                   TABLE_NAME: courses
                  COLUMN_NAME: tid
             ORDINAL_POSITION: 1
POSITION_IN_UNIQUE_CONSTRAINT: 1
      REFERENCED_TABLE_SCHEMA: studentscore
        REFERENCED_TABLE_NAME: teachers
       REFERENCED_COLUMN_NAME: tid
1 row in set (0.34 sec)
```

同样可以查看到，教师表 teachers 被 courses 表所引用，外键约束名称为 courses_ibfk_1，并且，课程表中授课教师 tid 关联到教师表 teachers 的主键 tid。

3）删除外键约束

当一个表中不需要外键约束时，就需要从表中将其删除。外键一旦删除，就会解除主表和从表间的关联关系。删除外键约束的语法格式如下：

```
ALTER TABLE <表名>
DROP FOREIGN KEY <外键约束名>;
```

3. 关联表操作

在没有添加外键约束时，关联表中的数据插入、更新和删除操作互不影响。而对于添加了外键约束的关联表而言，数据的插入、更新和删除操作就会受到一定的约束。

在添加外键约束时可以使用 ON DELETE 或 ON UPDATE 子句，用于设置主表中的数据被删除或修改时，从表对应数据的处理办法。比如，在创建表时添加外键约束并设置级联操作，语句如下：

```
CREATE TABLE <表名>(
...
[CONSTRAINT <外键约束名>] FOREIGN KEY  [<索引名>](<列名>)
REFERENCES <主表名>(<列名>)
[ON DELETE {RESTRICT | CASCADE | SET NULL | NO ACTION }]
[ON UPDATE {RESTRICT | CASCADE | SET NULL | NO ACTION }]
);
```

在修改表添加外键约束并设置级联操作，语句如下：

```
ALTER TABLE <表名>
ADD [CONSTRAINT <外键约束名>] FOREIGN KEY [<索引名>](<列名>)
REFERENCES <主表名>(<列名>)
[ON DELETE {RESTRICT | CASCADE | SET NULL | NO ACTION }]
[ON UPDATE {RESTRICT | CASCADE | SET NULL | NO ACTION }];
```

其中，ON DELETE 子句指定当主表中的记录被删除时，从表的记录怎样执行操作。如果省略 ON DELETE 子句，那么删除主表中的记录时，如果从表中有引用主表中相关联的数据，那么 MySQL 将拒绝删除。ON UPDATE 子句指定当主表中的记录更新时，从表中的记录怎样执行操作。如果省略 ON UPDATE 子句，那么当主表中的记录被更新时，如果从表中有引用相关联的数据，那么 MySQL 将拒绝更新。

此外，还可以定义 RESTRICT、CASCADE、SET NULL、NO ACTION、SET DEFAULT 等级联操作。

① CASCADE：如果在主表中删除或更新数据，会自动删除或更新从表中匹配到的相应的数据。

② RESTRICT：如果在主表中删除和更新数据，MySQL 拒绝删除或更新操作。

③ NO ACTION：NO ACTION 是标准 SQL 中的关键字，在 MySQL 中 NO ACTION 和 RESTRICT 的作用相同，都是在修改或者删除之前去检查从表中是否有对应的数据，如果有，拒绝删除或更新，此时将回滚对主表中行的删除或更新操作。

④ SET NULL：如果在主表中删除或更新数据，则会将从表中外键对应的字段设置为 NULL。若要执行此约束，外键列必须可为空值。

⑤ SET DEFAULT：如果在主表中删除或更新数据，则会将从表中外键对应的字段设置为默认值。若要执行此约束，所有外键列都必须有默认定义。如果某个列可为空值，并且未设置显式的默认值，则将使用 NULL 作为该列的隐式默认值。需要注意的是，SET DEFAULT 只是 MySQL 解析器认可，InnoDB 目前不支持。

1) 添加数据

一个具有外键约束的从表在插入数据时，外键字段的值会受主表数据的约束，保证从表插入的数据必须符合约束规范的要求。例如，课程表和教师表建立了约束关系，在课程

表中想要插入教师表中不存在的教师编号时,语句如下:

```
mysql> INSERT INTO `courses` VALUES
    -> (108,'软件测试',4,20120038);
ERROR 1452 (23000): Cannot add or update a child row: a foreign key constraint fails (`studentscore`.`courses`, CONSTRAINT `courses_ibfk_1` FOREIGN KEY (`tid`) REFERENCES `teachers` (`tid`))
```

这时,由于主表中并没有教师编号为 20120038 的教师信息,插入这条数据 MySQL 会报错。这说明,在给从表添加数据时,从表外键字段不能插入主表中不存在的数据。

2) 更新数据

尽管外键约束的主要目的是控制可以存储在从表中的数据,但它还可以控制对主表中数据的更改。对于建立外键约束的关联数据表来说,若对主表进行更新操作,从表将按照其建立外键约束时设置的 ON UPDATE 参数自动执行相应的操作。

比如,ON UPDATE CASCADE 操作允许执行级联更新,如果主表发生更新,则从表也会对相应的字段进行更新。ON UPDATE SET NULL 操作会将从表中相应的值重置为 NULL 值。ON UPDATE NO ACTION 或 UPDATE RESTRICT 操作将拒绝任何更新。限制更新或者级联更新是两种最常见的选项。

以建立了约束关系的课程表和教师表为例,如果在课程表中添加外键约束时设置了 ON UPDATE RESTRICT,那么修改教师表中的记录时(比如将教师表中的教师编号 20100004 修改为 20100014),语句如下:

```
mysql> UPDATE teachers
    -> SET tid = 20100014
    -> WHERE tid = 20100004;
ERROR 1451 (23000): Cannot delete or update a parent row: a foreign key constraint fails (`studentscore`.`courses`, CONSTRAINT `courses_ibfk_1` FOREIGN KEY (`tid`) REFERENCES `teachers` (`tid`))
```

这时,由于从表中有引用教师编号为 20100004 的记录,当主表中的记录被更新时,MySQL 将拒绝更新并报错。

如果在课程表中添加外键约束时设置的是 ON UPDATE CASCADE,那么将教师表中的教师编号 20100004 修改为 20100014,再重新执行语句如下:

```
mysql> UPDATE teachers
    -> SET tid = 20100014
    -> WHERE tid = 20100004;
Query OK, 1 row affected (0.07 sec)
Rows matched: 1  Changed: 1  Warnings: 0
```

这时,MySQL 提示教师表的更新操作成功,并匹配到一行记录并同步修改,查看课程表中的信息。

```
mysql> select * from courses;
+-----+------------------+--------------+----------+
| cid | cname            | credit_point | tid      |
+-----+------------------+--------------+----------+
| 101 | 数据库基础与应用 |      5       | 20070003 |
| 102 | PHP程序设计      |      3       | 20080009 |
| 103 | java程序设计     |      5       | 20030008 |
| 104 | python程序设计   |      3       | 20050011 |
| 105 | 计算机基础       |      3       | 19980002 |
| 106 | android程序设计  |      4       | 19990001 |
| 107 | 网页制作         |      4       | 20100014 |
+-----+------------------+--------------+----------+
7 rows in set (0.00 sec)
```

可以看到，课程表中引用了此教师编号的记录会进行相应的修改。

3）删除数据

对于已建立外键约束的关联数据表来说，若要对主表执行删除操作，从表将按照其建立外键约束时设置的 ON DELETE 参数自动执行相应的操作。

比如，ON DELETE CASCADE 表示，如果主表进行删除操作，而且从表中的外键字段有关联记录，MySQL 就会进行同步删除从表中的相关记录。如果使用 ON DELETE SET NULL 操作，表示当主表中的记录被删除时，MySQL 会将从表中的外键列值设置为 NULL，前提条件是从表中的外键列必须接受 NULL 值。如果使用 ON DELETE NO ACTION 或 ON DELETE RESTRICT 操作，则 MySQL 将拒绝删除。限制删除或者级联删除是两种最常见的选项。

以建立了约束关系的课程表和教师表为例，如果在课程表中添加外键约束时设置了 ON DELETE RESTRICT，当主表进行删除操作时，若从表中的外键字段有关联记录，就会阻止主表的删除操作。比如要将教师表中的教师编号为 20080009 的记录删除，语句如下：

```
mysql> DELETE FROM teachers
    -> WHERE tid = 20080009;
ERROR 1451 (23000): Cannot delete or update a parent row: a foreign key constraint fails ('studentscore'.'courses', CONSTRAINT 'courses_ibfk_1' FOREIGN KEY ('tid') REFERENCES 'teachers' ('tid') ON UPDATE CASCADE)
```

这时，由于从表中有引用教师编号为 20080009 的记录，当主表中的记录被删除时，MySQL 将报错。

如果在课程表中添加外键约束时设置的是 ON DELETE CASCADE，当主表进行删除操作时，若从表中的外键字段有关联记录，则从表中与之对应的记录也会被同步删除。重新执行语句如下：

```
mysql> DELETE FROM teachers
    -> WHERE tid = 20080009;
Query OK, 1 row affected (0.07 sec)
```

这时，MySQL 提示教师表的删除操作成功，并匹配到一行记录并同步删除，查看课程表中的信息。

```
mysql> select * from courses;
+-----+---------------------+--------------+----------+
| cid | cname               | credit_point | tid      |
+-----+---------------------+--------------+----------+
| 101 | 数据库基础与应用    |            5 | 20070003 |
| 103 | java 程序设计       |            5 | 20030008 |
| 104 | python 程序设计     |            3 | 20050011 |
| 105 | 计算机基础          |            3 | 19980002 |
| 106 | android 程序设计    |            4 | 19990001 |
| 107 | 网页制作            |            4 | 20100004 |
+-----+---------------------+--------------+----------+
6 rows in set (0.00 sec)
```

可以看到，课程表中引用了此教师编号的记录被同步删除。

6.2.3 任务实现

【例 6-2-1】 在创建表时添加添加外键约束。

在学生成绩管理数据库中新创建一个 resume 表，用来存放学生的简历信息，使用 CREATE TABLE 创建数据表 resume，表中加入四个字段 id、author_no、title、content，分别表示简历编号、作者编号、简历名称、简历内容，设置 id 为主键。添加外键约束，将 resume 表中作者编号 author_no 和学生表 students 中的学生学号 sid 关联起来。

```
mysql> CREATE TABLE 'resume' (
    -> 'id' INT NOT NULL AUTO_INCREMENT,
    -> 'author_no'  int(20) unsigned NOT NULL,
    -> 'title' varchar(50) NOT NULL UNIQUE,
    -> 'content' text DEFAULT NULL,
    -> PRIMARY KEY ('id'),
    -> FOREIGN KEY ('author_no') REFERENCES students('sid')
    -> );
Query OK, 0 rows affected (0.44 sec)
```

这里在 CREATE TABLE 语句中省略了关键字 CONSTRAINT，MYSQL 会自动生成一个外键约束的名称，FOREIGN KEY 关键字指明了从表中需要引用外键的字段是 author_no，REFERENCES 关键字指明了关联到主表 studtents 中的字段是学生学号 sid。

【例6-2-2】 在修改表时添加外键约束。

在学生成绩管理数据库中已有students表和courses表,创建学生成绩表scores,用来存放学生的分数信息,表中加入三个字段sid、cid、score,分别表示学生学号、课程编号、考试分数,将sid和cid作为联合主键。

```
mysql> CREATE TABLE `scores` (
    -> `sid` int(10) unsigned NOT NULL,
    -> `cid` int(10) unsigned NOT NULL,
    -> `score` tinyint unsigned DEFAULT NULL,
    -> PRIMARY KEY (`sid`,`cid`)
    -> ) ENGINE=InnoDB DEFAULT CHARSET=utf8;
Query OK, 0 rows affected (0.33 sec)
```

使用ALTER TABLE的方式给scores表添加两个外键约束,将scores表中的sid关联到students表中的sid。

```
mysql> ALTER TABLE `scores`
    -> ADD
    -> FOREIGN KEY (`sid`) REFERENCES `students` (`sid`);
Query OK, 0 rows affected (1.07 sec)
Records: 0  Duplicates: 0  Warnings: 0
```

将scores表中的cid关联到courses表中的cid。

```
mysql> ALTER TABLE `scores`
    -> ADD
    -> FOREIGN KEY (`cid`) REFERENCES `courses` (`cid`);
Query OK, 0 rows affected (1.12 sec)
Records: 0  Duplicates: 0  Warnings: 0
```

【例6-2-3】 查看表中的外键约束。

查看学生成绩管理数据库中resume表和scores表的外键约束信息。

① 使用SHOW CREATE TABLE查看resume表信息。

```
mysql> SHOW CREATE TABLE resume \G
*************************** 1. row ***************************
       Table: resume
Create Table: CREATE TABLE `resume` (
  `id` int(11) NOT NULL AUTO_INCREMENT,
  `author_no` int(20) unsigned NOT NULL,
  `title` varchar(50) NOT NULL,
  `content` text,
  PRIMARY KEY (`id`),
  UNIQUE KEY `title` (`title`),
```

```
  KEY 'author_no' ('author_no'),
  CONSTRAINT 'resume_ibfk_1' FOREIGN KEY ('author_no') REFERENCES 'students' ('sid')
) ENGINE=InnoDB DEFAULT CHARSET=utf8
1 row in set (0.00 sec)
```

可以看到,resume 表中有一个外键约束名为 resume_ibfk_1,此约束将 resume 表中的作者编号和学生表 students 中的学生学号关联起来。

② 使用 SHOW CREATE TABLE 查看 scores 表信息。

```
mysql> SHOW CREATE TABLE scores \G
*************************** 1. row ***************************
       Table: scores
Create Table: CREATE TABLE 'scores' (
  'sid' int(10) unsigned NOT NULL,
  'cid' int(10) unsigned NOT NULL,
  'score' tinyint(3) unsigned DEFAULT NULL,
  PRIMARY KEY ('sid','cid'),
  KEY 'cid' ('cid'),
  CONSTRAINT 'scores_ibfk_1' FOREIGN KEY ('sid') REFERENCES 'students' ('sid'),
  CONSTRAINT 'scores_ibfk_2' FOREIGN KEY ('cid') REFERENCES 'courses' ('cid')
) ENGINE=InnoDB DEFAULT CHARSET=utf8
1 row in set (0.00 sec)
```

可以看到,此表共有两个外键约束,分别命名为 scores_ibfk_1 和 scores_ibfk_2,将 scores 表分别与 students 表和 courses 表关联起来。

【例 6-2-4】 设置级联更新和级联删除。

在学生成绩管理数据库中重新创建 resume 表,表中加入四个字段 id、author_no、title、content,分别表示简历编号、作者编号、简历名称、简历内容。添加外键约束,将 resume 表中的作者编号和学生表 students 中的学生学号关联起来。要求如果 students 表中的学号发生更新,则 resume 表也会对相应的字段进行更新;如果 students 表进行删除操作,resume 表也会同步删除相关记录。

```
mysql> CREATE TABLE 'resume' (
    -> 'id' INT NOT NULL AUTO_INCREMENT,
    -> 'author_no'  int(20) unsigned NOT NULL,
    -> 'title' varchar(50) NOT NULL UNIQUE,
    -> 'content' text DEFAULT NULL,
    -> PRIMARY KEY ('id'),
    -> FOREIGN KEY ('author_no') REFERENCES students('sid')
    -> ON UPDATE CASCADE ON DELETE CASCADE
    -> );
Query OK, 0 rows affected (0.48 sec)
```

这里使用 ON UPDATE CASCADE 操作允许执行级联更新,如果主表 students 发生更新,则从表 resume 也会对相应的字段进行更新,ON DELETE CASCADE 表示级联删除,如果主表 students 进行删除操作,从表 resume 会同步删除相应的记录。使用 SHOW CREATE TABLE 查看建表信息。

```
mysql> SHOW CREATE TABLE resume \G
*************************** 1. row ***************************
       Table: resume
Create Table: CREATE TABLE 'resume' (
  'id' int(11) NOT NULL AUTO_INCREMENT,
  'author_no' int(20) unsigned NOT NULL,
  'title' varchar(50) NOT NULL,
  'content' text,
  PRIMARY KEY ('id'),
  UNIQUE KEY 'title' ('title'),
  KEY 'author_no' ('author_no'),
  CONSTRAINT 'resume_ibfk_1' FOREIGN KEY ('author_no') REFERENCES 'students' ('sid') ON DELETE CASCADE ON UPDATE CASCADE
) ENGINE=InnoDB DEFAULT CHARSET=utf8
1 row in set (0.00 sec)
```

可以看到自动生成外键约束名称为 resume_ibfk_1,将 resume 表和 students 表关联起来。

【例 6-2-5】 使用 ALTER TABLE 删除外键约束。

将学生成绩管理数据库中删除外键约束 resume_ibfk_1。

```
mysql> ALTER TABLE 'resume' DROP FOREIGN KEY 'resume_ibfk_1';
Query OK, 0 rows affected (0.14 sec)
Records: 0  Duplicates: 0  Warnings: 0
```

重新查看建表信息。

```
mysql> show create table resume \G
*************************** 1. row ***************************
       Table: resume
Create Table: CREATE TABLE 'resume' (
  'id' int(11) NOT NULL AUTO_INCREMENT,
  'author_no' int(20) unsigned NOT NULL,
  'title' varchar(50) NOT NULL,
  'content' text,
  PRIMARY KEY ('id'),
  UNIQUE KEY 'title' ('title'),
  KEY 'author_no' ('author_no')
) ENGINE=InnoDB DEFAULT CHARSET=utf8
1 row in set (0.00 sec)
```

可以看到，外键约束删除成功。外键一旦删除，就会解除主表和从表间的关联关系。

6.2.4 训练任务

在图书管理数据库表中创建外键约束，具体要求如下。

① 给借阅表 Borrow 添加外键约束，分别关联 Reader 表的 name 和 Book 表的 id。

② 设置级联删除，如果 Reader 表中删除了一位读者，同时删除 Borrow 表中的数据。

③ 设置级联更新，如果 Reader 表中更新了一位读者的姓名，同时更新 Borrow 表中的数据。

6.2.5 任务总结

外键的作用就是用来在两个表的数据之间建立连接，保证两个表中数据的一致性和完整性。设置外键约束至少要有两种表，被约束的表是从表，另一张表是主表。外键约束的主要目的是用来控制可以存储在从表中的数据，还可以控制对主表中数据的更改和删除。

本任务主要讲解了什么是外键约束、添加和删除外键约束的方法，以及外键约束下的数据表的关联操作。通过本任务的学习，读者应掌握外键约束的添加和使用，能够使用外键约束保证数据的完整性和一致性。

任务 3

视　　图

6.3.1 任务描述

数据库将数据存储在数据表中，如果一个查询或子查询的结果被经常使用，我们就可以将它保存为视图。视图一般用来简化复杂查询，通过数据库视图，我们只需使用简单的 SQL 语句，无须使用具有多个连接的复杂的 SQL 语句，让查询更简单高效。

视图还有助于限制对特定用户的数据访问，比如某个表中有敏感数据是不对开发者开放的，只能给出某几列的权限，这时也可以使用视图来简化操作。视图在数据库中的作用类似于窗口，用户可以通过这个窗口看到只对自己有用的数据，那些对用户没用，或者用户没有权限了解的信息，都可以直接屏蔽掉，这样既保障了数据的安全性，又大大提高了查询效率。

本任务主要在学生成绩管理数据库中创建视图来简化查询，首先进行视图相关知识点的讲解。

6.3.2 知识准备

1. 视图的概念

视图是基于 SQL 语句结果形成的一张虚拟表,表的结构和数据都依赖于基本表。虽然视图看起来和基本表一样,也包含行和列,但是它们不是基本表。基本表的内容是持久的,而视图的内容是在使用过程中动态产生的。

视图是一个虚拟表,也称为临时表。视图保存的是 SELECT 查询语句而不是数据本身,视图本身并不包含数据。数据库中只存放了视图的定义。视图中行和列的数据来自定义视图的查询中所使用的表,在使用视图时动态生成。也就是说,视图展示的数据都存放在定义视图查询所引用的真实表中。使用视图查询数据时,数据库会从真实表中取出对应的数据。因此,视图中的数据是依赖于真实表中的数据的,一旦真实表中的数据发生改变,显示在视图中的数据也会发生改变。

视图主要有以下几个作用。

(1) 简化查询语句

使用视图可以简化用户的查询操作,使查询更加快捷。视图作为一个访问接口,不管基本表的表结构或者关联条件、筛选条件有多复杂,对使用视图的用户来说已经是过滤好的复合条件的结果集,用户只需基于视图进行简单查询。

(2) 数据安全性

通过视图可以更方便地进行权限控制。使用视图的用户只能访问他们被允许查询的结果集。视图可以把要保密的数据对无权访问这些数据的用户隐藏起来,对数据提供一定程度的安全保护。

(3) 数据独立性

视图还能屏蔽真实表结构变化带来的影响,体现逻辑数据的独立性。一旦视图的结构确定了,可以屏蔽表结构变化对用户的影响。基本表增加列对视图没有影响,基本表修改列名则可以通过修改视图来解决,不会造成对访问者的影响。

2. 视图操作

1) 创建视图

创建视图的语句如下:

```
CREATE VIEW〈视图名称〉
AS
SELECT 语句;
```

其中,CREATE VIEW 表示创建一个视图,并将 SELECT 语句存放到此视图中。

比如,在学生成绩管理数据库中,要查询某个课程的授课教师姓名,由于课程表中只有教师编号,是没有教师姓名的,故需要先将课程表和教师表做连接,查询语句如下:

```
mysql> select cid,cname,credit_point,courses.tid,tname
    -> from courses inner join teachers
    -> on courses.tid = teachers.tid;
+-----+------------------+--------------+----------+--------+
| cid | cname            | credit_point | tid      | tname  |
+-----+------------------+--------------+----------+--------+
| 101 | 数据库基础与应用  |      5       | 20070003 | 欧巧曼 |
| 102 | PHP 程序设计     |      3       | 20080009 | 何寻文 |
| 103 | java 程序设计    |      5       | 20030008 | 易安翔 |
| 104 | python 程序设计  |      3       | 20050011 | 杨文康 |
| 105 | 计算机基础       |      3       | 19980002 | 宁光明 |
| 106 | android 程序设计 |      4       | 19990001 | 胡昊空 |
| 107 | 网页制作         |      4       | 20100004 | 赵思恩 |
+-----+------------------+--------------+----------+--------+
7 rows in set (0.00 sec)
```

如果需要查询"java 程序设计"这门课程的教师姓名，那么查询语句应该改为如下：

```
mysql> select cid,cname,credit_point,courses.tid,tname
    -> from courses inner join teachers
    -> on courses.tid = teachers.tid
    -> where cname="java 程序设计";
+-----+---------------+--------------+----------+--------+
| cid | cname         | credit_point | tid      | tname  |
+-----+---------------+--------------+----------+--------+
| 103 | java 程序设计 |      5       | 20030008 | 易安翔 |
+-----+---------------+--------------+----------+--------+
1 row in set (0.00 sec)
```

如果课程表和教师表的连接操作需要经常使用，那么就可以将这个连接查询存放在视图中，语句如下：

```
mysql> CREATE VIEW courses_teachers_join_view
    -> AS
    -> select cid,cname,credit_point,courses.tid,tname
    -> from courses inner join teachers
    -> on courses.tid = teachers.tid;
Query OK, 0 rows affected (0.12 sec)
```

上述语句表示创建了一个视图，视图名称为 courses_teachers_join_view，我们可以像操作基本表一样来操作视图，如查询视图中所有数据如下：

```
mysql> select * from courses_teachers_join_view;
+-----+-------+--------------+-----+-------+
| cid | cname | credit_point | tid | tname |
+-----+-------+--------------+-----+-------+
```

```
| 101 | 数据库基础与应用      |   5 | 20070003 | 欧巧曼 |
| 102 | PHP 程序设计         |   3 | 20080009 | 何寻文 |
| 103 | java 程序设计        |   5 | 20030008 | 易安翔 |
| 104 | python 程序设计      |   3 | 20050011 | 杨文康 |
| 105 | 计算机基础           |   3 | 19980002 | 宁光明 |
| 106 | android 程序设计     |   4 | 19990001 | 胡昊空 |
| 107 | 网页制作             |   4 | 20100004 | 赵思恩 |
+-----+----------------------+-----+----------+--------+
7 rows in set (0.00 sec)
```

同样，在视图中查询"java 程序设计"这门课程的教师姓名，查询语句如下：

```
mysql> select * from courses_teachers_join_view
    -> where cname="java 程序设计";
+-----+---------------+--------------+----------+--------+
| cid | cname         | credit_point | tid      | tname  |
+-----+---------------+--------------+----------+--------+
| 103 | java 程序设计 |            5 | 20030008 | 易安翔 |
+-----+---------------+--------------+----------+--------+
1 row in set (0.00 sec)
```

可以看到，使用视图查询与使用基本表查询结果一致。

2）查看视图

视图其实就是一张虚拟的表，它与普通的表有些不同，查看视图可以有以下三种写法。

① 查看视图的字段信息。可以直接使用 DESCRIBE 语句查看视图的字段信息，语法格式如下：

```
DESC 〈视图名称〉;
```

比如，查看视图 courses_teachers_join_view 的字段信息。

```
mysql> DESC courses_teachers_join_view;
+--------------+------------------+------+-----+---------+-------+
| Field        | Type             | Null | Key | Default | Extra |
+--------------+------------------+------+-----+---------+-------+
| cid          | int(10) unsigned | NO   |     | 0       |       |
| cname        | varchar(20)      | YES  |     | NULL    |       |
| credit_point | int(5)           | YES  |     | NULL    |       |
| tid          | int(10) unsigned | NO   |     | NULL    |       |
| tname        | varchar(20)      | YES  |     | NULL    |       |
+--------------+------------------+------+-----+---------+-------+
5 rows in set (0.00 sec)
```

这里显示了视图 courses_teachers_join_view 的字段信息，其属性含义如表 6-3-1 所示。

表 6-3-1　视图描述属性信息

属性名	属性含义
Field	表示字段名
Type	表示字段类型
NULL	表示该列是否可以存储 NULL 值
Key	表示该列是否有索引
Default	表示该列是否有默认值
Extra	表示获取到的与给定列相关的附加信息

② 查看视图的基本信息。MySQL 中使用 SHOW TABLE STATUS 语句，可以查看视图的基本信息，语法格式如下：

```
SHOW TABLE STATUS LIKE '<视图名称>';
```

比如，查看视图 courses_teachers_join_view 的基本信息。

```
mysql> SHOW TABLE STATUS LIKE 'courses_teachers_join_view' \G
*************************** 1. row ***************************
           Name: courses_teachers_join_view
         Engine: NULL
        Version: NULL
     Row_format: NULL
           Rows: NULL
 Avg_row_length: NULL
    Data_length: NULL
Max_data_length: NULL
   Index_length: NULL
      Data_free: NULL
 Auto_increment: NULL
    Create_time: NULL
    Update_time: NULL
     Check_time: NULL
      Collation: NULL
       Checksum: NULL
 Create_options: NULL
        Comment: VIEW
1 row in set (0.00 sec)
```

可以看出，表的说明项 Comment 的值为 VIEW，说明查询的 courses_teachers_join_view 是一个视图；而存储引擎、数据长度等信息都显示为 NULL，说明视图是虚拟表。此语句也可以写作如下：

```
SHOW TABLE STATUS WHERE comment='view';
```

③ 查看视图的定义。使用 SHOW CREATE VIEW 语句不仅可以查看创建视图时的定义语句,还可以查看视图的字符编码,语法格式如下:

```
SHOW CREATE VIEW <视图名称>;
```

比如,查看视图 courses_teachers_join_view 的定义。

```
mysql> SHOW CREATE VIEW courses_teachers_join_view \G
*************************** 1. row ***************************
                View: courses_teachers_join_view
         Create View: CREATE ALGORITHM = UNDEFINED DEFINER = 'root'@'localhost' SQL
SECURITY DEFINER VIEW 'courses_teachers_join_view' AS select 'courses'.'cid' AS 'cid',
'courses'.'cname' AS 'cname','courses'.'credit_point' AS 'credit_point','courses'.'tid' AS 'tid',
'teachers'.'tname' AS 'tname' from ('courses' join 'teachers' on(('courses'.'tid' = 'teachers'.
'tid')))
character_set_client: utf8
collation_connection: utf8_general_ci
1 row in set (0.00 sec)
```

可以看到,这里显示了视图的名称、视图的创建语句、字符编码等信息。

3) 修改视图

使用 ALTER VIEW 可以重新定义视图的查询语句,修改视图的语法格式如下:

使用 UPDATE
修改视图数据

```
ALTER VIEW <视图名称>
AS
SELECT 语句;
```

比如,将视图 courses_teachers_join_view 显示的数据中添加一个字段信息 tschool。

```
mysql> ALTER VIEW courses_teachers_join_view
    -> AS
    -> select cid,cname,credit_point,courses.tid,tname,tschool
    -> from courses inner join teachers
    -> on courses.tid = teachers.tid;
Query OK, 0 rows affected (0.14 sec)
```

重新查看视图字段信息。

```
mysql> desc courses_teachers_join_view;
+-------+--------------+------+-----+---------+-------+
| Field | Type         | Null | Key | Default | Extra |
+-------+--------------+------+-----+---------+-------+
| cid   | int(10) unsigned | NO |     | 0       |       |
| cname | varchar(20)  | YES  |     | NULL    |       |
```

```
| credit_point | int(5) | YES |   | NULL |   |
| tid | int(10) unsigned | NO |   | NULL |   |
| tname | varchar(20) | YES |   | NULL |   |
| tschool | varchar(20) | YES |   | NULL |   |
+--------------+------------------+-----+---+------+---+
6 rows in set (0.00 sec)
```

可以看到,添加字段 tschool 成功。

4)删除视图

使用 DROP VIEW 加上视图名称,就可以将视图删除掉。删除视图的语法格式如下:

```
DROP VIEW〈视图名称〉;
```

比如,删除视图 courses_teachers_join_view。

```
mysql> DROP VIEW courses_teachers_join_view ;
Query OK, 0 rows affected (0.00 sec)
```

重新查看数据库中的视图。

```
mysql> SHOW TABLE STATUS WHERE comment='view';
Empty set (0.00 sec)
```

可以看到,视图删除成功。

6.3.3 任务实现

【例 6-3-1】 使用 CREATE VIEW 创建视图。

在学生成绩管理数据库中,经常要进行学生成绩统计,所以需要将学生表、课程表和成绩表这三个表进行连接,将连接查询保存为视图。在此视图中查询"java 程序设计"这门课程的平均成绩。首先,需要创建视图。

```
mysql> CREATE VIEW sc_s_c_view
    -> AS
    -> SELECT sc.sid,s.sname,s.sgender,sc.cid,c.cname,sc.score
    -> FROM scores  sc
    -> INNER JOIN students s
    -> ON sc.sid = s.sid
    -> INNER JOIN courses c
    -> ON sc.cid = c.cid;
Query OK, 0 rows affected (0.12 sec)
```

接下来使用此视图中做查询操作。

```
mysql> select cname,avg(score)
    -> from sc_s_c_view
    -> where cname = "java 程序设计";
+----------------+------------+
| cname          | avg(score) |
+----------------+------------+
| java 程序设计  |    72.4286 |
+----------------+------------+
1 row in set (0.15 sec)
```

【例 6-3-2】 查看视图的详细信息。

查看学生成绩管理数据库中的视图 sc_s_c_view 的字段信息。

```
mysql> desc sc_s_c_view;
+---------+------------------+------+-----+---------+-------+
| Field   | Type             | Null | Key | Default | Extra |
+---------+------------------+------+-----+---------+-------+
| sid     | int(10) unsigned | NO   |     | NULL    |       |
| sname   | varchar(20)      | YES  |     | NULL    |       |
| sgender | varchar(10)      | YES  |     | NULL    |       |
| cid     | int(10) unsigned | NO   |     | NULL    |       |
| cname   | varchar(20)      | YES  |     | NULL    |       |
| score   | tinyint(3) unsigned | YES |    | NULL    |       |
+---------+------------------+------+-----+---------+-------+
6 rows in set (0.00 sec)
```

查看视图 sc_s_c_view 的 SQL 语句定义。

```
mysql> SHOW CREATE VIEW sc_s_c_view \G
*************************** 1. row ***************************
                View: sc_s_c_view
         Create View: CREATE ALGORITHM = UNDEFINED DEFINER = 'root'@'localhost' SQL
SECURITY DEFINER VIEW `sc_s_c_view` AS select `sc`.`sid` AS `sid`,`s`.`sname` AS `sname`,`s`.
`sgender` AS `sgender`,`sc`.`cid` AS `cid`,`c`.`cname` AS `cname`,`sc`.`score` AS `score` from
((`scores` `sc` join `students` `s` on((`sc`.`sid` = `s`.`sid`))) join `courses` `c` on((`sc`.`cid` = `c`.
`cid`)))
character_set_client: utf8
collation_connection: utf8_general_ci
1 row in set (0.00 sec)
```

可以看到视图的定义及编码信息。

【例 6-3-3】 使用 ALTER VIEW 修改视图。

在学生成绩管理数据库中修改视图 sc_s_c_view 的查询语句，删除字段 sgender，语句如下：

```
mysql> ALTER VIEW sc_s_c_view
    -> AS
    -> SELECT sc.sid,s.sname,sc.cid,c.cname,sc.score
    -> FROM scores   sc
    -> INNER JOIN students s
    -> ON sc.sid = s.sid
    -> INNER JOIN courses c
    -> ON sc.cid = c.cid;
Query OK, 0 rows affected (0.12 sec)
```

重新查看视图的字段信息。

```
mysql> desc sc_s_c_view;
+-------+---------------------+------+-----+---------+-------+
| Field | Type                | Null | Key | Default | Extra |
+-------+---------------------+------+-----+---------+-------+
| sid   | int(10) unsigned    | NO   |     | NULL    |       |
| sname | varchar(20)         | YES  |     | NULL    |       |
| cid   | int(10) unsigned    | NO   |     | NULL    |       |
| cname | varchar(20)         | YES  |     | NULL    |       |
| score | tinyint(3) unsigned | YES  |     | NULL    |       |
+-------+---------------------+------+-----+---------+-------+
5 rows in set (0.00 sec)
```

可以看到,视图修改成功。

【例6-3-4】 使用 DROP VIEW 删除视图。

在学生成绩管理数据库中删除视图 sc_s_c_view。

```
mysql> DROP VIEW sc_s_c_view ;
Query OK, 0 rows affected (0.00 sec)
```

重新查看数据库中的视图。

```
mysql> SHOW TABLE STATUS WHERE comment='view';
Empty set (0.00 sec)
```

可以看到,视图 sc_s_c_view 删除成功。

6.3.4 训练任务

在图书管理数据库中创建视图并简化查询。具体要求如下。

① 创建视图 view_borrow_reader_1,查询借阅《1984》这本书的读者信息。

② 使用视图 view_borrow_reader_2,查询显示所有19岁读者的借阅信息。

③ 创建视图 view_borrow_book,要求显示图书的借阅情况,包括图书编号、书名、库存

数、借阅次数字段。

6.3.5 任务总结

视图是一张虚拟表，与包含数据的表不一样，视图只包含使用时动态检索数据的查询。也就是说，视图保存的是 SELECT 查询语句而不是数据本身。需要注意的是，如果视图由复杂的多表查询所定义，那么查询视图需要花费一定的时间。此外，创建视图时不允许使用 ORDER BY、INTO 等子句。

本任务主要讲解了视图的概念以及作用、视图的基本操作及使用视图的注意事项等知识点。读者应掌握视图的创建和使用，能够运用视图简化查询。

任务 4
存 储 过 程

6.4.1 任务描述

在开发过程中，经常会遇到重复使用某一功能的情况。存储过程是数据库的一个重要的对象，可以用来封装 SQL 语句集，完成一些较复杂的业务逻辑，与程序设计语言中函数的封装方式十分相似。

MySQL 5.0 版本开始支持存储过程。存储过程是数据库 SQL 语言层面的代码封装与重用，将重复性很高的一些操作封装到一个存储过程中，简化了对这些 SQL 的调用。我们可以将编译好的一段或多段 SQL 语句放置数据库端的存储过程中，方便开发者直接调用，而且创建存储过程时会预先编译后保存，开发者后续的调用不需再次编译。

本任务主要在学生成绩管理数据库中创建存储过程来进行代码重用，在完成任务之前，首先进行存储过程相关知识点的讲解。

6.4.2 知识准备

1. 存储过程的概念

存储过程（Stored Procedure）是一种在数据库中存储复杂程序，以便外部程序调用的一种数据库对象。存储过程用来完成特定的功能，经编译创建并保存在数据库中，用户可通过指定存储过程的名字并给定参数来调用执行。我们也可以理解为，存储过程就是数据库中被编译的 SQL 函数，它和普通函数一样有函数体，有顺序结构、条件结构、循环结构，有参数，有输入输出等。

使用存储过程有哪些好处呢？

第一，存储过程有助于提高应用程序的性能。当我们创建的存储过程被编译之后，就存储在数据库中。MySQL 存储过程按需编译,在编译存储过程之后,将其放入缓存中。MySQL 为每个连接维护自己的存储过程高速缓存。如果应用程序在单个连接中多次使用存储过程,则会使用编译版本,提高了应用程序的性能。

第二,存储过程有助于减少应用程序和数据库服务器之间的流量,因为应用程序不必发送多个冗长的 SQL 语句,而只需要发送存储过程的名称和参数。此外,存储的程序对任何应用程序都是可重用的和透明的。

第三,存储过程将数据库接口提供给所有应用程序,以便开发人员不必开发存储过程中已支持的功能。数据库管理员也可以向访问数据库中存储过程的应用程序授予适当的权限,便于开发者或数据库管理员使用和维护。

第四,在实际开发中,我们可以通过直接修改存储过程的方式修改业务逻辑或 bug,而不用重启服务器。存储过程可以回传值,并可以接受参数,执行速度快,存储过程经过编译之后会比单独一条一条编译执行要快很多。

存储过程的缺点也很明显,主要有以下三个方面。

第一,如果使用大量存储过程,那么使用这些存储过程的每个连接的内存使用量将会大大增加。由于采用过程化编程,开发具有复杂业务逻辑的存储过程变得更加困难。如果在存储过程中过度使用大量逻辑操作,则 CPU 使用率也会增加。

第二,MySQL 不提供调试存储过程的功能。开发和维护存储过程并不容易,复杂业务处理的维护成本高。而且不是所有应用程序开发人员都具备数据库开发和维护的技能,这可能会导致应用程序开发和维护阶段的问题。

第三,当切换到其他厂商的数据库系统时,需要重写原有的存储过程。存储过程的性能调校与撰写,受限于各种数据库系统。由于不同数据库语法不一致,因此会导致不同数据库之间可移植性差。

2. 存储过程的创建和调用

存储过程就是具有名字的一段代码,用来完成一个特定的功能。它可以定义批量插入的语句,也可以定义一个接收不同条件的 SQL,创建的存储过程保存在数据库的数据字典中。接下来详细介绍存储过程的创建和调用。

1)创建存储过程

创建一个存储过程的语句如下:

```
DELIMITER //
CREATE PROCEDURE <存储过程名>(参数列表)
BEGIN
    sql 语句;
END //
DELIMITER ;
```

我们可以分为三个部分来理解上述存储过程的定义。

首先需要注意的是开头和结尾的"DELIMITER //"和"DELIMITER ;"两句,它与存储过程语法无关。这里 DELIMITER 是分隔符的意思,我们使用 DELIMITER 来声明语句结束符,比如:

```
DELIMITER //
```

表示以"//"作为语句结束符,语句结束符也可以自定义,比如:

```
DELIMITER $ $
```

则表示以＄＄作为语句结束符。如果以＄＄作为语句结束符,那么上述存储过程的定义语句就可以改为:

```
DELIMITER $ $
CREATE PROCEDURE〈存储过程名〉(参数列表)
BEGIN
    sql 语句;
END $ $
DELIMITER ;
```

为什么必须更改分隔符呢?这是因为 MySQL 默认以";"为分隔符,如果我们没有声明分隔符,那么编译器会把存储过程当成 SQL 语句进行处理,则存储过程的编译过程会报错。所以,要事先用 DELIMITER 关键字声明分隔符,将语句的结束符号改为"//"或"＄＄",这样 MySQL 才会将";"当作存储过程中的代码,而不会执行这些代码。需要注意的是,存储过程定义完成之后还要用 DELIMITER 语句把分隔符还原成分号";"。

接下来,使用 CREATE PROCEDURE 语句创建一个新的存储过程,并指定存储过程的名称。声明存储过程语句如下:

```
CREATE PROCEDURE〈存储过程名〉(参数列表)
```

这个参数列表详细描述应该写作:

```
CREATE PROCEDURE〈存储过程名〉( [ IN | OUT | INOUT ]〈参数名称〉〈参数类型〉)
```

参数列表中具体表示的含义我们在后面的小节中会详细讲到。

最后,BEGIN 和 END 之间的部分称为存储过程的主体。主体部分的开始与结束分别使用 BEGIN 与 END 进行标识。过程体格式如下:

```
BEGIN
    sql 语句;
END //
```

我们将 SQL 语句放在主体中以处理业务逻辑。存储过程体包含了在过程调用时必须执行的语句,比如 DDL 语句、DML 语句、声明变量的 DECLARE 语句、存储过程的控制语句等,在后面的章节中会详细讲到。

我们以查看学生成绩管理数据库中学生表的所有学生姓名操作为例,创建一个简单的存储过程 GetAllStudents。

```
mysql> DELIMITER //
mysql> CREATE PROCEDURE GetAllStudents()
    -> BEGIN
    -> SELECT sname FROM students;
    -> END //
Query OK, 0 rows affected (0.00 sec)
mysql> DELIMITER ;
```

在这里,我们定义的存储过程名字为 GetAllStudents,不带任何参数,在这个存储过程中,我们使用一个简单的 SELECT 语句来查询 students 表中的数据。这样,一个 MySQL 简单的存储过程就定义完成了。

2)调用存储过程

和程序设计语言中的函数一样,定义好了存储过程,我们需要进行调用才能执行它,MySQL 中调用存储过程的语句如下:

```
CALL <存储过程名>(参数列表);
```

那么调用刚才建好的存储过程 GetAllStudents。

```
mysql> CALL GetAllStudents();
+--------+
| sname  |
+--------+
| 刘洋   |
| 邓嘉蓉 |
| 葛佳音 |
| 索凤洋 |
| 曹宏美 |
| 郭思宇 |
| 周慧敏 |
+--------+
7 rows in set (0.00 sec)
Query OK, 0 rows affected (0.01 sec)
```

3. 存储过程的使用

1)查看存储过程

通过 SHOW PROCEDURE STATUS 和 SHOW CREATE PROCEDURE 语句,可以列出数据库中的存储过程并获取存储过程的详细信息。

① 使用 SHOW PROCEDURE STATUS 查看数据库中的存储过程。我们知道,如果

想要查看一个数据库里面有哪些表,一般采用 SHOW TABLES 进行查看。那么要查看某个数据库下面有哪些存储过程呢？我们也可以用 SHOW PROCEDURE 语句进行查看,语法格式如下：

```
SHOW PROCEDURE STATUS WHERE db='〈数据库名〉';
```

比如,查看 studentscore 数据库中定义了哪些存储过程。

```
mysql> SHOW PROCEDURE STATUS WHERE db='studentscore';
+------------+-------------+-----------+----------------+---------------------+---------------------+---------------+---------+--------------------+--------------------+--------------------+
| Db         | Name        | Type      | Definer        | Modified            | Created             | Security_type | Comment | character_set_client | collation_connection | Database Collation |
+------------+-------------+-----------+----------------+---------------------+---------------------+---------------+---------+--------------------+--------------------+--------------------+
|studentscore|GetAllStudents| PROCEDURE |root@localhost | 2021-06-09 20:44:13 | 2021-06-09 20:44:13 | DEFINER       |         | gbk                | gbk_chinese_ci     | utf8_general1      |
+------------+-------------+-----------+----------------+---------------------+---------------------+---------------+---------+--------------------+--------------------+--------------------+
1 row in set (0.016 sec)
```

这里查看到的存储过程信息可读性比较弱,加上"\G"来将查到的结构变成纵向。

```
mysql> SHOW PROCEDURE STATUS WHERE db='studentscore'\G
*************************** 1. row ***************************
                  Db: studentscore
                Name: GetAllStudents
                Type: PROCEDURE
             Definer: root@localhost
            Modified: 2021-06-09 19:41:44
             Created: 2021-06-09 19:41:44
       Security_type: DEFINER
             Comment:
character_set_client: utf8
collation_connection: utf8_general_ci
  Database Collation: utf8_general_ci
1 row in set (0.00 sec)
```

从上面可以看到刚才定义的存储过程 GetAllStudents,包括创建时间以及编码格式等信息。

② 使用 SHOW CREATE PROCEDURE 查看存储过程的定义。如果我们想知道某个存储过程的详细定义,可以使用语句如下：

```
SHOW CREATE PROCEDURE 〈存储过程名〉;
```

那么,查看存储过程 GetAllStudents 的详细定义的语句如下：

```
mysql> SHOW CREATE PROCEDURE GetAllStudents\G
*************************** 1. row ***************************
       Procedure: GetAllStudents
```

```
    sql_mode:
  STRICT_TRANS_TABLES,NO_AUTO_CREATE_USER,NO_ENGINE_SUBSTITUTION
  Create Procedure: CREATE DEFINER='root'@'localhost' PROCEDURE
  'GetAllStudents'()
BEGIN
    select sname from students;
END
character_set_client: utf8
 collation_connection: utf8_general_ci
   Database Collation: utf8_general_ci
1 row in set (0.00 sec)
```

2）修改存储过程

存储过程创建完成后，如果想要修改，可以使用 ALTER PROCEDURE 语句进行修改，语法格式如下：

```
ALTER PROCEDURE〈存储过程名〉[characteristic……];
```

其中 characteristic 表示特性选项，用于指定修改存储过程的哪个部分，其取值具体如下表 6-4-1 所示。

表 6-4-1　存储过程特征选项

特征选项	特征值描述
CONTAINS SQL	表示子程序包含除读或写数据的 SQL 语句
NO SQL	表示子程序中不包含 SQL 语句
READS SQL DATA	表示子程序中包含读数据的语句
MODIFIES SQL DATA	表示子程序中包含写数据的语句
SQL SECURITY {DEFINER | INVOKER}	指明谁有权限来执行，其中 DEFINER 表示只有定义者自己才能够执行，INVOKER 表示调用者可以执行
COMMENT '注释内容'	表示给存储过程添加注释

这里以修改存储过程 GetAllStudents 的权限和添加注释为例，语句如下：

```
mysql> ALTER PROCEDURE GetAllStudents
    -> SQL SECURITY INVOKER
    -> COMMENT '获取学生信息';
Query OK, 0 rows affected (0.22 sec)
```

再查看 GetAllStudents 存储过程的状态。

```
mysql> SHOW PROCEDURE STATUS WHERE db='studentscore' \G
*************************** 1. row ***************************
                  Db: studentscore
                Name: GetAllStudents
                Type: PROCEDURE
             Definer: root@localhost
            Modified: 2021-06-09 19:57:36
             Created: 2021-06-09 19:41:44
       Security_type: INVOKER
             Comment: 获取学生信息
character_set_client: utf8
collation_connection: utf8_general_ci
  Database Collation: utf8_general_ci
1 row in set (0.00 sec)
```

可以看到,通过 ALTER PROCEDURE 语句,我们修改了 GetAllStudents 存储过程的权限为调用者可以执行,并添加了注释信息。

3) 删除存储过程

删除一个存储过程比较简单,和删除表一样,语句如下:

```
DROP PROCEDURE IF EXISTS〈存储过程名〉;
```

那么从数据库中删除之前创建的存储过程 GetAllStudents,可以使用语句如下:

```
mysql> DROP PROCEDURE IF EXISTS GetAllStudents;
Query OK, 0 rows affected (0.01 sec)
```

然后使用 SHOW PROCEDURE STATUS 查看数据库中的存储过程,发现删除成功。

```
mysql> SHOW PROCEDURE STATUS WHERE db='studentscore';
Empty set (0.14 sec)
```

4. 变量的定义和赋值

一般程序设计语言中都会用到变量,用来保存数据处理过程中的值。在 MySQL 存储过程中也可以声明和使用变量,接下来进行详细的讲解。

1) 变量定义

在存储过程的过程体中,我们可以声明局部变量,用来临时保存一些值,变量的值可以在存储过程执行期间更改。声明变量的语法格式如下:

```
DECLARE 变量名 datatype [DEFAULT 默认值];
```

首先,在 DECLARE 关键字后面要指定变量名,变量名必须遵循 MySQL 表列名称的命名规则。datatype 表示数据类型,我们需要指定变量的数据类型及其大小。变量可以是任何 MySQL 数据类型,如 INT、VARCHAR、DATETIME 等。使用 DEFAULT 子句给变量提供一个默认值,这个默认值可以是常数,也可以指定为一个表达式。如果没有 DEFAULT 子句,则变量初始值为 NULL。

需要注意的是,DECLARE 仅被用在 BEGIN … END 语句块里,而且一定要放在存储过程体的开始。我们也可以在一个存储过程中定义多个变量,语法格式如下:

```
CREATE PROCEDURE〈存储过程名〉(参数列表)
BEGIN
    DECLARE 变量名1  datatype [DEFAULT 默认值];
    DECLARE 变量名2  datatype [DEFAULT 默认值];
    DECLARE……;
    sql 语句;
END
```

例如,我们可以在存储过程中声明一个名为 num 的变量,数据类型为 INT,默认值为 100,语句如下:

```
DECLARE num INT DEFAULT 100;
```

MySQL 允许使用单个 DECLARE 语句声明共享相同数据类型的两个或多个变量,例如:

```
DECLARE x, y INT DEFAULT 0;
```

则表示声明了两个整数变量 x 和 y,其默认值均设置为 0。

2) 变量赋值

我们知道,变量的值可以在存储过程执行期间更改。那么变量声明了,怎么修改呢?在常规情况下,我们通常用 SET 语句给变量赋值,语法格式如下:

```
SET 变量名 = 值;
```

比如,编写存储过程 GetStudentInfo,定义一个变量表示学号,根据学号查找此学生的姓名和专业,语句如下:

```
mysql> DELIMITER //
mysql> CREATE PROCEDURE GetStudentInfo()
    -> BEGIN
    -> DECLARE stu_id INT UNSIGNED DEFAULT 0;
    -> SET stu_id = 1933062301;
    -> SELECT sname,smajor FROM students where sid = stu_id;
    -> END //
Query OK, 0 rows affected (0.00 sec)
mysql> DELIMITER ;
```

注意:变量必须先声明后才能使用它。上述代码中首先使用 DECLARE 语句声明变量 stu_id,然后使用 SET 语句直接给变量 stu_id 赋值,调用存储过程。

```
mysql> CALL GetStudentInfo();
+-------+--------+
| sname | smajor |
+-------+--------+
| 刘洋  | 软件技术 |
+-------+--------+
1 row in set (0.00 sec)
```

除了使用 SET 语句赋值，变量还可以通过 SELECT INTO 的方式赋值，也就是使用 SELECT INTO 语句将查询的结果分配给一个变量，重新定义 GetStudentInfo 存储过程。

```
mysql> DELIMITER //
mysql> CREATE PROCEDURE GetStudentInfo()
    -> BEGIN
    -> DECLARE total_students INT DEFAULT 0;
    -> SELECT count(*) INTO total_students FROM students;
    -> SELECT total_students;
    -> END //
Query OK, 0 rows affected (0.14 sec)
mysql> DELIMITER ;
```

在这个存储过程中，首先声明一个名为 total_students 的变量，并将其值初始化为 0。然后，使用 SELECT INTO 语句来分配值给 total_students 变量，这个值是从 studentscore 数据库 students 表中计算的学生数量。最后使用 SELECT 语句将变量输出。调用此存储过程。

```
mysql> CALL GetStudentInfo();
+----------------+
| total_students |
+----------------+
|              7 |
+----------------+
1 row in set (0.14 sec)
```

3) MySQL 中的变量

MySQL 中使用的变量可以大致分为三类：系统变量、用户变量、局部变量，下面进行详细讲解。

（1）系统变量

系统变量一般用两个@符号表示，根据作用域不同系统变量又可以分为全局变量与会话变量。

① 全局变量(@@global)：全局变量是 MySQL 数据库内置的变量，其作用域为整个服务器，也就是针对于所有连接(会话)有效。MySQL 启动的时候由服务器自动将全局变量

初始化为默认值。

② 会话变量(@@session)：服务器为每一个连接的客户端提供了系统变量,作用域为当前客户端与数据库服务器端的一次连接当中。如果连接断开,那么会话变量全部丢失。会话变量是由系统提供的,只在当前会话(连接)中有效。在每次建立一个新的连接的时候,由 MySQL 来初始化,MySQL 会将当前所有全局变量的值复制一份作为会话变量。

全局变量与会话变量的区别是,对全局变量的修改会影响到整个服务器,但是对会话变量的修改只会影响到当前的会话。

查看系统变量的语句如下：

```
SHOW [GLOBAL | SESSION] VARIABLES [LIKE '匹配模式' | WHERE 条件表达式];
```

比如,我们想查看全局变量中变量名有 char 的记录,语句如下：

```
mysql> SHOW GLOBAL  VARIABLES like '%char%';
+--------------------------+--------------------------------------------------------+
| Variable_name            | Value                                                  |
+--------------------------+--------------------------------------------------------+
| character_set_client     | latin1                                                 |
| character_set_connection | latin1                                                 |
| character_set_database   | latin1                                                 |
| character_set_filesystem | binary                                                 |
| character_set_results    | latin1                                                 |
| character_set_server     | latin1                                                 |
| character_set_system     | utf8                                                   |
| character_sets_dir       | c:\wamp64\bin\mysql\mysql5.7.14\share\charsets\        |
+--------------------------+--------------------------------------------------------+
8 rows in set, 1 warning (0.09 sec)
```

查看指定的系统变量的值,语句如下：

```
select @@[global | session].变量名;
```

给系统变量赋值的语句如下：

```
SET  @@[global | session].变量名 = 值;
```

如果没有显式声明 global 还是 session,默认后者。

比如,我们要查看全局变量 character_set_client 的值,语句如下：

```
mysql> select @@global.character_set_client;
+-------------------------------+
| @@global.character_set_client |
+-------------------------------+
| latin1                        |
+-------------------------------+
1 row in set (0.00 sec)
```

查看会话变量 character_set_client 的值,语句如下：

```
mysql> select @@session.character_set_client;
+--------------------------------+
| @@session.character_set_client |
+--------------------------------+
| utf8                           |
+--------------------------------+
1 row in set (0.00 sec)
```

（2）用户变量

用户变量是指用户自定义的会话变量，用一个@符号表示，它也是当前会话（连接）有效。也就是说，在当前连接中声明的用户变量，在连接断开的时候就会失效。

查看用户变量的语句如下：

SELECT @变量名；

给用户变量赋值的语句如下：

SET @变量名 = 值；

比如，我们可以直接声明一个用户变量并查看，语句如下：

```
mysql> SET @user_name = 'sunny';
Query OK, 0 rows affected (0.10 sec)
mysql> select @user_name;
+------------+
| @user_name |
+------------+
| sunny      |
+------------+
1 row in set (0.00 sec)
```

需要注意的是，用户变量随处可以定义，不定义也可以直接使用，不需要提前声明，默认值为 NULL。

（3）局部变量

我们前一小节在存储过程中用 DECLARE 语句声明的变量都属于局部变量。局部变量只在存储过程内部使用，在过程体外是没有意义的。当存储过程体处理完后，局部变量就会超出其作用域，在其他代码块中无法访问。

对于局部变量的使用，需要注意的是：局部变量使用 DECLARE 语句进行声明，而且必须先声明后使用。局部变量的作用范围在 BEGIN 与 END 之间。局部变量可以通过 SET 赋值，也可以通过 SELECT INTO 的方式赋值。如果需要查看局部变量的值，可以使用 SELECT 语句。

5. 存储过程的参数

在前面的章节中我们讲解了声明存储过程，语句如下：

CREATE PROCEDURE 〈存储过程名〉(参数列表)

这个参数列表详细描述应该写作：

CREATE PROCEDURE 〈存储过程名〉([IN | OUT | INOUT]〈参数名称〉〈参数类型〉)

本节将详细讲解存储过程中参数的使用。存储过程可以有 0 个或多个参数，多个参数则需要用","分隔开。如果括号内什么都不定义，就说明该存储过程没有参数。我们在前面创建的存储过程都是没有参数的。实际上，存储过程根据需要可能会有 IN 输入参数、OUT 输出参数、INOUT 输入输出参数这三种模式。

① IN 输入参数：是定义传入参数的关键字，表示调用者向存储过程传入值，该参数的值必须在调用存储过程时指定。另外，即使在存储过程中更改了 IN 参数的值，在存储过程结束后仍保留其原始值。

② OUT 输出参数：是定义输出参数的关键字，表示存储过程向调用者返回值。我们可以在存储过程中更改 OUT 参数的值，并将其更改后的新值传递回调用者。存储过程在启动时无法访问 OUT 参数的初始值。

③ INOUT 输入输出参数：是定义一个出入参数都可以的参数。既表示调用者向存储过程传入值，又表示存储过程向调用者返回值。

1) IN 输入参数

重新定义存储过程 GetStudentInfo，将学号的值作为 IN 输入参数传入，根据学号查找此学生的姓名和专业，语句如下：

```
mysql> DELIMITER //
mysql> CREATE PROCEDURE GetStudentInfo(IN stu_id INT UNSIGNED)
    -> BEGIN
    -> SELECT sname,smajor FROM students WHERE sid = stu_id;
    -> END //
Query OK, 0 rows affected (0.02 sec)
mysql> DELIMITER ;
```

在上面的定义的存储过程中，参数列表中传入了一个 IN 输入参数，命名为 stu_id，类型为 INT 型。此参数的值必须在调用存储过程时指定，指定的值可以是常量，也可以是变量。比如，传入字面量值进行调用，语句如下：

```
mysql> CALL GetStudentInfo(1933062301);
+-------+-----------+
| sname | smajor    |
+-------+-----------+
| 刘洋  | 软件技术  |
+-------+-----------+
1 row in set (0.06 sec)
```

或者，也可以传入变量的值进行调用，这时变量必须在存储过程之外使用，我们需要使用用户变量，也就是加上一个@符号，因为用户变量不定义也可以直接使用，所以可以用 SET 语句直接给用户变量赋值，语句如下：

```
mysql> SET @stu_id = 1933062301;
Query OK, 0 rows affected (0.00 sec)
mysql> CALL GetStudentInfo(@stu_id);
+-------+-----------+
| sname | smajor    |
+-------+-----------+
| 刘洋  | 软件技术  |
+-------+-----------+
1 row in set (0.10 sec)

Query OK, 0 rows affected (0.12 sec)
```

注意：这里 stu_id 和@stu_id 是两个不同的变量，stu_id 只能在存储过程中使用，是局部变量；而@stu_id 是用户变量，可以在多个块之间使用。

2）OUT 输出参数

IN 类型参数一般只用于传入，在调用存储过程中一般不作修改和返回，如果调用存储过程中需要修改和返回值，可以使用 OUT 类型参数，下面举例说明。

比如，定义一个新的存储过程 GetStudentName，将学号的值作为 IN 输入参数传入，根据学号查找此学生的姓名并返回，语句如下：

```
DELIMITER //
CREATE PROCEDURE GetStudentName(
  IN stu_id INT UNSIGNED,
  OUT stu_name VARCHAR(20))
BEGIN
  SELECT sname INTO stu_name FROM students WHERE sid = stu_id;
END //
DELIMITER ;
```

在这个存储过程中，根据学号值查找到学生姓名，并将学生姓名 sname 的值存入了 OUT 参数 stu_name 中。如果我们要想在存储过程之外获取到学生姓名这个值，语句如下：

```
mysql> SET @stu_id = 1933062301;
Query OK, 0 rows affected (0.00 sec)
mysql> CALL GetStudentName(@stu_id,@stu_name);
Query OK, 1 row affected (0.00 sec)
```

```
mysql> SELECT @stu_name;
+-----------+
| @stu_name |
+-----------+
| 刘洋      |
+-----------+
1 row in set (0.00 sec)
```

注意：这里@stu_name 是一个新的用户变量，用户变量不定义也可以直接使用，这时@stu_name 里面存放的就是存储过程 OUT 参数 stu_name 的值，这样我们就实现了存储过程向调用者返回值。

3) INOUT 输入输出参数

INOUT 参数是 IN 和 OUT 参数的组合。调用程序可以向存储过程传递参数，并且存储过程可以修改参数值并将新值传递回调用程序。下面定义一个存储过程 SetStudentCount，用来修改学生总数。语句如下：

```
mysql> DELIMITER //
mysql> CREATE PROCEDURE SetStudentCount(
    ->   IN increase INT,
    ->   INOUT total_students INT)
    -> BEGIN
    ->   SELECT count(*) INTO total_students FROM students;
    ->   SET total_students = total_students + increase;
    -> END //
Query OK, 0 rows affected (0.09 sec)
mysql> DELIMITER ;
```

这个存储过程接收一个 IN 参数（increase）和一个 INOUT 参数（total_students），在存储过程中，通过 increase 参数的值增加学生总数。语句如下：

```
mysql> CALL SetStudentCount(10,@total_students);
Query OK, 1 row affected (0.08 sec)
mysql> SELECT @total_students;
+-----------------+
| @total_students |
+-----------------+
|              17 |
+-----------------+
1 row in set (0.00 sec)
```

以上就是定义存储过程时可以使用的三种参数，一般来说，需要输入值使用 IN 参数，需要返回值使用 OUT 参数,既要传入值又要返回值可以使用 INOUT 参数。需要注意的是,如果存储过程没有参数,也必须在过程名后面写上小括号。此外,必须确保参数的名字和列的名字不同,否则在过程体中,参数名会被当作列名来处理。

6. 存储过程的控制语句

在存储过程中可以使用流程控制语句来控制程序的流程。MySQL 中流程控制语句有:IF 语句、CASE 语句、WHILE 语句、REPEAT 语句、LOOP 语句、LEAVE 语句和 ITERATE 语句等。下面将详细讲解这些流程控制语句。

1) IF 语句

IF 语句用来进行条件判断,根据是否满足条件来执行不同的语句,是流程控制中最常用的判断语句。IF 语句有三种常见的写法,分别是单分支语句、双分支语句以及多分支语句。

存储过程中的 IF 语句

(1) 单分支 IF 语句

最简单的单分支 IF 语句的语法格式如下:

```
IF 条件表达式 THEN
    语句列表
END IF;
```

其意思是,如果条件表达式的值为 TRUE,执行相应的 SQL 语句列表。下面举例说明。

定义一个存储过程 GetScoreLevel,用来判定学生各门课程的平均成绩是否大于等于 60 分,如果结果为 TRUE,则设定此学生的成绩等级为及格。语句如下:

```
mysql> DELIMITER //
mysql> CREATE PROCEDURE GetScoreLevel(
    ->     IN stu_id INT UNSIGNED,
    ->     OUT stu_level VARCHAR(20))
    -> BEGIN
    ->     DECLARE avg_score INT DEFAULT 0;
    ->     SELECT AVG(score) INTO avg_score FROM scores
    ->     WHERE sid = stu_id;
    ->     IF avg_score >= 60 THEN
    ->         SET stu_level = "及格";
    ->     END IF;
    -> END //
Query OK, 0 rows affected (0.06 sec)
mysql> DELIMITER ;
```

这个存储过程中定义了一个 IN 输入参数 stu_id 表示学生学号,一个 OUT 输出参数 stu_level 用来表示学生成绩等级。语句如下:

```
mysql> SET @stu_id = 1933062301;
Query OK, 0 rows affected (0.00 sec)
mysql> CALL GetScoreLevel(@stu_id,@stu_level);
Query OK, 1 row affected (0.00 sec)
mysql> SELECT @stu_level;
+------------+
| @stu_level |
+------------+
| 及格       |
+------------+
1 row in set (0.00 sec)
```

可以看到学号为 1933062301 的同学，其平均分是大于等于 60 分的，所以学生成绩等级 @stu_level 的值是及格。但如果我们换一个同学，比如学号为 1933062304，语句如下：

```
mysql> SET @stu_id = 1933062304;
Query OK, 0 rows affected (0.00 sec)
mysql> CALL GetScoreLevel(@stu_id,@stu_level);
Query OK, 1 row affected (0.00 sec)
mysql> SELECT @stu_level;
+------------+
| @stu_level |
+------------+
| NULL       |
+------------+
1 row in set (0.00 sec)
```

可以看到，学号为 1933062304 的这位同学平均分是低于 60 分的，但这时 @stu_level 的值为 NULL。如果我们需要指明学生平均成绩小于 60 分设置成绩等级为不及格，需要使用双分支 IF-ELSE 语句。

（2）双分支 IF 语句

双分支 IF 语句的语法格式如下：

```
IF 条件表达式 THEN
    语句列表
ELSE
    语句列表
END IF;
```

这个语句表示，如果条件判断语句的值为 TRUE，执行相应的 SQL 语句列表。如果条件判断语句的值为 FALSE，则 ELSE 子句的语句列表被执行。重新定义 GetScoreLevel，语句如下：

```
mysql> DELIMITER //
mysql> CREATE PROCEDURE GetScoreLevel(
    ->     IN stu_id INT UNSIGNED,
    ->     OUT stu_level VARCHAR(20))
    -> BEGIN
    ->     DECLARE avg_score INT DEFAULT 0;
    ->     SELECT AVG(score) INTO avg_score FROM scores
    ->     WHERE sid = stu_id;
    ->     IF avg_score >= 60 THEN
    ->         SET stu_level = "及格";
    ->     ELSE
    ->         SET stu_level = "不及格";
    ->     END IF;
    -> END //
Query OK, 0 rows affected (0.00 sec)
mysql> DELIMITER ;
```

调用存储过程查看学号为 1933062304 同学成绩的等级，语句如下：

```
mysql> SET @stu_id = 1933062304;
Query OK, 0 rows affected (0.00 sec)
mysql> CALL GetScoreLevel(@stu_id,@stu_level);
Query OK, 1 row affected (0.00 sec)
mysql> SELECT @stu_level;
+------------+
| @stu_level |
+------------+
| 不及格     |
+------------+
1 row in set (0.00 sec)
```

这时会发现，该同学的成绩等级变成了不及格。如果我们想对学生成绩多设置几个等级，比如大于等于 80 分设置为 A 等，60~80 分之间为 B 等，60 分以下为 C 等，像这样有多个选项需要处理，这时就需要用到多分支语句。

(3) 多分支 IF 语句

多分支 IF 语句的语法格式如下：

```
IF 条件表达式 THEN
    语句列表
ELSEIF 条件表达式 THEN
    语句列表
...
ELSE
    语句列表
END IF;
```

重新定义 GetScoreLevel，编写多分支语句，如果平均分大于等于 80 分，设置 stu_level 为 A 等，平均分在 60~80 分之间，设置 stu_level 为 B 等，60 分以下设置 stu_level 为 C 等，语句如下：

```
mysql> DELIMITER //
mysql> CREATE PROCEDURE GetScoreLevel(
    ->   IN stu_id INT UNSIGNED,
    ->   OUT stu_level VARCHAR(20))
    -> BEGIN
    ->   DECLARE avg_score INT DEFAULT 0;
    ->   SELECT AVG(score) INTO avg_score FROM scores
    ->   WHERE sid = stu_id;
    ->   IF avg_score >= 80 THEN
    ->     SET stu_level = "A等";
    ->   ELSEIF avg_score >= 60 THEN
    ->     SET stu_level = "B等";
    ->   ELSE
    ->     SET stu_level = "C等";
    ->   END IF;
    -> END //
Query OK, 0 rows affected (0.00 sec)
mysql> DELIMITER ;
```

调用存储过程，语句如下：

```
mysql> SET @stu_id = 1933062301;
Query OK, 0 rows affected (0.00 sec)
mysql> CALL GetScoreLevel(@stu_id,@stu_level);
Query OK, 1 row affected (0.00 sec)
mysql> SELECT @stu_level;
+------------+
| @stu_level |
+------------+
| A等        |
+------------+
1 row in set (0.00 sec)
```

读者可传入不同学号调用此存储过程，验证不同学生的平均成绩等级。

2）CASE 语句

CASE 语句也是用来进行条件判断的，它提供了多个条件进行选择，可以实现比 IF 语句更复杂的条件判断。简单 CASE 语句的语法格式如下：

存储过程中的 CASE 语句

```
CASE case_value
  WHEN when_value_1 THEN
```

```
        语句列表
    WHEN when_value_2 THEN
        语句列表
    ...
    ELSE
        语句列表
END CASE;
```

其中,case_value 参数表示条件判断的变量,可以是任何有效的表达式。我们将 case_value 的值与每个 WHEN 子句中的 when_value 值进行比较,如果 case_value 和 when_value_n 的值相等,则执行相应的 WHEN 分支中的语句列表。如果 WHEN 子句中所有 when_value 值与 case_value 的值都不匹配,则 ELSE 子句中的语句列表将被执行。需要注意的是,ELSE 子句是可选的。但是如果省略了 ELSE 子句,并且找不到匹配项,MySQL 将引发错误。CASE 语句都要使用 END CASE 结束。

简单 CASE 语句仅允许将表达式的值与一组不同的值进行匹配。为了执行更复杂的匹配,我们可以使用可搜索 CASE 语句。该形式的语法格式如下:

```
CASE
    WHEN search_condition THEN
        语句列表
    WHEN search_condition THEN
        语句列表
    ...
    ELSE 语句列表
END CASE;
```

其中,search_condition 参数表示条件判断语句。可搜索 CASE 语句类似于 IF 语句,但是它的构造更加可读。与简单 CASE 语句不同的是,该语句中的 WHEN 语句将被逐个执行,如果某个 search_condition 表达式为真,则执行对应 THEN 关键字后面的语句列表。如果没有一个 WHEN 语句的表达式为真,则 ELSE 子句里的语句被执行。接下来举例说明。

重新定义 GetScoreLevel,编写 CASE 语句。如果平均分大于等于 80 分,设置 stu_level 为 A 等;平均分在 60~80 分之间,设置 stu_level 为 B 等;60 分以下设置 stu_level 为 C 等。语句如下:

```
mysql> DELIMITER //
mysql> CREATE PROCEDURE GetSCoreLevel(
    ->     IN stu_id INT UNSIGNED,
    ->     OUT stu_level VARCHAR(20))
    -> BEGIN
    ->     DECLARE avg_score INT DEFAULT 0;
```

```
    -> SELECT AVG(score) INTO avg_score FROM scores
    -> WHERE sid = stu_id;
    -> CASE
    ->     WHEN avg_score >= 80 THEN
    ->         SET stu_level = "A等";
    ->     WHEN avg_score < 80 AND avg_score >= 60 THEN
    ->         SET stu_level = "B等";
    ->     ELSE
    ->         SET stu_level = "C等";
    -> END CASE;
    -> END //
Query OK, 0 rows affected (0.03 sec)
mysql> DELIMITER ;
```

调用存储过程,语句如下:

```
mysql> SET @stu_id = 1933062302;
Query OK, 0 rows affected (0.00 sec)
mysql> CALL GetSCoreLevel(@stu_id,@stu_level);
Query OK, 1 row affected (0.00 sec)
mysql> SELECT @stu_level;
+------------+
| @stu_level |
+------------+
| B等        |
+------------+
1 row in set (0.10 sec)
```

可以看到,可搜索 CASE 语句类似于 IF 语句的写法,读者可传入不同学号调用此存储过程,验证不同学生的平均成绩等级。

3) WHILE 语句

如果要重复执行一个 SQL 代码块,MySQL 提供了三种循环语句:WHILE、REPEAT 和 LOOP。下面逐一进行讲解。

WHILE 语句的语法形式如下:

利用循环语句求解水仙花数

```
WHILE search_condition DO
    语句列表
END WHILE;
```

其中,search_condition 参数表示循环执行的条件表达式,如果表达式结果为 TRUE,将执行 WHILE 和 END WHILE 之间的语句列表,直到表达式结果为 FALSE,循环结束。接下来进行举例说明。

定义一个存储过程 GetLoopNumbers,用来查看循环变量的取值,语句如下:

```
mysql> DELIMITER //
mysql> CREATE PROCEDURE GetLoopNumbers()
    -> BEGIN
    ->   DECLARE x INT ;
    ->   DECLARE string_value VARCHAR(20) ;
    ->   SET x = 1;
    ->   SET string_value = "";
    ->   WHILE x <= 5 DO
    ->     SET string_value = CONCAT(string_value,x," ");
    ->     SET x = x + 1;
    ->   END WHILE;
    ->   SELECT string_value;
    -> END //
Query OK, 0 rows affected (0.14 sec)
mysql> DELIMITER ;
```

在上述存储过程中,有三个关键语句,首先设置变量 x 的初值为 1,当 x≤5 时,将每个 x 的值用空格连接起来,然后通过"x=x+1;"语句将循环变量 x 转到下一项,这三个关键语句保证了循环的正常退出。调用存储过程,语句如下:

```
mysql> CALL GetLoopNumbers();
+--------------+
| string_value |
+--------------+
| 1 2 3 4 5    |
+--------------+
1 row in set (0.00 sec)
```

可以看到,此循环语句将变量 x 的取值从 1 到 5,并连成字符串输出。

4) REPEAT 语句

REPEAT 语句也是有条件控制的循环语句,和 WHILE 语句不同的是,REPEAT 语句是在每次语句执行完毕之后,才会对条件表达式 search_condition 进行判断。如果表达式返回值为 TRUE,则循环结束,否则继续执行循环中的语句列表。REPEAT 语句的语法格式如下:

```
REPEAT
    语句列表
    UNTIL search_condition
END REPEAT;
```

重新定义存储过程 GetLoopNumbers,可用 REPEAT 语句实现,语句如下:

```
mysql> DELIMITER //
mysql> CREATE PROCEDURE GetLoopNumbers()
    -> BEGIN
    ->   DECLARE x INT ;
    ->   DECLARE string_value VARCHAR(20) ;
    ->   SET x = 1;
    ->   SET string_value = "";
    ->   REPEAT
    ->     SET string_value = CONCAT(string_value,x," ");
    ->     SET x = x + 1;
    ->     UNTIL x > 5
    ->   END REPEAT;
    ->   SELECT string_value;
    -> END //
Query OK, 0 rows affected (0.00 sec)
mysql> DELIMITER ;
```

在上述存储过程中，首先设置变量 x 的初值为 1，先不做任何判断，直接将第一个 x 的值用空格连接起来，然后通过"x=x+1;"语句将循环变量 x 转到下一项，直到 x>5 条件成立的时候循环退出。调用存储过程，语句如下：

```
mysql> CALL GetLoopNumbers();
+--------------+
| string_value |
+--------------+
| 1 2 3 4 5    |
+--------------+
1 row in set (0.00 sec)
Query OK, 0 rows affected (0.01 sec)
```

由此可见，使用 REPEAT 语句和 WHILE 语句执行的结果一样，但执行流程却是不一样的。WHILE 语句是先判断条件后执行，REPEATE 语句先执行一次后再判断条件，也就是不管条件真假至少执行一次。

5) LOOP 语句

LOOP 语句可以使某些特定的语句重复执行，LOOP 语句实现了一个简单的循环，并不进行条件判断。LOOP 语句的语法格式如下：

```
label:LOOP
    语句列表
END LOOP;
```

其中，label 是标记循环开始，表示循环执行 LOOP 中的语句列表。LOOP 一般用于实现简单的死循环，也就是说，LOOP 语句本身没有停止循环的语句，必须使用 LEAVE 语句才能停止循环,跳出循环过程。

6) LEAVE 语句和 ITERATE 语句

LEAVE 语句主要用于跳出循环控制,可以和 LOOP 语句一起使用,用来终止循环,其语法格式如下:

```
LEAVE label;
```

其中,label 参数是 LOOP 循环的标记,用来标记退出标记所指的循环。重新定义存储过程 GetLoopNumbers,用 LOOP 语句和 LEAVE 语句实现,语句如下:

```
mysql> DELIMITER //
mysql> CREATE PROCEDURE GetLoopNumbers()
    -> BEGIN
    -> DECLARE x INT;
    -> DECLARE string_value VARCHAR(20);
    -> SET x = 1;
    -> SET string_value = "";
    -> loop_label_name:LOOP
    ->   IF x > 5 THEN
    ->     LEAVE loop_label_name;
    ->   END IF;
    ->   SET string_value = CONCAT(string_value,x," ");
    ->   SET x = x + 1;
    -> END LOOP;
    -> SELECT string_value;
    -> END //
Query OK, 0 rows affected (0.00 sec)
mysql> DELIMITER ;
```

在上述存储过程中,首先设置变量 x 的初值为 1,直接设置一个 LOOP 循环,标记为 loop_label_name,在循环将每一个 x 的值用空格连接起来,然后通过"x=x+1;"语句将循环变量 x 转到下一项。在上面的代码中,我们需要用到 LEAVE 语句将循环退出,也就是用 IF 语句进行条件判断,如果 x>5 条件成立的时候就使用 LEAVE 语句退出循环。调用存储过程,语句如下:

```
mysql> CALL GetLoopNumbers();
+--------------+
| string_value |
+--------------+
| 1 2 3 4 5    |
+--------------+
1 row in set (0.00 sec)
Query OK, 0 rows affected (0.01 sec)
```

和 LEAVE 语句比较相似的还有一个 ITERATE 语句,ITERATE 语句主要用来跳出本次循环,直接进入下一次循环。ITERATE 语句的基本语法格式如下:

```
ITERATE label;
```

其中,label 参数表示循环的标记,ITERATE 语句必须跟在循环标志前面。重新定义存储过程 GetLoopNumbers,用来构造具有偶数数字的字符串,语句如下:

```
mysql> DELIMITER //
mysql> CREATE PROCEDURE GetLoopNumbers()
    -> BEGIN
    ->   DECLARE x INT;
    ->   DECLARE string_value VARCHAR(20);
    ->   SET x = 1;
    ->   SET string_value = "";
    ->   loop_label_name:LOOP
    ->     IF x > 10 THEN
    ->       LEAVE loop_label_name;
    ->     END IF;
    ->     SET x = x + 1;
    ->     IF (x % 2) THEN
    ->       ITERATE loop_label_name;
    ->     ELSE
    ->       SET string_value = CONCAT(string_value,x," ");
    ->     END IF;
    ->   END LOOP;
    ->   SELECT string_value;
    -> END //
Query OK, 0 rows affected (0.11 sec)
mysql> DELIMITER ;
```

在上述存储过程中,如果循环变量 x 不能被 2 整除,也就是 x 是奇数的情况下,使用 ITERATE 语句跳出本次循环,提前进入下一次循环。如果 x 的值是偶数,则 ELSE 语句中的块将使用偶数构建字符串。另外,如果 x 的值大于 10,由 LEAVE 语句将循环结束。调用存储过程,语句如下:

```
mysql> CALL GetLoopNumbers();
+--------------+
| string_value |
+--------------+
| 2 4 6 8 10   |
+--------------+
1 row in set (0.00 sec)
Query OK, 0 rows affected (0.01 sec)
```

需要注意的是,LEAVE 语句和 ITERATE 语句都用来跳出循环语句,但两者的功能是不一样的。LEAVE 语句是跳出整个循环,然后执行循环后面的程序。而 ITERATE 语句是跳出本次循环,然后进入下一次循环。使用这两个语句时一定要区分清楚。

6.4.3 任务实现

【例 6-4-1】 使用 CREATE PROCEDURE 创建存储过程。

在学生成绩管理数据库中定义一个存储过程 GetStudentAvgScore,将学生姓名的值作为 IN 输入参数传入,根据学生姓名求出此学生的平均成绩并返回,调用此存储过程查看结果。编写此存储过程的语句如下:

```
mysql> DELIMITER //
mysql> CREATE PROCEDURE GetStudentAvgScore(
    -> IN stu_name_in VARCHAR(20),
    -> OUT avg_score DECIMAL(5,2))
    -> BEGIN
    -> SELECT avg(sc.score) INTO avg_score
    -> FROM scores  sc
    -> INNER JOIN students s
    -> ON sc.sid = s.sid
    -> WHERE s.sname = stu_name_in;
    -> END //
Query OK, 0 rows affected (0.00 sec)
mysql> DELIMITER ;
```

这里先用 dilimiter 修改分隔符为"//",然后使用 CREATE PROCEDURE 表示创建存储过程 GetStudentAvgScore。这里定义了两个参数:IN 表示输入参数,命名为 stu_name_in,用来表示学生姓名,类型为 varchar;OUT 表示输出参数,命名为 avg_score,用于表示待计算的平均成绩,类型为 DECIMAL(5,2)。这里输入输出参数的类型要与它所表示的字段类型一致。接下来把查询语句直接放到 BEGIN END 语句块之间。

注意:查询语句修改了两处地方。第一处需要修改 s.sname = stu_name_in,表明查找的学生姓名是由输入参数 stu_name_in 给定的值来决定的。第二处修改了 select 展示的字段,这里添加了 INTO avg_score,表明将计算后的平均成绩放在了输出变量 avg_score 中。最后用 dilimiter 把分隔符还原成分号。调用此存储过程的语句如下:

```
mysql> CALL GetStudentAvgScore("刘洋",@avg_score);
Query OK, 1 row affected, 1 warning (0.00 sec)
mysql> SELECT @avg_score;
+------------+
| @avg_score |
+------------+
|      89.67 |
```

```
+------------------+
1 row in set (0.00 sec)
```

可以看到，调用存储过程时传入学生名字刘洋，然后调用存储过程并将执行完的结果存放到用户变量@avg_score中，之后使用 SELECT 语句查看此变量，就得到了该学生的平均成绩。

【例 6-4-2】 使用 SHOW PROCEDURE 查看存储过程。

使用 SHOW PROCEDURE STATUS 语句查看 studentscore 数据库中的所有存储过程。语句如下：

```
mysql> SHOW PROCEDURE STATUS WHERE db='studentscore' \G
*************************** 1. row ***************************
                  Db: studentscore
                Name: GetLoopNumbers
                Type: PROCEDURE
             Definer: root@localhost
            Modified: 2021-06-09 21:19:13
             Created: 2021-06-09 21:19:13
       Security_type: DEFINER
             Comment:
character_set_client: utf8
collation_connection: utf8_general_ci
  Database Collation: utf8_general_ci
*************************** 2. row ***************************
                  Db: studentscore
                Name: GetSCoreLevel
                Type: PROCEDURE
             Definer: root@localhost
            Modified: 2021-06-09 21:02:46
             Created: 2021-06-09 21:02:46
       Security_type: DEFINER
             Comment:
character_set_client: utf8
collation_connection: utf8_general_ci
  Database Collation: utf8_general_ci
*************************** 3. row ***************************
                  Db: studentscore
                Name: GetStudentAvgScore
                Type: PROCEDURE
             Definer: root@localhost
            Modified: 2021-06-09 21:58:40
             Created: 2021-06-09 21:58:40
```

```
        Security_type: DEFINER
              Comment:
character_set_client: utf8
collation_connection: utf8_general_ci
    Database Collation: utf8_general_ci
*************************** 4. row ***************************
                  Db: studentscore
                Name: GetStudentInfo
                Type: PROCEDURE
             Definer: root@localhost
            Modified: 2021-06-09 20:32:24
             Created: 2021-06-09 20:32:24
       Security_type: DEFINER
              Comment:
character_set_client: utf8
collation_connection: utf8_general_ci
    Database Collation: utf8_general_ci
*************************** 5. row ***************************
                  Db: studentscore
                Name: GetStudentName
                Type: PROCEDURE
             Definer: root@localhost
            Modified: 2021-06-09 20:41:44
             Created: 2021-06-09 20:41:44
       Security_type: DEFINER
              Comment:
character_set_client: utf8
collation_connection: utf8_general_ci
    Database Collation: utf8_general_ci
*************************** 6. row ***************************
                  Db: studentscore
                Name: SetStudentCount
                Type: PROCEDURE
             Definer: root@localhost
            Modified: 2021-06-09 20:44:13
             Created: 2021-06-09 20:44:13
       Security_type: DEFINER
              Comment:
character_set_client: utf8
collation_connection: utf8_general_ci
    Database Collation: utf8_general_ci
6 rows in set (0.00 sec)
```

【例6-4-3】 使用 DROP PROCEDURE 删除存储过程。

删除学生成绩管理数据库中的存储过程 GetStudentAvgScore,语句如下：

```
mysql> DROP PROCEDURE IF EXISTS GetStudentAvgScore;
Query OK, 0 rows affected (0.00 sec)
```

重新查看数据库中的存储过程：

```
mysql> SHOW PROCEDURE STATUS WHERE db='studentscore';
```

Db	Name	Type	Definer	Modified	Created	Security_type	Comment	character_set_client	collation_connection	Database Collation
studentscore	GetLoopNumbers	PROCEDURE	root@localhost	2021-06-09 21:19:13	2021-06-09 21:19:13	DEFINER		gbk	gbk_chinese_ci	utf8_general1
studentscore	GetSCoreLevel	PROCEDURE	root@localhost	2021-06-09 21:02:46	2021-06-09 21:02:46	DEFINER		gbk	gbk_chinese_ci	utf8_general1
studentscore	GetStudentInfo	PROCEDURE	root@localhost	2021-06-09 20:32:24	2021-06-09 20:32:24	DEFINER		gbk	gbk_chinese_ci	utf8_general1
studentscore	GetStudentName	PROCEDURE	root@localhost	2021-06-09 20:41:44	2021-06-09 20:41:44	DEFINER		gbk	gbk_chinese_ci	utf8_general1
studentscore	GetStudentCount	PROCEDURE	root@localhost	2021-06-09 20:44:13	2021-06-09 20:44:13	DEFINER		gbk	gbk_chinese_ci	utf8_general1

5 row in set (0.016 sec)

可以看到,存储过程删除成功。

6.4.4 训练任务

在图书管理数据库中创建和调用存储过程。

① 定义的存储过程名字为 GetAllReaders,不带任何参数,用来查看读者表中所有读者姓名,调用此存储过程查看结果。

② 定义一个存储过程 GetReaderName,将读者编号的值作为 IN 输入参数传入,根据读者编号查找此读者的姓名并返回。调用此存储过程查看结果。

③ 查看图书管理数据库中的存储过程。

④ 删除存储过程 GetAllReaders。

6.4.5 任务总结

存储过程是事先经过编译并存储在数据库中的一段 SQL 语句的集合,调用存储过程可以简化开发人员的很多工作,减少数据和应用服务器之间的传输,提高了数据处理效率,简化了 SQL 语句的调用。

本任务主要讲解了存储过程的概念、存储过程的创建及调用、查看存储过程、删除存储过程的方法,以及存储过程中的变量使用和流程控制。通过本任务的学习,读者应掌握存储过程的编写和调用,能够使用存储过程提高数据库效率。

任务 5

触 发 器

6.5.1 任务描述

在实际开发项目时,我们经常会遇到以下情况:在学生表中添加一条关于学生的记录时,学生的总数就需要同时改变;增加一条成绩记录时,需要检查成绩是否符合范围要求;删除一条学生信息时,需要在数据库存档表中保留一个备份副本。虽然上述情况实现的业务逻辑不同,但是它们都需要在数据表发生更改时,自动进行一些处理,这时我们可以使用触发器来进行处理。

MySQL 5.0 版本开始支持触发器。触发器和存储过程一样,都是 MySQL 中的一段数据库存储类型程序,是 MySQL 中管理数据的有力工具。不同的是,执行存储过程要使用 CALL 语句来调用,而触发器的执行不需要使用 CALL 语句调用,也不需要手工启动,而是通过对数据表的相关操作来触发并执行。触发器与数据表关系密切,主要用于保护表中的数据,特别是当有多个表具有一定的相互联系的时候,触发器能够让不同的表保持数据的一致性。

本任务主要是在学生成绩管理数据库使用触发器来保持数据完整性。在完成任务之前,首先进行触发器相关知识点的讲解。

6.5.2 知识准备

1. 触发器的概念

触发器(TRIGGER)是在数据库中执行任何 DML 操作时可以触发的脚本或指令集。触发器是一种特殊类型的存储过程,但不同于存储过程,触发器主要是通过事件进行触发而被执行的,而存储过程可以通过存储过程名字直接调用。

当对某一数据表进行诸如 INSERT、UPDATE 和 DELETE 这些操作时,MySQL 就会自动执行触发器所定义的 SQL 语句,从而确保对数据的处理必须符合由这些 SQL 语句所定义的规则。触发器是与表有关的数据库对象,在满足定义条件时触发,并执行触发器中定义的语句集合。我们也可以简单理解为:当我们执行一条 SQL 语句,这条 SQL 语句的执行会自动去触发执行其他的 SQL 语句。

触发器的主要作用就是能够实现由主键和外键所不能保证的复杂的参照完整性和数据的一致性。因为触发器的执行是自动的,当对触发器相关表的数据作出相应的修改后立即执行。触发器可以实施比 FOREIGN KEY 约束、CHECK 约束更为复杂的检查和操作,可以实现表数据的级联更改,在一定程度上保证了数据的完整性。

触发器也有一些明显的缺点。由于触发器是一个隐藏的存储过程,它不需要参数,也不需要显式调用,往往在人不知情的情况下已经做了很多操作,从这个角度来说,无形中增加了系统的复杂性,非 DBA 人员理解数据库就会有困难,因为触发器不执行根本感觉不到它的存在。而且,使用触发器实现的业务逻辑在出现问题时很难进行定位,特别是涉及多个触发器的情况下,会使后期维护变得困难。此外,MySQL 触发器基于行触发,在需要变动整个数据集而数据量又较大时,触发器的执行效率就会非常低。

2. 触发器的创建和执行

1) 创建触发器

创建触发器的语法格式如下:

```
CREATE TRIGGER 〈触发器名称〉〈触发时间〉〈触发事件〉
ON〈表名〉
FOR EACH ROW
BEGIN
    语句列表
END
```

说明:

① 〈触发器名称〉:这里使用 CREATE TRIGGER 语句创建一个新的触发器,并指定触发器名称。

② 〈触发时间〉:可以取值为 BEFORE 或者 AFTER,表示触发器被触发的时间,表明触发器是在激活它的语句之前或之后触发。BEFORE 表示在检查约束前触发,AFTER 表示在检查约束后触发。一般来说,若希望验证新数据是否满足条件,则使用 BEFORE 选项;若希望在激活触发器的语句执行之后完成几个或更多的改变,则通常使用 AFTER 选项。

③ 〈触发事件〉:可以取值为 INSERT、UPDATE 或者 DELETE 三种,表示触发器是因为什么事件被触发。在 MySQL 中,只有执行 INSERT、UPDATE 和 DELETE 操作时才能激活触发器,其他 SQL 语句则不会激活触发器。三种触发器的执行时间如下。

INSERT:将新行插入表时激活触发器,可能通过 INSERT 语句触发。

UPDATE:更改表中某一行数据时激活触发器,可能通过 UPDATE 语句触发。

DELETE:删除表中某一行数据时激活触发器,可能通过 DELETE、REPLACE 语句触发。

④ ON〈表名〉:指明触发器是绑定在哪张数据表上的,当在这个表上执行 INSERT、UPDATE 和 DELETE 操作的时候就导致触发器被激活。

⑤ FOR EACH ROW:表示任何一条记录上的操作满足触发事件都会触发该触发器,也就是说触发器的触发频率是针对每一行数据触发一次。

⑥ BEGIN...END 语句块:BEGIN...END 语句块定义了触发器一旦被触发会执行的 SQL 语句列表。一般来说,当触发器主体中有多个语句时使用 BEGIN 和 END 子句;对于单个 SQL 语句,不需要加 BEGIN 和 END 直接编写也可以。

⑦ NEW 与 OLD 关键字：MySQL 中定义了 NEW 和 OLD 关键字，用来引用触发器中发生变化的记录内容，使用方法如下：

```
NEW.columnName
```

或者：

```
OLD.columnName
```

其中，OLD 或 NEW 代表的是触发了触发器的那一行数据，columnName 为相应数据表某一列名，有如下三种情况：

a. 在触发目标上执行 INSERT 操作后会有一个新行，如果在触发事件中需要用到这个新行的变量，可以用 NEW 关键字表示，NEW 用来表示将要（BEFORE）或已经（AFTER）插入的新数据；

b. 在触发目标上执行 UPDATE 操作后原记录是旧行，新记录是新行，可以使用 OLD 和 NEW 关键字来分别操作，OLD 用来表示将要或已经被修改的原数据，NEW 用来表示将要或已经修改为的新数据；

c. 在触发目标上执行 DELETE 操作后会有一个旧行，如果在触发事件中需要用到这个旧行的变量，可以用 OLD 关键字表示，OLD 用来表示将要或已经被删除的原数据。

另外，OLD 是只读的，而 NEW 则可以在触发器中使用 SET 赋值，这样不会再次触发触发器，造成循环调用。

在 MySQL 5.7.2 之前的版本中，对同一个表，相同触发时间、相同触发事件只能定义一个触发器，也就是每个表最多可以定义六个触发器，如表 6-5-1 所示。

表 6-5-1 MySQL 六种触发器

触发器类型	说明
BEFORE INSERT	在将数据插入表中之前激活触发器
AFTER INSERT	将数据插入表后激活触发器
BEFORE UPDATE	在更新表中的数据之前激活触发器
AFTER UPDATE	更新表中的数据后激活触发器
BEFORE DELETE	在从表中删除数据之前激活触发器
AFTER DELETE	从表中删除数据后激活触发器

注：MySQL 5.7.2+版本解决了这样的限制，并允许为表中的相同事件和动作时间创建多个触发器，当事件发生时，触发器将依次激活。

以学生成绩管理数据库中成绩表的修改操作为例，创建一个 AFTER UPDATE 型触发器。如果成绩表中的分数有修改，就会自动往 scores_log 表里插入一条记录，记下成绩修改的详细信息。

首先，在数据库中新创建一个表 scores_log，用来存放学生成绩修改的日志信息，表中加入六个字段 id、sid、cid、old_score、new_score、update_time，分别表示日志编号、学生学号、课程编号、原来的分数、修改后的分数、修改时间，并设置 id 作为主键，语句如下：

```
mysql> CREATE TABLE 'scores_log' (
    -> 'id'   INT AUTO_INCREMENT PRIMARY KEY ,
    -> 'sid' int(10) unsigned NOT NULL,
    -> 'cid' int(10) unsigned NOT NULL,
    ->   'old_score' tinyint unsigned DEFAULT NULL,
    ->   'new_score' tinyint unsigned DEFAULT NULL,
    -> 'update_time' datetime NOT NULL
    -> ) ENGINE = InnoDB DEFAULT CHARSET = utf8;
Query OK, 0 rows affected (0.44 sec)
```

接下来创建一个触发器，用来实现当更新某个学生的成绩之后，自动在 scores_log 表中新增一条修改成绩的记录，语句如下：

```
mysql> DELIMITER //
mysql> CREATE TRIGGER after_scores_update AFTER UPDATE
    -> ON scores
    -> FOR EACH ROW
    -> BEGIN
    ->  insert  into 'scores_log'
    -> ('sid','cid','old_score','new_score','update_time')
    -> values
    -> (OLD.sid,OLD.cid,OLD.score,NEW.score,NOW());
    -> END //
Query OK, 0 rows affected (0.19 sec)
mysql> DELIMITER ;
```

这里先用 DELIMITER 修改分隔符为"//"，然后使用 CREATE TRIGGER 语句创建名为 after_scores_update 的触发器。设置触发时间为 AFTER，触发事件为 UPDATE，表名为 scores，这三个信息指明了什么情况下触发器会被触发。也就是当 scores 中更新记录之后，触发器会被触发。FOR EACH ROW 表示触发器会对每一行数据的插入都执行相应的触发器动作。

接下来定义触发器被触发时要做的动作。当 scores 表中成绩更新时，我们会往 scores_log 表中插入一条数据，并把 INSERT 语句放到 BEGIN END 语句块之间。

```
insert  into 'scores_log'
('sid','cid','old_score','new_score','update_time')
values
(OLD.sid,OLD.cid,OLD.score,NEW.score,NOW());
```

这条 INSERT 语句表示将学生学号、课程编号、原来的分数、修改后的分数，以及更新数据的时间,插入 scores_log 表。这里用 OLD 指代原来的那条记录,NEW 指代新插入的那条记录。最后用 DELIMITER 把分隔符还原成分号,这样触发器就创建完成了。

2)执行触发器

触发器在创建完成后,若要执行触发器,则需要让触发器指定的数据表执行对应的操作。刚刚创建的 after_scores_update 是一个 UPDATE 型触发器,也就是当我们向 scores 中更新数据之后,触发器会被触发执行。如果我们使用 UPDATE 语句修改一条学生成绩,语句如下:

```
mysql> UPDATE scores
    -> SET score = 100
    -> WHERE sid = "1933062301" AND cid = "101";
Query OK, 1 row affected (0.22 sec)
Rows matched: 1  Changed: 1  Warnings: 0
```

可以看到我们将学号为 1933062301 的学生所修课程编号为 101 这门课程的成绩修改为 100,这里显示更新操作成功。

```
mysql> select * from scores
    -> WHERE sid = "1933062301" AND cid = "101";
+------------+-----+-------+
| sid        | cid | score |
+------------+-----+-------+
| 1933062301 | 101 |   100 |
+------------+-----+-------+
1 row in set (0.00 sec)
```

那么成绩表中更新了学生成绩,触发器有没有被触发呢?接下来查看 scores_log 表中的记录。

```
mysql> select * from scores_log;
+----+------------+-----+-----------+-----------+---------------------+
| id | sid        | cid | old_score | new_score | update_time         |
+----+------------+-----+-----------+-----------+---------------------+
|  1 | 1933062301 | 101 |        80 |       100 | 2021-06-10 10:04:51 |
+----+------------+-----+-----------+-----------+---------------------+
1 row in set (0.00 sec)
```

可以看到,成绩更新的信息已经自动写入了 scores_log 表,触发器中的动作被成功执行。

3. 触发器操作

1)查看触发器

查看触发器可以有两种方法。

① 使用 SHOW TRIGGERS 查看数据库中的触发器。

语法格式如下：

```
SHOW TRIGGERS [ IN〈数据库名称〉]
[LIKE'匹配模式'| | WHERE 条件表达式];
```

比如，查看学生成绩管理数据库中的所有触发器，语句如下：

```
mysql> SHOW TRIGGERS IN studentscore \G
*************************** 1. row ***************************
             Trigger: after_scores_update
               Event: UPDATE
               Table: scores
           Statement: BEGIN
 insert   into 'scores_log'
 ('sid','cid','old_score','new_score','update_time')
 values
 (OLD.sid,OLD.cid,OLD.score,NEW.score,NOW());
END
              Timing: AFTER
             Created: 2021-06-10 10:03:01.25
            sql_mode:
 STRICT_TRANS_TABLES,NO_AUTO_CREATE_USER,NO_ENGINE_SUBSTITUTION
             Definer: root@localhost
character_set_client: utf8
 collation_connection: utf8_general_ci
   Database Collation: utf8_general_ci
1 row in set (0.00 sec)
```

可以查看到我们刚才创建的 after_scores_update 触发器。它是一个 UPDATE 型触发器，触发事件绑定在 scores 表中，触发器的激活时间是 AFTER。还可以查看到触发器的其他详细信息，比如调用触发器时要执行的语句以及触发器的创建时间等。

② 通过库 information_schema 的 triggers 表查看。

所有触发器信息都存储在 information_schema 数据库下的 triggers 表中，可以使用 SELECT 语句查询。如果触发器过多，最好通过 TRIGGER_NAME 字段指定查询，语法格式如下：

```
SELECT * FROM INFORMATION_SCHEMA.TRIGGERS
WHERE TRIGGER_NAME = "〈触发器名称〉";
```

比如，查看学生成绩管理数据库 after_scores_update 触发器的详细信息，语句如下：

```
mysql> SELECT * FROM INFORMATION_SCHEMA.TRIGGERS
   -> where  TRIGGER_NAME = "after_scores_update" \G
*************************** 1. row ***************************
        TRIGGER_CATALOG: def
```

```
                TRIGGER_SCHEMA: studentscore
                  TRIGGER_NAME: after_scores_update
            EVENT_MANIPULATION: UPDATE
          EVENT_OBJECT_CATALOG: def
           EVENT_OBJECT_SCHEMA: studentscore
            EVENT_OBJECT_TABLE: scores
                  ACTION_ORDER: 1
              ACTION_CONDITION: NULL
              ACTION_STATEMENT: BEGIN
insert   into 'scores_log'
('sid','cid','old_score','new_score','update_time')
values
(OLD.sid,OLD.cid,OLD.score,NEW.score,NOW());
END
            ACTION_ORIENTATION: ROW
                 ACTION_TIMING: AFTER
    ACTION_REFERENCE_OLD_TABLE: NULL
    ACTION_REFERENCE_NEW_TABLE: NULL
      ACTION_REFERENCE_OLD_ROW: OLD
      ACTION_REFERENCE_NEW_ROW: NEW
                       CREATED: 2021-06-10 10:03:01.25
                      SQL_MODE:
STRICT_TRANS_TABLES,NO_AUTO_CREATE_USER,NO_ENGINE_SUBSTITUTION
                       DEFINER: root@localhost
          CHARACTER_SET_CLIENT: utf8
          COLLATION_CONNECTION: utf8_general_ci
            DATABASE_COLLATION: utf8_general_ci
1 row in set (0.05 sec)
```

同样可以查询到触发器 after_scores_update 的详细信息。

2）删除触发器

不需要某个触发器时，一定要将这个触发器删除，以免造成意外操作。要删除指定数据库中已存在的触发器，语法格式如下：

```
DROP TRIGGER [IF EXISTS] [〈数据库名〉.]〈触发器名称〉
```

比如，我们要删除触发器 after_scores_update，语句如下：

```
mysql> DROP TRIGGER IF EXISTS after_scores_update;
Query OK, 0 rows affected (0.00 sec)
```

再使用 SHOW TRIGGERS 查看，语句如下：

```
mysql> SHOW TRIGGERS IN studentscore;
Empty set (0.00 sec)
```

可以看到，触发器删除成功。需要注意的是，当删除数据表时，也会同时删除该表上创建的触发器。

6.5.3 任务实现

【例 6-5-1】 创建 BEFORE INSERT 型触发器。

在学生成绩管理数据库创建一个 BEFORE INSERT 型触发器，用来实现当向成绩表中录入新的学生成绩时，将录入的分数自动加上 10 分的附加分。创建触发器语句如下：

```
mysql> DELIMITER //
mysql> CREATE TRIGGER before_scores_insert BEFORE INSERT
    -> ON scores
    -> FOR EACH ROW
    -> BEGIN
    ->   set NEW.score = NEW.score + 10;
    -> END //
Query OK, 0 rows affected (0.21 sec)
mysql> DELIMITER ;
```

先使用 CREATE TRIGGER 语句创建名为 before_scores_insert 的触发器。设置触发时间为 BEFORE，触发事件为 INSERT，表名为 scores，指明当 scores 中插入新的记录之前触发器会被触发。FOR EACH ROW 表示触发器会对每一行数据的插入都执行相应的触发器动作。BEGIN...END 语句块之间定义了触发器被触发时要做的动作，这里设置当 scores 表中插入新的记录时，将成绩加上 10 分的附加分。接下来在 scores 表中插入一条新记录，语句如下：

```
mysql> INSERT INTO 'scores' VALUES
    -> (1933062301,104,60);
Query OK, 1 row affected (0.14 sec)
```

这时，查看触发器是否被执行，查看 scores 表中信息。

```
mysql> select * from scores
    -> WHERE sid = "1933062301" AND cid = "104";
+------------+-----+-------+
| sid        | cid | score |
+------------+-----+-------+
| 1933062301 | 104 |    70 |
+------------+-----+-------+
1 row in set (0.00 sec)
```

可以看到，成绩自动修改为 70 分，触发器被成功执行。

【例 6-5-2】 创建 AFTER DELETE 型触发器。

在学生成绩管理数据库创建一个 AFTER DELETE 型触发器,用来实现当向学生表中删除一条学生记录时,将删除的学生信息自动备份到 student_del 表中。首先创建 student_del 表,用来存放删除学生的信息,包含 id、学号、姓名、班级、性别、专业、出生日期、删除时间等信息,建表的语句如下:

```
mysql> CREATE TABLE `students_del` (
    -> `id`  INT AUTO_INCREMENT PRIMARY KEY ,
    -> `sid` int(20) unsigned NOT NULL ,
    -> `sname` varchar(20) DEFAULT NULL,
    -> `sclass` int(10) DEFAULT NULL,
    -> `sgender` varchar(10) DEFAULT NULL,
    -> `smajor` varchar(20) DEFAULT NULL,
    -> `sbirth` date DEFAULT NULL,
    -> `del_time` datetime NOT NULL
    -> ) ENGINE=InnoDB DEFAULT CHARSET=utf8;
Query OK, 0 rows affected (0.42 sec)
```

接下来定义触发器,实现删除学生信息时的自动备份,语句如下:

```
mysql> DELIMITER //
mysql> CREATE TRIGGER after_students_delete AFTER DELETE
    -> ON students
    -> FOR EACH ROW
    -> BEGIN
    -> insert into `students_del`
    -> (`sid`,`sname`,`sclass`,`sgender`,`smajor`,`sbirth`,`del_time`)
    -> values
    -> (OLD.sid,OLD.sname,OLD.sclass,OLD.sgender,OLD.smajor,OLD.sbirth,NOW());
    -> END //
Query OK, 0 rows affected (0.19 sec)
mysql> DELIMITER ;
```

这里使用 CREATE TRIGGER 语句创建名为 after_students_delete 的触发器。设置触发时间为 AFTER,触发事件为 DELETE,表名为 students,指明当 students 中删除一条记录之后触发器会被触发。FOR EACH ROW 表示触发器会对每一行数据的插入都执行相应的触发器动作。BEGIN END 语句块之间定义了触发器被触发时要做的动作,这里设置当 students 表中删除一条学生记录时,将删除的记录插入 students_del 表中。接下来在 students 表中删除一条记录,语句如下:

```
mysql> DELETE FROM students
    -> WHERE sid = 1933062301;
Query OK, 1 row affected (0.05 sec)
```

查看 students_del 表中信息。

```
mysql> select * from students_del;
+----+------------+-------+--------+---------+--------+------------+---------------------+
| id | sid        | sname | sclass | sgender | smajor | sbirth     | del_time            |
+----+------------+-------+--------+---------+--------+------------+---------------------+
|  1 | 1933062301 | 刘洋  | 1901   | 女      | 软件技术 | 2000-07-09 | 2021-06-10 18:03:09 |
+----+------------+-------+--------+---------+--------+------------+---------------------+
1 row in set (0.00 sec)
```

可以看到，被删除的记录自动插入 students_del 表中，触发器执行成功。

【例 6-5-3】 使用 SHOW TRIGGERS 查看触发器。

查看在学生成绩管理数据库中创建的触发器，语句如下：

```
mysql> SHOW TRIGGERS IN studentscore \G
*************************** 1. row ***************************
             Trigger: before_scores_insert
               Event: INSERT
               Table: scores
           Statement: BEGIN
  set NEW.score = NEW.score + 10;
END
              Timing: BEFORE
             Created: 2021-06-10 17:14:16.66
            sql_mode: 
    STRICT_TRANS_TABLES,NO_AUTO_CREATE_USER,NO_ENGINE_SUBSTITUTION
             Definer: root@localhost
character_set_client: utf8
collation_connection: utf8_general_ci
  Database Collation: utf8_general_ci
*************************** 2. row ***************************
             Trigger: after_students_delete
               Event: DELETE
               Table: students
           Statement: BEGIN
  insert  into 'students_del'
 ('sid','sname','sclass','sgender','smajor','sbirth','del_time')
  values
  (OLD.sid,OLD.sname,OLD.sclass,OLD.sgender,OLD.smajor,OLD.sbirth,NOW());
END
              Timing: AFTER
             Created: 2021-06-10 17:52:25.66
            sql_mode: 
    STRICT_TRANS_TABLES,NO_AUTO_CREATE_USER,NO_ENGINE_SUBSTITUTION
             Definer: root@localhost
```

```
 character_set_client: utf8
 collation_connection: utf8_general_ci
   Database Collation: utf8_general_ci
2 rows in set (0.11 sec)
```

可以看到目前数据库中有两个触发器，分别是 before_scores_insert 和 after_students_delete。

【例 6-5-4】 使用 DROP TRIGGER 删除触发器。

删除学生成绩管理数据库中的 before_scores_insert 和 after_students_delete 触发器，语句如下：

```
mysql> DROP TRIGGER IF EXISTS before_scores_insert;
Query OK, 0 rows affected (0.02 sec)
mysql> DROP TRIGGER IF EXISTS after_students_delete;
Query OK, 0 rows affected (0.00 sec)
```

重新查看数据库中的触发器。

```
mysql> SHOW TRIGGERS IN studentscore;
Empty set (0.00 sec)
```

可以看到，数据库中的触发器删除成功。

6.5.4 训练任务

在图书管理数据库的图书表上创建触发器并执行。

① 创建一个 AFTER UPDATE 触发器，一旦在图书表中有修改动作，就会自动往 reader_log 表里插入当前时间。

② 创建一个 AFTER DELETE 触发器，删除一条读者信息时，删除其借阅表上的对应记录。

6.5.5 任务总结

触发器是 MySQL 用来保证数据完整性的一种方法。触发器的执行不是由程序调用，也不是手工启动，而是由事件来触发，比如当对一个表进行 DML 操作时就会激活它执行。触发器与数据表关系密切，主要用于保护表中的数据，特别是当有多个表具有一定的相互联系的时候，触发器能够让不同的表保持数据的一致性。

本任务主要讲解了触发器的概念、触发器与存储过程的区别、触发器的创建和执行、查看触发器、删除触发器等知识点。通过本任务的学习，读者应掌握触发器的应用场景，能够通过创建触发器维护数据的完整性。

任务6
事 务

6.6.1 任务描述

生活中我们经常遇见银行转账的场景。要完成 A 转账到 B,需要两个步骤:从 A 账户减去相应金额;给 B 账户加上相应金额。这两步要么一起成功,要么全部失败,否则就会造成数据不一致。比如 A 的钱少了,但 B 的钱没增加,或者 A 的扣款失败,B 的钱却增加了。所以需要一种机制来保证这一操作过程中每一步的正确性,当其中任意操作失败时应该将已经进行的操作回滚,保证整体都失败,此时这些被绑定的一连串操作便形成了事务。

事务是 MySQL 中处理数据的一种方式,主要用在数据完整性高、数据之间依赖性大的情况。数据库引入事务的主要目的是,事务会把数据库会从一种一致状态转换到另一种一致状态,数据库提交工作时可以确保要么所有修改都保存,要么所有修改都不保存。数据库使用事务机制来保证即使机器出故障的情况下,数据仍然是正确的。

在 MySQL 中使用事务需要 MySQL 中的存储引擎支持,只有使用了 Innodb 数据库引擎的数据库或表才支持事务。事务处理可以用来维护数据库的完整性,保证成批的 SQL 语句要么全部执行,要么全部不执行。事务一般用来管理 INSERT、UPDATE、DELETE 语句。

本任务主要通过事务来管理学生成绩管理数据库中的多条数据操作。在完成任务之前,首先进行事务相关知识点的讲解。

6.6.2 知识准备

1. 事务的基本概念和 ACID 特性

首先了解事务的基本概念。到底什么是事务?

事务是数据库管理系统执行过程中的一个逻辑单位,在执行 SQL 语句的时候,某些业务要求一系列操作必须全部执行,而不能仅执行一部分,这种把多条语句作为一个整体进行操作的功能,被称为数据库事务。

事务是并发控制的基本单位,由一个有限的数据库操作序列构成。事务机制可以用来维护数据库的完整性,保证成批的 SQL 语句要么全部执行,要么全部不执行。事务可以确保该事务范围内的所有操作都可以全部成功或者全部失败,如果事务失败,那么效果就和没有执行这些 SQL 一样,不会对数据库数据有任何改动。事务的使用是数据库管理系统区别文件系统的重要特征之一。

一般来说,事务需要满足以下四个条件:原子性(Atomicity)、一致性(Consistency)、隔离性(Isolation)、持久性(Durability),我们习惯称之为 ACID 特性,下面分别对这四个特性进行解释。

1) 原子性

原子性是指事务开始后的所有操作,要么全部做完,要么全部不做,不可能结束在中间某个环节。如果事务在执行过程中发生错误,会被回滚(Rollback)到事务开始前的状态,就像这个事务从来没有执行过一样。比如前面提到的银行转账的场景,张三需要给李四转账100元,此时我们需要把这段业务逻辑看成一个事务,这个事务有两个步骤:张三账户减少100元;李四账户增加100元。如果张三扣了100元,但李四没增加100元,张三不干;或者张三没扣100元,李四平白无故增加100元,银行亏本。因此,张三扣100元,李四增加100元,整个交易必须是一个整体,要么全做,要么全不做——这就是事务的原子性。也就是说,事务是一个不可分割的整体,就像化学中原子是物质构成的基本单位一样,事务也具有原子性。

2) 一致性

一致性是指事务将数据库从一种状态转变为另一种一致的状态,事务开始前和结束后,数据库的完整性约束没有被破坏。比如,张三给李四的银行转账过程中,在交易前和交易后,钱的总量是不变的,整个交易前后,钱都是一致的状态,这是交易过程的一致性。

3) 隔离性

数据库允许多个并发事务同时对其数据进行读写和修改,隔离性可以防止多个事务并发执行时由于交叉执行而导致数据的不一致。事务的隔离性要求每个读写事务的对象对其他事务的操作对象能互相分离,即该事务提交前对其他事务不可见。

我们也可以理解为,多个事务并发访问时,事务之间是隔离的,一个事务不应该影响其他事务运行效果。也就是说,在并发环境中,当不同的事务同时操纵相同的数据时,每个事务都有各自的完整数据空间,由并发事务所作的修改必须与任何其他并发事务所作的修改隔离。例如,张三给李四转账100元的过程中,如果这时王五要给张三转账200元,那么这两个转账交易同时发生时,在一个交易执行之前,不能受到另外一个交易的影响。

MySQL通过锁机制来保证事务的隔离性。事务隔离分为四个不同级别,包括读未提交(read uncommitted)、读提交(read committed)、可重复读(repeatable read)和串行化(serializable)。

4) 持久性

持久性是指事务一旦提交,则其结果就是永久性的。即使发生宕机的故障,数据库也能将数据恢复。也就是说事务完成后,事务对数据库的所有更新将被保存到数据库,不能回滚。假如两个人去银行转账,交易结束后,过了一段时间,因为机器故障,两个人的交易不生效了,这显然是不合理的,所以持久性是说整个交易过程一旦结束,无论出现任何情况,交易应该是永久生效的。这只是从事务本身的角度来保证,排除MySQL本身发生的故障。MySQL使用redo log来保证事务的持久性。

2. 事务的开启、提交和回滚操作

MySQL 中，我们通过 START TRANSACTION 语句来开始一个事务，通过 COMMIT 语句提交事务，通过 ROLLBACK 语句回滚事务。还是以银行转账操作为例，我们先进行建库和建表的准备操作，语句如下：

```
mysql> CREATE DATABASE demo DEFAULT CHARSET=utf8 COLLATE=utf8_general_ci;
Query OK, 1 row affected (0.00 sec)
mysql> use demo;
Database changed
```

在数据库 demo 中创建一个 account 表，并插入数据，语句如下：

```
mysql> CREATE TABLE account(
    ->     id INT AUTO_INCREMENT PRIMARY KEY,
    ->     custname VARCHAR(50),
    ->     balance DECIMAL(20,2)
    -> );
Query OK, 0 rows affected (0.35 sec)

mysql> INSERT INTO account(custname, balance) VALUES
    ->     ('张三', 800),
    ->     ('李四', 800),
    ->     ('王五', 500),
    ->     ('赵六', 300);
Query OK, 4 rows affected (0.06 sec)
Records: 4  Duplicates: 0  Warnings: 0

mysql> select * from account;
+----+----------+--------+
| id | custname | balance |
+----+----------+--------+
|  1 | 张三     | 800.00 |
|  2 | 李四     | 800.00 |
|  3 | 王五     | 500.00 |
|  4 | 赵六     | 300.00 |
+----+----------+--------+
4 rows in set (0.00 sec)
```

1) 开启事务

对于单条 SQL 语句，数据库系统自动将其作为一个事务执行，这种事务被称为隐式事务。也就是说，单条语句被回车执行后，该语句便生效了，变更会保存在 MySQL 的文件中，无法撤销。

如果要手动把多条 SQL 语句作为一个事务执行，我们需要使用 START TRANSACTION

语句开启一个事务,这种事务被称为显式事务,显式地开启一个事务的语句如下:

```
mysql> START TRANSACTION;
Query OK, 0 rows affected (0.00 sec)
```

以银行转账业务为例,假设从张三账户给李四账户转账 100 元,第一步是将张三账户余额减去 100 元,语句如下:

```
mysql> UPDATE account SET balance = balance - 100 WHERE custname = '张三';
Query OK, 1 row affected (0.02 sec)
Rows matched: 1  Changed: 1  Warnings: 0
```

第二步,将李四账户余额加上 100 元,语句如下:

```
mysql> UPDATE account SET balance = balance + 100 WHERE custname = '李四';
Query OK, 1 row affected (0.00 sec)
Rows matched: 1  Changed: 1  Warnings: 0
```

但这个时候,如果新打开一个 Command Line Client,重新连接数据库会发现,此时查询到张三和李四的账户余额钱数都未改变。

```
mysql> select * from account;
+----+----------+---------+
| id | custname | balance |
+----+----------+---------+
|  1 | 张三     |  800.00 |
|  2 | 李四     |  800.00 |
|  3 | 王五     |  500.00 |
|  4 | 赵六     |  300.00 |
+----+----------+---------+
4 rows in set (0.00 sec)
```

这说明,数据库中这两条 UPDATE 语句并未生效。这是因为,开启事务后,默认的自动提交是失效的,此时事务并未提交,所以数据更新操作并没有生效,张三和李四的钱数没有改变。接下来讲解如何手动提交事务。

2) 提交事务

提交事务是指让数据库中进行的所有修改成为永久性的,可以使用 COMMIT 语句进行事务提交,如果将上一小节的事务提交,语句如下:

```
mysql> COMMIT;
Query OK, 0 rows affected (0.16 sec)
```

这时,再重新连接数据库会发现,此时查询到张三和李四的钱数已经完成更改,张三账户减少了 100 元,李四账户增加了 100 元。

```
mysql> select * from account;
+----+----------+---------+
| id | custname | balance |
+----+----------+---------+
|  1 | 张三     |  700.00 |
|  2 | 李四     |  900.00 |
|  3 | 王五     |  500.00 |
|  4 | 赵六     |  300.00 |
+----+----------+---------+
4 rows in set (0.00 sec)
```

实际上，MySQL 默认情况下开启了一个自动提交的模式，通过查询@@autocommit 变量来查看当前是否是自动提交的状态，默认值是 1，表示是自动提交的。而当显式声明开始一个事务时，autocommit 就会自动变成关闭状态，也就是 autocommit 的值会从默认值 1 变成 0，开启事务后查看 autocommit 的值。

```
mysql> SELECT @@autocommit;
+--------------+
| @@autocommit |
+--------------+
|            0 |
+--------------+
1 row in set (0.00 sec)
```

此外，除了通过 START TRANSACTION 语句显式声明开始一个事务让 autocommit 自动变成关闭状态，也可以手动修改 autocommit 的值为 0，语句如下：

```
mysql> SET autocommit = 0;
Query OK, 0 rows affected (0.00 sec)
mysql> SELECT @@autocommit;
+--------------+
| @@autocommit |
+--------------+
|            0 |
+--------------+
1 row in set (0.00 sec)
```

如果使用此命令禁止自动提交，事务则在用户本次对数据进行操作时自动开启，在用户执行 COMMIT 命令时提交。

3) 回滚事务

ROLLBACK 语句用于回滚结束用户的事务，表示撤销正在进行的所有未提交的修改。这里需要注意的是，无论是否为自动提交模式，语句执行后都会生效。区别在于，非自动提交模式下(autocommit 值为 0)，没提交的那些操作是可以回滚的，一旦提交后便不可撤销了。以转账操作为例,下面的 SQL 语句展示了启动事务后但是事务未提交之前,用 ROLLBACK 进行回滚到初始状态。这是 account 表的初始状态。

```
mysql> select * from account;
+----+----------+---------+
| id | custname | balance |
+----+----------+---------+
|  1 | 张三     |  800.00 |
|  2 | 李四     |  800.00 |
|  3 | 王五     |  500.00 |
|  4 | 赵六     |  300.00 |
+----+----------+---------+
4 rows in set (0.00 sec)
```

首先开启一个事务,并执行更新操作。

```
mysql> START TRANSACTION;
Query OK, 0 rows affected (0.00 sec)

mysql> UPDATE account SET balance = balance - 100 WHERE custname = '张三';
Query OK, 1 row affected (0.00 sec)
Rows matched: 1  Changed: 1  Warnings: 0

mysql> select * from account;
+----+----------+---------+
| id | custname | balance |
+----+----------+---------+
|  1 | 张三     |  700.00 |
|  2 | 李四     |  800.00 |
|  3 | 王五     |  500.00 |
|  4 | 赵六     |  300.00 |
+----+----------+---------+
4 rows in set (0.00 sec)

mysql> UPDATE account SET balance = balance + 100 WHERE custname = '李四';
Query OK, 1 row affected (0.00 sec)
Rows matched: 1  Changed: 1  Warnings: 0

mysql> select * from account;
+----+----------+---------+
| id | custname | balance |
+----+----------+---------+
|  1 | 张三     |  700.00 |
|  2 | 李四     |  900.00 |
|  3 | 王五     |  500.00 |
|  4 | 赵六     |  300.00 |
+----+----------+---------+
4 rows in set (0.00 sec)
```

这时并未提交事务,如果使用 ROLLBACK 语句进行回滚,语句如下:

```
mysql> ROLLBACK;
Query OK, 0 rows affected (0.15 sec)
```

那么,这时重新查询账户信息可以看到,张三和李四的账户余额回到了初始状态。

```
mysql> select * from account;
+----+----------+--------+
| id | custname | balance|
+----+----------+--------+
| 1  | 张三     | 800.00 |
| 2  | 李四     | 800.00 |
| 3  | 王五     | 500.00 |
| 4  | 赵六     | 300.00 |
+----+----------+--------+
4 rows in set (0.00 sec)
```

需要特别注意的是,当用户执行 START TRANSACTION 命令时开启事务,自动提交关闭(autocommit 的值变成 0),事务操作中可随时调用 ROLLBACK 进行回滚,但一旦执行了 COMMIT 操作便不能回滚了。

从用户执行 START TRANSACTION 命令到用户执行 COMMIT 命令之间的一系列操作为一个完整的事务周期,若不执行 COMMIT 命令,系统则默认事务回滚。

3. 事务的保存点

我们从前面的小节中了解到,如果事务在执行的过程中遇到错误,可以使用 ROLLBACK 语句使事务回滚到起点,将事务中对数据库的所有已完成的操作全部撤销,回到事务开始时的状态,这里的操作指对数据库的更新操作。

当条件回滚只影响事务的一部分时,事务不需要全部撤销已执行的操作,我们可以让事务回滚到指定位置,此时,需要在事务中设定保存点(SAVEPOINT),保存点所在位置之前的事务语句不用回滚,即保存点之前的操作被视为有效的。SAVEPOINT 相关的 SQL 语句如下。

1) 创建保存点

```
SAVEPOINT <保存点名>;
```

保存点的创建通过 SAVEPOINT 语句来实现,表示在事务中创建一个保存点,一个事务中可以有多个 SAVEPOINT。

2) 删除保存点

```
RELEASE SAVEPOINT <保存点名>;
```

该语句用来删除一个事务的保存点,当指定的保存点不存在时,执行该语句会抛出一个异常。

3) 回滚到保存点

```
ROLLBACK TO <保存点名>;
```

该语句用来将事务回滚到某个保存点,同时,系统将清除自事务起点到某个保存点所做的所有的数据修改,并且释放由事务控制的资源。因此,这条语句也标志着事务的结束。

为了更好地理解,还是以银行转账操作为例,下面是 account 表的初始状态。

```
mysql> select * from account;
+----+----------+---------+
| id | custname | balance |
+----+----------+---------+
|  1 | 张三     |  800.00 |
|  2 | 李四     |  800.00 |
|  3 | 王五     |  500.00 |
|  4 | 赵六     |  300.00 |
+----+----------+---------+
4 rows in set (0.00 sec)
```

在事务开始时,可以通过 SAVEPOINT 保存这个初始状态,这个保存点命名为 initial_save。

```
mysql> START TRANSACTION;
Query OK, 0 rows affected (0.00 sec)

mysql> SAVEPOINT initial_save;
Query OK, 0 rows affected (0.00 sec)
```

将张三账户减去 100 元,在 UPDATE 语句执行后,事务中的临时数据如下,保存这个转出成功的状态,命名为 out_complete。

```
mysql> UPDATE account SET balance = balance - 100 WHERE custname = '张三';
Query OK, 1 row affected (0.00 sec)
Rows matched: 1  Changed: 1  Warnings: 0

mysql> select * from account;
+----+----------+---------+
| id | custname | balance |
+----+----------+---------+
|  1 | 张三     |  700.00 |
|  2 | 李四     |  800.00 |
|  3 | 王五     |  500.00 |
|  4 | 赵六     |  300.00 |
+----+----------+---------+
4 rows in set (0.00 sec)
```

```
mysql> SAVEPOINT out_complete;
Query OK, 0 rows affected (0.00 sec)
```

然后，执行给李四的账户增加 100 元的操作，这是转账成功的状态，语句如下：

```
mysql> UPDATE account SET balance = balance + 100 WHERE custname = '李四';
Query OK, 1 row affected (0.00 sec)
Rows matched: 1  Changed: 1  Warnings: 0

mysql> select * from account;
+----+----------+--------+
| id | custname | balance|
+----+----------+--------+
| 1  | 张三     | 700.00 |
| 2  | 李四     | 900.00 |
| 3  | 王五     | 500.00 |
| 4  | 赵六     | 300.00 |
+----+----------+--------+
4 rows in set (0.00 sec)
```

如果这时，想回退到张三转出成功，但李四还没收到的状态，也就是回退到保存点 out_complete，那么语句如下：

```
mysql> ROLLBACK TO out_complete;
Query OK, 0 rows affected (0.00 sec)

mysql> select * from account;
+----+----------+--------+
| id | custname | balance|
+----+----------+--------+
| 1  | 张三     | 700.00 |
| 2  | 李四     | 800.00 |
| 3  | 王五     | 500.00 |
| 4  | 赵六     | 300.00 |
+----+----------+--------+
4 rows in set (0.00 sec)
```

如果需要回退到起始状态，那么语句如下：

```
mysql> ROLLBACK TO initial_save;
Query OK, 0 rows affected (0.18 sec)
```

```
mysql> select * from account;
+----+----------+---------+
| id | custname | balance |
+----+----------+---------+
|  1 | 张三     |  800.00 |
|  2 | 李四     |  800.00 |
|  3 | 王五     |  500.00 |
|  4 | 赵六     |  300.00 |
+----+----------+---------+
4 rows in set (0.00 sec)
```

在MySQL中，保存点SAVEPOINT属于事务控制处理部分。利用SAVEPOINT可以回滚指定部分事务，从而使事务处理更加灵活和精细。

4. 事务的四种隔离级别

在事务的四个特性中，事务的隔离性涉及并发控制。对于两个并发执行的事务，如果涉及操作同一条记录的时候，可能会发生问题。因为并发操作会带来数据的不一致性，包括脏读、不可重复读、幻读等。数据库系统提供了四种隔离级别，我们可以有针对性地选择事务的隔离级别，避免数据不一致的问题。

SQL标准定义了四种隔离级别，包括读未提交（read uncommitted）、读提交（read committed）、可重复读（repeatable read）和串行化（serializable），分别对应可能出现的数据不一致的情况，接下来依次介绍这四种隔离级别。

1) READ UNCOMMITTED(读未提交)

该隔离级别的事务会读到其他未提交事务的数据，此现象也称之为脏读。脏读是指一个事务读取了另一个事务没有提交的数据，或者读取一个事务未提交的更新。

2) READ COMMITTED(读提交)

一个事务可以读取另一个已提交的事务，多次读取会造成不一样的结果，此现象称为不可重复读问题，不可重复读是指同一个事务中两次读取同一行记录的值有可能不一样。该隔离级别是Oracle和SQL Server的默认隔离级别。

3) REPEATABLE READ(可重复读)

该隔离级别是MySQL默认的隔离级别，在同一个事务里，select的结果是事务开始时时间点的状态，因此，同样的select操作读到的结果会是一致的。但是会有幻读现象，幻读是两次读取的结果包含的行记录不一样。幻读和不可重复读的区别是：幻读是读取的行记录数不一样，不可重复读是读取的行记录的值是不一样的。MySQL的InnoDB引擎可以通过next-key locks机制来避免幻读。

4) SERIALIZABLE(串行化)

在该隔离级别下事务都是串行顺序执行的，MySQL数据库的InnoDB引擎会给读操作隐式加一把读共享锁，从而避免了脏读、不可重复读和幻读问题，但所有的事务都是串行执行的,这样也会导致数据库性能差。

SQL 实现了四个标准的隔离级别,每一种级别都规定了一个事务中所做的修改,哪些在事务内和事务间是可见的,哪些是不可见的。低级别的隔离级一般支持更高的并发处理,并拥有更低的系统开销。一般说来,事务隔离级别越高,数据库性能越差。表 6-6-1 列出了各个隔离级别下产生的数据一致性问题。

表 6-6-1 各个隔离级别下产生的一些问题

隔离级别	脏读	不可重复读	幻读
读未提交(READ UNCOMMITTED)	可以出现	可以出现	可以出现
读提交(READ COMMITTED)	不允许出现	可以出现	可以出现
可重复读(REPEATABLE READ)	不允许出现	不允许出现	可以出现
串行化(SERIALIZABLE)	不允许出现	不允许出现	不允许出现

我们可以通过@@tx_isolation 来查看事务的隔离级别。语句如下:

```
mysql> SELECT @@tx_isolation;
+-----------------+
| @@tx_isolation  |
+-----------------+
| REPEATABLE-READ |
+-----------------+
```

可以看到,MySQL 默认的隔离级别是 REPEATABLE READ(可重复读)。

6.6.3 任务实现

【例 6-6-1】 使用 COMMIT 语句提交事务。

在学生成绩管理数据库中,如果学生表中修改了学生的学号信息,那么成绩表中的记录需要同步修改,否则就会造成数据的不一致,将这两个操作作为一个事务进行管理。首先使用 START TRANSACTION 语句开启事务。

```
mysql> START TRANSACTION;
Query OK, 0 rows affected (0.00 sec)
```

在 students 表中修改学生 1933062301 的学号为 2033062301。语句如下:

```
mysql> UPDATE students
    -> SET sid = 2033062301
    -> WHERE sid = 1933062301;
Query OK, 1 row affected (0.13 sec)
Rows matched: 1  Changed: 1  Warnings: 0
```

在 scores 表中同步修改学号信息。语句如下：

```
mysql> UPDATE scores
    -> SET sid = 2033062301
    -> WHERE sid = 1933062301;
Query OK, 3 rows affected (0.00 sec)
Rows matched: 3  Changed: 3  Warnings: 0
```

最后用 COMMIT 语句提交事务。

```
mysql> COMMIT;
Query OK, 0 rows affected (0.06 sec)
```

【例 6-6-2】 使用 ROLLBACK 语句回滚事务。

在学生成绩管理数据库中，利用事务实现删除 students 表中的记录，发现误删之后使用 ROLLBACK 回滚取消删除。使用 START TRANSACTION 语句开始事务。

```
mysql> START TRANSACTION;
Query OK, 0 rows affected (0.00 sec)
```

在事务中执行操作，删除 scores 表中的记录。

```
mysql> DELETE FROM students
    -> WHERE sid = 2033062301;
Query OK, 1 row affected (0.11 sec)
```

这时，使用 select 查询该生记录。

```
mysql> select * from students where sid = 2033062301;
Empty set (0.03 sec)
```

可以看到记录被删除。如果这时我们发现进行了误删操作，那么就可以使用 ROLLBACK 语句回滚事务。

```
mysql> ROLLBACK;
Query OK, 0 rows affected (0.14 sec)
```

重新使用 select 查询该生记录。

```
mysql> select * from students where sid = 2033062301;
+------------+-------+--------+---------+----------+------------+---------------+
| sid        | sname | sclass | sgender | smajor   | sbirth     | credit_points |
+------------+-------+--------+---------+----------+------------+---------------+
| 2033062301 | 刘洋  | 191    | 女      | 软件技术 | 2000-09-01 |            12 |
+------------+-------+--------+---------+----------+------------+---------------+
1 row in set (0.00 sec)
```

可以看到，删除操作被撤销。

【例 6-6-3】 使用 SAVEPOINT 回滚到保存点。

在学生成绩管理数据库中,利用事务实现在 students 表中添加一条新记录,更新学号为 1933062302 的学生姓名,然后删除学号为 2033062301 的记录,记下每一步操作步骤,让事务可以分步撤销,查看学生初始记录。

```
mysql> select * from students;
+------------+--------+--------+---------+----------+------------+---------------+
| sid        | sname  | sclass | sgender | smajor   | sbirth     | credit_points |
+------------+--------+--------+---------+----------+------------+---------------+
| 1933062302 | 邓嘉蓉 | 1901   | 女      | 软件技术 | 2000-04-02 | 12            |
| 1933062303 | 葛佳音 | 1901   | 女      | 软件技术 | 2000-07-09 | 12            |
| 1933062304 | 索凤洋 | 1901   | 女      | 软件技术 | 2000-08-04 | 12            |
| 1933062305 | 曹宏美 | 1901   | 女      | 软件技术 | 2000-11-07 | 12            |
| 1933062306 | 郭思宇 | 1901   | 女      | 软件技术 | 2000-03-21 | 12            |
| 1933062307 | 周慧敏 | 1901   | 女      | 软件技术 | 2000-03-23 | 12            |
| 2033062301 | 刘洋   | 191    | 女      | 软件技术 | 2000-09-01 | 12            |
+------------+--------+--------+---------+----------+------------+---------------+
7 rows in set (0.00 sec)
```

首先使用 START TRANSACTION 语句开始事务,并使用保存点记下初始状态。

```
mysql> START TRANSACTION;
Query OK, 0 rows affected (0.00 sec)
mysql> SAVEPOINT initial;
Query OK, 0 rows affected (0.00 sec)
```

接下来在事务中执行操作,将添加新记录、更新学生成绩、删除一条记录作为一个事务进行处理,并为每一步设置保存点。

```
mysql> INSERT INTO 'students' VALUES
    -> (1933062308,'顾晓曼',192,'女','软件技术','2002-03-21',12);
Query OK, 1 row affected (0.11 sec)
mysql> SAVEPOINT ins;
Query OK, 0 rows affected (0.03 sec)

mysql> UPDATE students SET sname = '乔月月'
    -> WHERE sid = 1933062302;
Query OK, 1 row affected (0.12 sec)
Rows matched: 1  Changed: 1  Warnings: 0
mysql> SAVEPOINT upd;
Query OK, 0 rows affected (0.02 sec)

mysql> DELETE FROM students WHERE sid = 2033062301;
Query OK, 1 row affected (0.00 sec)
mysql> SAVEPOINT del;
Query OK, 0 rows affected (0.00 sec)
```

可以看到,我们设置了三个保存点,分别是 ins、upd、del。这时查看表的临时状态。

```
mysql> select * from students;
+------------+--------+--------+---------+---------+------------+---------------+
| sid        | sname  | sclass | sgender | smajor  | sbirth     | credit_points |
+------------+--------+--------+---------+---------+------------+---------------+
| 1933062302 | 乔月月 | 1901   | 女      | 软件技术 | 2000-04-02 |            12 |
| 1933062303 | 葛佳音 | 1901   | 女      | 软件技术 | 2000-07-09 |            12 |
| 1933062304 | 索凤洋 | 1901   | 女      | 软件技术 | 2000-08-04 |            12 |
| 1933062305 | 曹宏美 | 1901   | 女      | 软件技术 | 2000-11-07 |            12 |
| 1933062306 | 郭思宇 | 1901   | 女      | 软件技术 | 2000-03-21 |            12 |
| 1933062307 | 周慧敏 | 1901   | 女      | 软件技术 | 2000-03-23 |            12 |
| 1933062308 | 顾晓曼 |  192   | 女      | 软件技术 | 2002-03-21 |            12 |
+------------+--------+--------+---------+---------+------------+---------------+
7 rows in set (0.00 sec)
```

如果我们想让事务回到保存点 upd 的状态,也就是执行完更新操作,但还未执行删除操作的状态。使用 ROLLBACK TO 返回到此保存点,语句如下:

```
mysql> ROLLBACK TO upd;
Query OK, 0 rows affected (0.09 sec)
```

重新查看 students 状态信息。

```
mysql> select * from students;
+------------+--------+--------+---------+---------+------------+---------------+
| sid        | sname  | sclass | sgender | smajor  | sbirth     | credit_points |
+------------+--------+--------+---------+---------+------------+---------------+
| 1933062302 | 乔月月 | 1901   | 女      | 软件技术 | 2000-04-02 |            12 |
| 1933062303 | 葛佳音 | 1901   | 女      | 软件技术 | 2000-07-09 |            12 |
| 1933062304 | 索凤洋 | 1901   | 女      | 软件技术 | 2000-08-04 |            12 |
| 1933062305 | 曹宏美 | 1901   | 女      | 软件技术 | 2000-11-07 |            12 |
| 1933062306 | 郭思宇 | 1901   | 女      | 软件技术 | 2000-03-21 |            12 |
| 1933062307 | 周慧敏 | 1901   | 女      | 软件技术 | 2000-03-23 |            12 |
| 1933062308 | 顾晓曼 | 1901   | 女      | 软件技术 | 2002-03-21 |            12 |
| 2033062301 | 刘洋   |  192   | 女      | 软件技术 | 2000-09-01 |            12 |
+------------+--------+--------+---------+---------+------------+---------------+
8 rows in set (0.00 sec)
```

可以看到,事务仅仅撤销了删除操作,回退到了更新成功的状态。

6.6.4 训练任务

利用事务实现在图书管理数据库中借阅表中添加一条新记录,读者表中更新赵鑫的邮箱为 zhaoxin@qq.com,然后在图书表中删除一条记录;记下每一步操作步骤,让事务可以分步撤销。

① 使用 START TRANSACTION 语句开始事务。

② 在事务中执行操作,将添加新记录、更新读者信息、删除一条记录作为一个事务进行处理。

③ 为每一步设置保存点,让事务可以回滚到任意一个保存点的状态。

④ 使用 COMMIT 语句提交事务。

6.6.5 任务总结

MySQL 每一步操作都可看成一个原子操作。默认情况下,autocommit 是开启状态,单个 SQL 语句都是执行后自动提交,语句产生的效果被记录保存了下来。开启事务后,语句不会自动提交,需要手动调用 COMMIT 来让效果被持久化。

我们可以通过 START TRANSACTION 开启一个事务,MySQL 中的事务从第一个可执行的 SQL 语句开始,并在使用 COMMIT 提交或 ROLLBACK 语句回滚时结束。事务有两种执行结果:提交事务后,所有修改都将成功;或者,当事务回滚时,所有修改都将被撤销。此外,还可以通过保存点让事务回滚到指定位置。

本任务主要讲解了事务的概念以及 ACID 特性、事务的基本操作、保存点的设置、事务的四种隔离级别等知识点。通过本任务的学习,读者应掌握事务的应用场景,了解事务不同隔离级别的特点,具备运用事务解决实际需求的能力。

项目 7　数据库综合实训

实训 1
数 据 库 操 作

【实训目的】

掌握数据库层面的 SQL 命令及操作。

① 查看 DBMS 中有哪些数据库。

```
show  databases;
```

② 创建数据库。

```
create database〈数据库名称〉;
```

③ 进入数据库。

```
use〈数据库名称〉;
```

④ 查看目前所在的数据库。

```
select database();
```

⑤ 删除数据库。

```
drop database〈数据库名称〉;
```

【实训内容】

① 查看 DBMS 中有哪些数据库。

② 创建一个新的数据库,命名为 test_db1。

③ 进入此数据库 test_db1。

④ 查看当前所在数据库。

⑤ 删除数据库 test_db1。

【实训步骤】

① 打开 MySQL 控制台，建立数据库连接，使用 SHOW DATABASES 查看 DBMS 中有哪些数据库，结果如下：

```
mysql> SHOW DATABASES;
+--------------------+
| Database           |
+--------------------+
| information_schema |
| employees          |
| mysql              |
| performance_schema |
| studentscore       |
| sys                |
+--------------------+
6 rows in set (0.00 sec)
```

② 创建一个名为 test_db1 的数据库。

```
mysql> CREATE DATABASE test_db1;
Query OK, 1 row affected (0.04 sec)
```

重新查看数据库，语句如下：

```
mysql> SHOW DATABASES;
+--------------------+
| Database           |
+--------------------+
| information_schema |
| employees          |
| mysql              |
| performance_schema |
| studentscore       |
| sys                |
| test_db1           |
+--------------------+
7 rows in set (0.00 sec)
```

可以看到，数据库 test_db1 创建成功。

③ 使用 USE 命令进入数据库。

```
mysql> USE test_db1;
Database changed
```

④ 使用 SELECT DATABASE()命令查看当前所在数据库。

```
mysql> SELECT DATABASE();
+------------+
| database() |
+------------+
| test_db1   |
+------------+
1 row in set (0.00 sec)
```

可以看到,当前所在的数据库是 test_db1。

⑤ 使用 DROP DATABASE 删除此数据库,语句如下:

```
mysql> DROP DATABASE test_db1;
Query OK, 0 rows affected (0.41 sec)
```

重新使用 SHOW DATABASES 查看 DBMS 中有哪些数据库,结果如下:

```
mysql> SHOW DATABASES;
+--------------------+
| Database           |
+--------------------+
| information_schema |
| employees          |
| mysql              |
| performance_schema |
| studentscore       |
| sys                |
+--------------------+
6 rows in set (0.00 sec)
```

可以看到,数据库 test_db1 删除成功。使用 SELECT DATABASE()命令查看当前所在数据库,语句如下:

```
mysql> SELECT DATABASE();
+------------+
| database() |
+------------+
| NULL       |
+------------+
1 row in set (0.00 sec)
```

可以看到,此时并未进入任何数据库。

实训 2

数 据 表 操 作

【实训目的】

掌握 MySQL 中的常用数据类型,熟练掌握数据表层面的 SQL 命令及操作。

① MySQL 中的常用数据类型。

MySQL 的数据类型大致分为三类:

a. 数值类型(INTEGER、INT、DECIMAL、NUMERIC、FLOAT、DOUBLE 等)。

b. 日期时间类型(DATE、TIME、YEAR、DATETIME、TIMESTAMP 等)。

c. 字符类型(CHAR、VARCHAR、BINARY、VARBINARY、BLOB、TEXT、ENUM 等)。

② 创建数据表。

```
create Table 〈表名〉(
    〈字段名 1〉〈数据类型〉,
〈字段名 2〉〈数据类型〉,
…
〈字段名n〉〈数据类型〉
);
```

③ 查看某个数据表的详细字段信息。

```
DESC 〈表名〉;
```

④ 查看数据库中所有的表。

```
SHOW TABLES;
```

⑤ 删除数据表。

```
DROP table 〈表名〉;
```

【实训内容】

① 创建一个新的数据库,命名为 demo。在此数据库中创建一个简单的 user 表,表中加入三个字段 name、age、tel,分别表示用户姓名、年龄、电话号码,并设置字段的数据类型。创建完成后查看 user 表的详细字段信息。

② 在数据库 demo 中创建一个 employee 数据表,表中加入六个字段 empId、empName、gender、jobTitle、salary、hire_date,分别表示员工编号、员工姓名、性别、职位、薪水、入职日期等,并设置字段的数据类型。创建完成后查看 employee 表的详细字段信息。

③ 查看数据库 demo 中所有的数据表。

④ 删除 demo 数据库中的 user 表和 employee 表。

【实训步骤】

① 打开 MySQL 控制台,建立数据库连接,创建一个新的数据库 demo 并设置默认编码。

```
mysql> CREATE DATABASE demo DEFAULT CHARSET=utf8 COLLATE=utf8_general_ci;
Query OK, 1 row affected (0.00 sec)
```

使用 USE 命令进入数据库。

```
mysql> USE demo;
Database changed
```

在此数据库中创建一个表,表名为 user,在此表中加入三个字段 name、age、tel,分别表示用户姓名、年龄、电话号码,语句如下:

```
mysql> CREATE TABLE user(
    ->    name varchar(20),
    ->    age int,
    ->    tel varchar(20)
    -> );
Query OK, 0 rows affected (0.52 sec)
```

使用 DESC 命令查看 user 表的详细字段信息。

```
mysql> DESC user;
+-------+-------------+------+-----+---------+-------+
| Field | Type        | Null | Key | Default | Extra |
+-------+-------------+------+-----+---------+-------+
| name  | varchar(20) | YES  |     | NULL    |       |
| age   | int(11)     | YES  |     | NULL    |       |
| tel   | varchar(20) | YES  |     | NULL    |       |
+-------+-------------+------+-----+---------+-------+
3 rows in set (0.04 sec)
```

② 在 demo 数据库中创建一个 employee 数据表,表中加入六个字段 empId、empName、gender、jobTitle、salary、hire_date,分别表示员工编号、员工姓名、性别、职位、薪水、入职日期,语句如下:

```
mysql> CREATE TABLE employee (
    ->    empId INT,
    ->    empName varchar(50),
    ->    gender ENUM("男","女"),
    ->    jobTitle varchar(50),
    ->    salary DOUBLE,
```

```
    -> hire_date DATE
    -> );
Query OK, 0 rows affected (0.30 sec)
```

使用 DESC 命令查看 employee 表的详细字段信息。

```
mysql> DESC employee;
+-----------+---------------+------+-----+---------+-------+
| Field     | Type          | Null | Key | Default | Extra |
+-----------+---------------+------+-----+---------+-------+
| empId     | int(11)       | YES  |     | NULL    |       |
| empName   | varchar(50)   | YES  |     | NULL    |       |
| gender    | enum('男','女')| YES  |     | NULL    |       |
| jobTitle  | varchar(50)   | YES  |     | NULL    |       |
| salary    | double        | YES  |     | NULL    |       |
| hire_date | date          | YES  |     | NULL    |       |
+-----------+---------------+------+-----+---------+-------+
6 rows in set (0.12 sec)
```

③ 查看数据库 demo 中所有的数据表。

```
mysql> SHOW TABLES;
+----------------+
| Tables_in_demo |
+----------------+
| employee       |
| user           |
+----------------+
2 rows in set (0.00 sec)
```

④ 删除 demo 数据库中的 user 表，语句如下：

```
mysql> DROP TABLE user;
Query OK, 0 rows affected (0.20 sec)
```

删除 demo 数据库中的 employee 表，SQL 语句如下：

```
mysql> DROP TABLE employee;
Query OK, 0 rows affected (0.36 sec)
```

重新查看数据库 demo 中的数据表。

```
mysql> SHOW tables;
Empty set (0.00 sec)
```

可以看到，demo 数据库中的表全部删除成功。

实训 3

数据的插入操作

【实训目的】

掌握 INSERT 语句插入数据的方法。理解表的四种约束，包括非空约束、默认约束、主键约束、唯一约束等，学会给表中的字段添加约束。

① 往数据表中插入数据。

```
INSERT INTO <表名>(<字段名1>,<字段名2>,…,<字段名N>)
values (<值1>,<值2>,…,<值N>);
```

② 设置自动增长。若某一列是数值类型的，使用关键字 AUTO_INCREMENT 来完成值的自动增长。

③ 设置表的约束。

  a. 非空约束(NOT NULL)。

  b. 默认约束(DEFAULT)。

  c. 主键约束(PRIMARY KEY)。

  d. 唯一约束(UNIQUE)。

【实训内容】

① 在数据库 demo 中创建 user 表，表中加入三个字段 name、age、tel，分别表示用户姓名、年龄、电话号码，并设置字段的数据类型，向 user 表中插入如下数据。

```
+--------+-----+-------------+
| name   | age | tel         |
+--------+-----+-------------+
| 王晓天 | 20  | 13459392049 |
| 李东东 | 21  | 13678438347 |
| 邱平平 | 18  | NULL        |
+--------+-----+-------------+
```

② 重新创建 user 表，表中加入四个字段 uid、name、age、tel，分别表示用户编号、用户姓名、年龄、电话号码，设置用户编号 uid 为自动增长，name 默认值为匿名，age 非空，tel 必须唯一，主键设置为 uid。创建 user 表完成后查看表的详细字段信息，并向 user 表中添加如下数据。

```
+-----+--------+-----+-------------+
| uid | name   | age | tel         |
+-----+--------+-----+-------------+
| 1   | 王晓天 | 20  | 13459392049 |
| 2   | 李东东 | 21  | 13678438347 |
| 3   | 匿名   | 20  | 13567789544 |
+-----+--------+-----+-------------+
```

③ 在数据库 demo 中创建一个 employee 数据表，表中加入六个字段 empId、empName、gender、jobTitle、salary、hire_date，分别表示员工编号、员工姓名、性别、职位、薪水、入职日期。设置员工编号 empId 为自动增长，empName、gender、jobTitle、salary、hire_date 为非空，hire_date 默认值为 2018-01-01。创建 employee 表完成后查看表的详细字段信息，并向 employee 表中添加如下数据。

```
+-------+---------+--------+----------------+--------+------------+
| empId | empName | gender | jobTitle       | salary | hire_date  |
+-------+---------+--------+----------------+--------+------------+
|     1 | 邱子石  | 男     | 软件开发工程师 |   9560 | 2019-10-12 |
|     2 | 张信瑞  | 女     | 测试工程师     |   5500 | 2018-12-02 |
|     3 | 萧正业  | 男     | 软件开发工程师 |   9800 | 2019-05-08 |
|     4 | 侯鹏运  | 男     | 软件开发工程师 |   9200 | 2020-08-31 |
|     5 | 周媛雪  | 女     | 文案           |   5000 | 2021-07-18 |
|     6 | 沈备    | 男     | 产品经理       |   7000 | 2020-09-13 |
|     7 | 任凯康  | 男     | 设计           |   6000 | 2018-10-19 |
|     8 | 孙安康  | 男     | 销售员         |   5000 | 2019-10-12 |
|     9 | 王云溪  | 女     | 销售员         |   6000 | 2019-11-23 |
|    10 | 张小雨  | 女     | 销售员         |   8500 | 2017-10-04 |
|    11 | 赵棠离  | 女     | 测试工程师     |   6500 | 2018-10-05 |
|    12 | 李甜甜  | 女     | 软件开发工程师 |   8800 | 2019-05-31 |
+-------+---------+--------+----------------+--------+------------+
```

【实训步骤】

① 在数据库 demo 中创建 user 表，语句如下：

```
CREATE TABLE user(
    name varchar(20),
    age int,
    tel varchar(20)
);
```

使用 INSERT 语句插入第一条数据，语句如下：

```
mysql> INSERT into user(name,age,tel)
    -> values
    -> ("王晓天",20,"13459392049");
Query OK, 1 row affected (0.16 sec)
```

使用 INSERT 语句插入第二条数据，语句如下：

```
mysql> INSERT into user
    -> values
    -> ("李东东",21,"13678438347");
Query OK, 1 row affected (0.14 sec)
```

注意：如果向表中的全部字段添加值，插入数据时可以省略字段名。

插入第三条数据，语句如下：

```
mysql> INSERT into user(name,age)
    -> values
    -> ("邱平平",18);
Query OK, 1 row affected (0.16 sec)
```

注意：这里插入第三条数据时，是向表中指定字段添加值，只需要在 INSERT 语句中指定要插入值的部分字段，其他字段的值为表定义时的默认值，这里为 NULL。

② 重新创建 user 表，表中加入四个字段 uid、name、age、tel，分别表示用户编号、用户姓名、年龄、电话号码，设置用户编号 uid 为自动增长，设置为主键，name 默认值为匿名，age 非空，tel 必须唯一，创建表的语句如下：

```
mysql> DROP TABLE user;
Query OK, 0 rows affected (0.53 sec)
mysql> CREATE TABLE user(
    ->     uid int AUTO_INCREMENT,
    ->     name varchar(20) DEFAULT '匿名',
    ->     age int NOT NULL,
    ->     tel varchar(20) UNIQUE,
    ->     PRIMARY KEY(uid)
    -> );
Query OK, 0 rows affected (0.47 sec)
```

创建完成后查看 user 表的详细字段信息。

```
mysql> DESC user;
+-------+-------------+------+-----+---------+----------------+
| Field | Type        | Null | Key | Default | Extra          |
+-------+-------------+------+-----+---------+----------------+
| uid   | int(11)     | NO   | PRI | NULL    | auto_increment |
| name  | varchar(20) | YES  |     | 匿名    |                |
| age   | int(11)     | NO   |     | NULL    |                |
| tel   | varchar(20) | YES  | UNI | NULL    |                |
+-------+-------------+------+-----+---------+----------------+
4 rows in set (0.16 sec)
```

可以看到，uid 已经设置成主键，并设置为自动增长列，name 添加了默认约束，默认值为匿名；age 添加了非空约束，tel 添加了唯一约束。接下来插入第一条和第二条数据，语句如下：

```
mysql> INSERT into user(name,age,tel)
    -> values
    -> ("王晓天",20,"13459392049"),
    -> ("李东东",21,"13678438347");
```

```
Query OK, 2 rows affected (0.13 sec)
Records: 2  Duplicates: 0  Warnings: 0
```

这里注意,多条数据的插入可以用逗号隔开。此外,由于设置了 uid 为自动增长,所以插入数据时不需要设置 uid 的值。由于 name 设置了默认约束,因此第三条数据插入的语句如下:

```
mysql> INSERT into user(age,tel)
    -> values
    -> (20,"13567789544");
Query OK, 1 row affected (0.11 sec)
```

需要注意的是,由于 age 设置了非空约束,所以如果某条数据中没有给 age 取值,则插入不成功。

```
mysql> INSERT into user(name,tel)
    -> values
    -> ("王丽","13768594356");
ERROR 1364 (HY000): Field 'age' doesn't have a default value
```

提示:Field 'age' doesn't have a default value,表示 age 必须非空。

此外,由于 tel 设置了唯一约束,因此如果某条数据中 tel 的取值重复了,插入也会不成功。

```
mysql> INSERT into user(name,age,tel)
    -> values
    -> ("张芳",18,"13459392049");
ERROR 1062 (23000): Duplicate entry '13459392049' for key 'tel'
```

提示:电话号码 13459392049 取值重复,不满足唯一约束,所以插入不成功。

③ 在数据库 demo 中创建一个 employee 数据表,表中加入六个字段 empId、empName、gender、jobTitle、salary、hire_date,分别表示员工编号、员工姓名、性别、职位、薪水、入职日期。设置员工编号 empId 为自动增长,empName、gender、jobTitle、salary、hire_date 为非空,hire_date 默认值为 2018-01-01。创建表的语句如下:

```
mysql> CREATE TABLE employee (
    -> empId INT AUTO_INCREMENT PRIMARY KEY,
    -> empName varchar(50)  NOT NULL,
    -> gender ENUM("男","女") NOT NULL ,
    -> jobTitle varchar(50) NOT NULL,
    -> salary DOUBLE NOT NULL,
    -> hire_date DATE NOT NULL DEFAULT "2018-01-01"
    -> );
Query OK, 0 rows affected (0.26 sec)
```

查看 employee 表的详细字段信息。

```
mysql> DESC employee;
+-----------+---------------+------+-----+------------+----------------+
| Field     | Type          | Null | Key | Default    | Extra          |
+-----------+---------------+------+-----+------------+----------------+
| empId     | int(11)       | NO   | PRI | NULL       | auto_increment |
| empName   | varchar(50)   | NO   |     | NULL       |                |
| gender    | enum('男','女')| NO   |     | NULL       |                |
| jobTitle  | varchar(50)   | NO   |     | NULL       |                |
| salary    | double        | NO   |     | NULL       |                |
| hire_date | date          | NO   |     | 2018-01-01 |                |
+-----------+---------------+------+-----+------------+----------------+
6 rows in set (0.00 sec)
```

向 employee 表中添加多条数据。

```
mysql> INSERT into 'employee'
    -> ('empName','gender','jobTitle','salary','hire_date')
    -> values
    -> ('邱予石','男','软件开发工程师',9560,'2019-10-12'),
    -> ('张信瑞','女','测试工程师',5500,'2018-12-2'),
    -> ('萧正业','男','软件开发工程师',9800,'2019-5-8'),
    -> ('侯鹏运','男','软件开发工程师',9200,'2020-8-31'),
    -> ('周嫒雪','女','文案',5000,'2021-7-18'),
    -> ('沈备','男','产品经理',7000,'2020-9-13'),
    -> ('任凯康','男','设计',6000,'2018-10-19'),
    -> ('孙安康','男','销售员',5000,'2019-10-12'),
    -> ('王云溪','女','销售员',6000,'2019-11-23'),
    -> ('张小雨','女','销售员',8500,'2017-10-4'),
    -> ('赵棠离','女','测试工程师',6500,'2018-10-5'),
    -> ('李甜甜','女','软件开发工程师',8800,'2019-5-31');
Query OK, 12 rows affected (0.18 sec)
Records: 12  Duplicates: 0  Warnings: 0
```

实训 4

简单数据查询

【实训目的】

掌握使用 SELECT 语句进行简单数据查询的方法。

① SELECT 语句基本用法。

a. 查询表中所有数据。
```
SELECT * FROM <表名>;
```
b. 查询表中某几列数据。
```
SELECT <字段名1>,<字段名2>,…<字段名N> FROM <表名>;
```

② 通过 WHERE 子句过滤。
```
SELECT <字段名1>,<字段名2>,…<字段名N> FROM <表名>
[WHERe <条件表达式>];
```

WHERE 子句是用来设定查询条件的。<条件表达式>类似于程序语言中的 if 条件，一般格式为：字段名运算符值。其中，运算符可以是＝、!＝、>、<、>＝、<＝等。此外，还可以使用 AND、OR、NOT 来指定一个或多个条件。

③ 使用 ORDER BY 对返回结果进行排序。
```
SELECT <字段名1>,<字段名2>,…<字段名N> FROM <表名>
[WHERE <条件表达式>]
[ORdER BY <字段名> [ASC|DESC] ];
```

其中，ASC 表示升序排列，DESC 表示降序排列，默认值为 ASC。

④ LIMIT 限制返回结果的数量。
```
SELECT <字段名1>,<字段名2>,…<字段名N> FROM <表名>
[WHERE <条件表达式>]
[LIMIt <数量>];
```

⑤ LIKE 进行字符串搜索过滤。
```
SELECT <字段名1>,<字段名2>,…<字段名N> FROM <表名>
[WHERE 字段名 like <匹配模式>]
```

LIKE 操作符用于在 WHERE 子句中搜索列中的指定模式。在 SQL 中，通配符经常与 LIKE 操作符一起使用。通配符可用于替代字符串中的任何其他字符，常用的通配符是"％"和"_"，其中"％"用来表示替代 0 个或多个字符，"_"则用来替代一个字符。

⑥ BETWEEN AND 进行范围查询，用来判断字段的值是否在指定范围内。

⑦ IN 运算符用来判断字段的值是否位于给出的列表中。

【实训内容】
① 进入 demo 数据库，查看 demo 数据库中的 employee 表中的所有数据。
② 显示查看 employee 表中每个员工的员工编号、员工姓名、职位、薪水信息。
③ 查找职位是软件开发工程师的所有员工信息。
④ 查找姓名为周媛雪的员工详细信息。
⑤ 查找所有薪水小于 6 000 元的员工信息。
⑥ 查找职位是软件开发工程师并且薪水大于 9 500 元的员工信息。

⑦ 查找职位是软件产品经理或者薪水等于 5 000 元的员工信息。
⑧ 查找职位不是软件开发工程师的所有员工信息。
⑨ 将员工信息按照工资从高到低进行排序。
⑩ 查询员工表中工资最高的三位员工信息。
⑪ 查询员工表中姓周的员工信息。
⑫ 查询员工表中名字以"康"结尾的员工信息。
⑬ 查找薪水在 6 000～8 000 元之间的所有员工信息。
⑭ 查找 2017～2019 年入职的所有员工信息。
⑮ 查找职位是文案、产品经理或者设计的所有员工信息。

【实训步骤】

① 进入 demo 数据库，查看 demo 数据库中 employee 表的所有数据。

```
mysql> USE demo;
Database changed
mysql> SELECT * FROM employee;
+-------+---------+--------+----------------+--------+------------+
| empId | empName | gender | jobTitle       | salary | hire_date  |
+-------+---------+--------+----------------+--------+------------+
|     1 | 邱子石  | 男     | 软件开发工程师 |   9560 | 2019-10-12 |
|     2 | 张信瑞  | 女     | 测试工程师     |   5500 | 2018-12-02 |
|     3 | 萧正业  | 男     | 软件开发工程师 |   9800 | 2019-05-08 |
|     4 | 侯鹏运  | 男     | 软件开发工程师 |   9200 | 2020-08-31 |
|     5 | 周媛雪  | 女     | 文案           |   5000 | 2021-07-18 |
|     6 | 沈备    | 男     | 产品经理       |   7000 | 2020-09-13 |
|     7 | 任凯康  | 男     | 设计           |   6000 | 2018-10-19 |
|     8 | 孙安康  | 男     | 销售员         |   5000 | 2019-10-12 |
|     9 | 王云溪  | 女     | 销售员         |   6000 | 2019-11-23 |
|    10 | 张小雨  | 女     | 销售员         |   8500 | 2017-10-04 |
|    11 | 赵棠离  | 女     | 测试工程师     |   6500 | 2018-10-05 |
|    12 | 李甜甜  | 女     | 软件开发工程师 |   8800 | 2019-05-31 |
+-------+---------+--------+----------------+--------+------------+
12 rows in set (0.00 sec)
```

② 显示查看 employee 表中每个员工的员工编号、员工姓名、职位、薪水信息。

```
mysql> SELECT empId as '员工编号',empName as '员工姓名',
    -> jobTitle as '职位',salary as '薪水' from employee;
+----------+----------+----------------+--------+
| 员工编号 | 员工姓名 | 职位           | 薪水   |
+----------+----------+----------------+--------+
|        1 | 邱子石   | 软件开发工程师 |   9560 |
|        2 | 张信瑞   | 测试工程师     |   5500 |
```

```
|   3 | 萧正业   | 软件开发工程师 | 9800 |
|   4 | 侯鹏运   | 软件开发工程师 | 9200 |
|   5 | 周媛雪   | 文案           | 5000 |
|   6 | 沈备     | 产品经理       | 7000 |
|   7 | 任凯康   | 设计           | 6000 |
|   8 | 孙安康   | 销售员         | 5000 |
|   9 | 王云溪   | 销售员         | 6000 |
|  10 | 张小雨   | 销售员         | 8500 |
|  11 | 赵棠离   | 测试工程师     | 6500 |
|  12 | 李甜甜   | 软件开发工程师 | 8800 |
```

12 rows in set (0.00 sec)

③ 查找职位是软件开发工程师的所有员工信息。

```
mysql> SELECT * FROM employee where jobTitle = '软件开发工程师';
+-------+---------+--------+----------------+--------+------------+
| empId | empName | gender | jobTitle       | salary | hire_date  |
+-------+---------+--------+----------------+--------+------------+
|     1 | 邱子石  | 男     | 软件开发工程师 |   9560 | 2019-10-12 |
|     3 | 萧正业  | 男     | 软件开发工程师 |   9800 | 2019-05-08 |
|     4 | 侯鹏运  | 男     | 软件开发工程师 |   9200 | 2020-08-31 |
|    12 | 李甜甜  | 女     | 软件开发工程师 |   8800 | 2019-05-31 |
+-------+---------+--------+----------------+--------+------------+
```

4 rows in set (0.00 sec)

④ 查找姓名为周媛雪的员工详细信息。

```
mysql> SELECT * FROM employee where empName = '周媛雪';
+-------+---------+--------+----------+--------+------------+
| empId | empName | gender | jobTitle | salary | hire_date  |
+-------+---------+--------+----------+--------+------------+
|     5 | 周媛雪  | 女     | 文案     |   5000 | 2021-07-18 |
+-------+---------+--------+----------+--------+------------+
```

1 row in set (0.00 sec)

⑤ 查找所有薪水小于6 000元的员工信息。

```
mysql> SELECT * FROM employee where salary < 6000;
+-------+---------+--------+------------+--------+------------+
| empId | empName | gender | jobTitle   | salary | hire_date  |
+-------+---------+--------+------------+--------+------------+
|     2 | 张信瑞  | 女     | 测试工程师 |   5500 | 2018-12-02 |
|     5 | 周媛雪  | 女     | 文案       |   5000 | 2021-07-18 |
|     8 | 孙安康  | 男     | 销售员     |   5000 | 2019-10-12 |
+-------+---------+--------+------------+--------+------------+
```

3 rows in set (0.00 sec)

⑥ 查找职位是软件开发工程师并且薪水大于 9 500 元的员工信息。

```
mysql> SELECT * FROM employee
    -> WHERE jobTitle = '软件开发工程师' AND salary > 9500;
+-------+---------+--------+------------------+--------+------------+
| empId | empName | gender | jobTitle         | salary | hire_date  |
+-------+---------+--------+------------------+--------+------------+
|     1 | 邱予石  | 男     | 软件开发工程师   |   9560 | 2019-10-12 |
|     3 | 萧正业  | 男     | 软件开发工程师   |   9800 | 2019-05-08 |
+-------+---------+--------+------------------+--------+------------+
2 rows in set (0.00 sec)
```

⑦ 查找职位是软件产品经理或者薪水等于 5 000 元的员工信息。

```
mysql> SELECT * FROM employee
    -> WHERE jobTitle = '产品经理' OR salary = 5000;
+-------+---------+--------+----------+--------+------------+
| empId | empName | gender | jobTitle | salary | hire_date  |
+-------+---------+--------+----------+--------+------------+
|     5 | 周媛雪  | 女     | 文案     |   5000 | 2021-07-18 |
|     6 | 沈备    | 男     | 产品经理 |   7000 | 2020-09-13 |
|     8 | 孙安康  | 男     | 销售员   |   5000 | 2019-10-12 |
+-------+---------+--------+----------+--------+------------+
3 rows in set (0.00 sec)
```

⑧ 查找职位不是软件开发工程师的所有员工信息。

```
mysql> SELECT * FROM employee
    -> WHERE  NOT jobTitle = '软件开发工程师';
+-------+---------+--------+--------------+--------+------------+
| empId | empName | gender | jobTitle     | salary | hire_date  |
+-------+---------+--------+--------------+--------+------------+
|     2 | 张信瑞  | 女     | 测试工程师   |   5500 | 2018-12-02 |
|     5 | 周媛雪  | 女     | 文案         |   5000 | 2021-07-18 |
|     6 | 沈备    | 男     | 产品经理     |   7000 | 2020-09-13 |
|     7 | 任凯康  | 男     | 设计         |   6000 | 2018-10-19 |
|     8 | 孙安康  | 男     | 销售员       |   5000 | 2019-10-12 |
|     9 | 王云溪  | 女     | 销售员       |   6000 | 2019-11-23 |
|    10 | 张小雨  | 女     | 销售员       |   8500 | 2017-10-04 |
|    11 | 赵棠离  | 女     | 测试工程师   |   6500 | 2018-10-05 |
+-------+---------+--------+--------------+--------+------------+
8 rows in set (0.00 sec)
```

⑨ 使用 ORDER BY 将员工信息按照工资从高到低进行排序。

```
mysql> SELECT * FROM employee order by salary desc;
+-------+---------+--------+------------------+--------+------------+
| empId | empName | gender | jobTitle         | salary | hire_date  |
+-------+---------+--------+------------------+--------+------------+
|     3 | 萧正业  | 男     | 软件开发工程师   |   9800 | 2019-05-08 |
|     1 | 邱子石  | 男     | 软件开发工程师   |   9560 | 2019-10-12 |
|     4 | 侯鹏运  | 男     | 软件开发工程师   |   9200 | 2020-08-31 |
|    12 | 李甜甜  | 女     | 软件开发工程师   |   8800 | 2019-05-31 |
|    10 | 张小雨  | 女     | 销售员           |   8500 | 2017-10-04 |
|     6 | 沈备    | 男     | 产品经理         |   7000 | 2020-09-13 |
|    11 | 赵棠离  | 女     | 测试工程师       |   6500 | 2018-10-05 |
|     7 | 任凯康  | 男     | 设计             |   6000 | 2018-10-19 |
|     9 | 王云溪  | 女     | 销售员           |   6000 | 2019-11-23 |
|     2 | 张信瑞  | 女     | 测试工程师       |   5500 | 2018-12-02 |
|     5 | 周嫒雪  | 女     | 文案             |   5000 | 2021-07-18 |
|     8 | 孙安康  | 男     | 销售员           |   5000 | 2019-10-12 |
+-------+---------+--------+------------------+--------+------------+
12 rows in set (0.00 sec)
```

⑩ 查询员工表中工资最高的三位员工信息。

```
mysql> SELECT * FROM employee order by salary desc limit 3;
+-------+---------+--------+------------------+--------+------------+
| empId | empName | gender | jobTitle         | salary | hire_date  |
+-------+---------+--------+------------------+--------+------------+
|     3 | 萧正业  | 男     | 软件开发工程师   |   9800 | 2019-05-08 |
|     1 | 邱子石  | 男     | 软件开发工程师   |   9560 | 2019-10-12 |
|     4 | 侯鹏运  | 男     | 软件开发工程师   |   9200 | 2020-08-31 |
+-------+---------+--------+------------------+--------+------------+
3 rows in set (0.10 sec)
```

⑪ 查询员工表中姓周的员工信息。

```
mysql> SELECT * FROM employee where empName like '周%';
+-------+---------+--------+----------+--------+------------+
| empId | empName | gender | jobTitle | salary | hire_date  |
+-------+---------+--------+----------+--------+------------+
|     5 | 周嫒雪  | 女     | 文案     |   5000 | 2021-07-18 |
+-------+---------+--------+----------+--------+------------+
1 row in set (0.16 sec)
```

⑫ 查询员工表中名字以"康"结尾的员工信息。

```
mysql> SELECT * FROM employee where empName like '%康';
+-------+---------+--------+----------+--------+------------+
| empId | empName | gender | jobTitle | salary | hire_date  |
+-------+---------+--------+----------+--------+------------+
|     7 | 任凯康  | 男     | 设计     |   6000 | 2018-10-19 |
|     8 | 孙安康  | 男     | 销售员   |   5000 | 2019-10-12 |
+-------+---------+--------+----------+--------+------------+
2 rows in set (0.00 sec)
```

⑬ 查找薪水在 6 000～8 000 元之间的所有员工信息。

```
mysql> SELECT * FROM employee where salary >= 6000 AND salary <= 8000;
+-------+---------+--------+--------------+--------+------------+
| empId | empName | gender | jobTitle     | salary | hire_date  |
+-------+---------+--------+--------------+--------+------------+
|     6 | 沈备    | 男     | 产品经理     |   7000 | 2020-09-13 |
|     7 | 任凯康  | 男     | 设计         |   6000 | 2018-10-19 |
|     9 | 王云溪  | 女     | 销售员       |   6000 | 2019-11-23 |
|    11 | 赵棠离  | 女     | 测试工程师   |   6500 | 2018-10-05 |
+-------+---------+--------+--------------+--------+------------+
4 rows in set (0.00 sec)
```

这里，也可以使用 BETWEEN AND 来编写 SQL 语句。

```
mysql> SELECT * FROM employee WHERE salary BETWEEN 6000 AND 8000;
+-------+---------+--------+--------------+--------+------------+
| empId | empName | gender | jobTitle     | salary | hire_date  |
+-------+---------+--------+--------------+--------+------------+
|     6 | 沈备    | 男     | 产品经理     |   7000 | 2020-09-13 |
|     7 | 任凯康  | 男     | 设计         |   6000 | 2018-10-19 |
|     9 | 王云溪  | 女     | 销售员       |   6000 | 2019-11-23 |
|    11 | 赵棠离  | 女     | 测试工程师   |   6500 | 2018-10-05 |
+-------+---------+--------+--------------+--------+------------+
4 rows in set (0.00 sec)
```

可以看到，结果完全相同。

⑭ 查找 2017～2019 年入职的所有员工信息。

```
mysql> SELECT * FROM employee WHERE year(hire_date) BETWEEN 2017 AND 2019;
+-------+---------+--------+------------------+--------+------------+
| empId | empName | gender | jobTitle         | salary | hire_date  |
+-------+---------+--------+------------------+--------+------------+
|     1 | 邱子石  | 男     | 软件开发工程师   |   9560 | 2019-10-12 |
|     2 | 张信瑞  | 女     | 测试工程师       |   5500 | 2018-12-02 |
|     3 | 萧正业  | 男     | 软件开发工程师   |   9800 | 2019-05-08 |
|     7 | 任凯康  | 男     | 设计             |   6000 | 2018-10-19 |
|     8 | 孙安康  | 男     | 销售员           |   5000 | 2019-10-12 |
|     9 | 王云溪  | 女     | 销售员           |   6000 | 2019-11-23 |
|    10 | 张小雨  | 女     | 销售员           |   8500 | 2017-10-04 |
|    11 | 赵棠离  | 女     | 测试工程师       |   6500 | 2018-10-05 |
|    12 | 李甜甜  | 女     | 软件开发工程师   |   8800 | 2019-05-31 |
+-------+---------+--------+------------------+--------+------------+
9 rows in set (0.10 sec)
```

⑮ 查找职位是文案、产品经理或者设计的所有员工信息。

```
mysql> SELECT * FROM employee
    -> WHERE jobTitle = "文案" or jobTitle = "产品经理" or jobTitle = "设计";
+-------+---------+--------+----------+--------+------------+
| empId | empName | gender | jobTitle | salary | hire_date  |
+-------+---------+--------+----------+--------+------------+
|     5 | 周媛雪  | 女     | 文案     |   5000 | 2021-07-18 |
|     6 | 沈备    | 男     | 产品经理 |   7000 | 2020-09-13 |
|     7 | 任凯康  | 男     | 设计     |   6000 | 2018-10-19 |
+-------+---------+--------+----------+--------+------------+
3 rows in set (0.00 sec)
```

这里，也可以使用 IN 来编写 SQL 语句。

```
mysql> SELECT * FROM employee WHERE jobTitle IN ("文案","产品经理","设计");
+-------+---------+--------+----------+--------+------------+
| empId | empName | gender | jobTitle | salary | hire_date  |
+-------+---------+--------+----------+--------+------------+
|     5 | 周媛雪  | 女     | 文案     |   5000 | 2021-07-18 |
|     6 | 沈备    | 男     | 产品经理 |   7000 | 2020-09-13 |
|     7 | 任凯康  | 男     | 设计     |   6000 | 2018-10-19 |
+-------+---------+--------+----------+--------+------------+
3 rows in set (0.02 sec)
```

可以看到，结果完全相同。

实训 5

数据的修改删除操作

【实训目的】

掌握数据的修改及删除操作。

① 使用 UPDATE 语句修改数据表数据。

```
UPDATE〈表名〉
SET〈字段名 1〉=〈新值 1〉,〈字段名 2〉=〈新值 2〉
[WHERE〈条件表达式〉];
```

UPDATE 语句可以同时更新一个或多个字段，并且可以在 WHERE 子句中指定任何条件。

② 使用 DELETE 语句删除 MySQL 数据表中的记录。

```
DELETE FROM〈表名〉
[WHERE〈条件表达式〉];
```

这里，我们可以在 WHERE 子句中指定任何条件。如果没有指定 WHERE 子句，那么表中的所有记录将被删除。

【实训内容】

① 修改 employee 表中销售员的工资为 10 000 元。

② 将员工萧正业的职位调整为项目经理，并且工资改为 20 000 元。

③ 在 employee 表中删除员工任凯康的记录。

④ 删除 employee 表中所有员工。

【实训步骤】

① 修改 employee 表中销售员的工资为 10 000 元。

```
mysql> UPDATE employee set salary = 10000
    -> WHERE jobTitle = "销售员";
Query OK, 3 rows affected (0.16 sec)
Rows matched: 3  Changed: 3  Warnings: 0
```

重新查询销售员的工资。

```
mysql> SELECT * FROM employee
    -> WHERE jobTitle = "销售员";
+-------+---------+--------+---------+--------+------------+
| empId | empName | gender | jobTitle| salary | hire_date  |
+-------+---------+--------+---------+--------+------------+
|   8   | 孙安康  |  男    | 销售员  | 10000  | 2019-10-12 |
|   9   | 王云溪  |  女    | 销售员  | 10000  | 2019-11-23 |
|  10   | 张小雨  |  女    | 销售员  | 10000  | 2017-10-04 |
+-------+---------+--------+---------+--------+------------+
3 rows in set (0.00 sec)
```

可以看到，所有销售员的工资调整成为 10 000 元。

② 将员工萧正业的职位调整为项目经理，并且工资改为 20 000 元。

```
mysql> UPDATE employee set jobTitle = "项目经理", salary = 20000
    -> WHERE empName = "萧正业";
Query OK, 1 row affected (0.13 sec)
Rows matched: 1  Changed: 1  Warnings: 0
```

这里同时更新了两个字段，重新查询萧正业的员工信息。

```
mysql> SELECT * FROM employee
    -> WHERE empName = "萧正业";
+-------+---------+--------+----------+--------+------------+
| empId | empName | gender | jobTitle | salary | hire_date  |
+-------+---------+--------+----------+--------+------------+
|   3   | 萧正业  |  男    | 项目经理 | 20000  | 2019-05-08 |
+-------+---------+--------+----------+--------+------------+
1 row in set (0.00 sec)
```

③ 在 employee 表中删除员工任凯康的记录。

```
mysql> DELETE FROM employee
    -> WHERE empName = "任凯康";
Query OK, 1 row affected (0.16 sec)
```

重新查询员工信息。

```
mysql> SELECT * FROM employee;
+-------+---------+--------+------------------+--------+------------+
| empId | empName | gender | jobTitle         | salary | hire_date  |
+-------+---------+--------+------------------+--------+------------+
|     1 | 邱子石  | 男     | 软件开发工程师   |   9560 | 2019-10-12 |
|     2 | 张信瑞  | 女     | 测试工程师       |   5500 | 2018-12-02 |
|     3 | 萧正业  | 男     | 项目经理         |  20000 | 2019-05-08 |
|     4 | 侯鹏运  | 男     | 软件开发工程师   |   9200 | 2020-08-31 |
|     5 | 周媛雪  | 女     | 文案             |   5000 | 2021-07-18 |
|     6 | 沈备    | 男     | 产品经理         |   7000 | 2020-09-13 |
|     8 | 孙安康  | 男     | 销售员           |  10000 | 2019-10-12 |
|     9 | 王云溪  | 女     | 销售员           |  10000 | 2019-11-23 |
|    10 | 张小雨  | 女     | 销售员           |  10000 | 2017-10-04 |
|    11 | 赵棠离  | 女     | 测试工程师       |   6500 | 2018-10-05 |
|    12 | 李甜甜  | 女     | 软件开发工程师   |   8800 | 2019-05-31 |
+-------+---------+--------+------------------+--------+------------+
11 rows in set (0.00 sec)
```

可以看到，员工任凯康的记录已经被成功删除。

④ 删除 employee 表中所有员工。

```
mysql> DELETE FROM employee;
Query OK, 11 rows affected (0.11 sec)
```

重新查询员工信息。

```
mysql> SELECT * FROM employee;
Empty SET (0.00 sec)
```

可以看到，如果省略了 WHERE 子句，所有的记录都会被删除。

实训 6

数据的聚合处理

【实训目的】

掌握使用聚合函数进行数据查询的方法，MySQL 中常用的聚合方式有：

① 使用 COUNT 函数对结果进行计数。

② 使用 DISTINCT 统计唯一值。

③ 使用 GROUP BY 子句进行数据整合。

④ 使用 MAX 函数和 MIN 函数求最大值最小值。

⑤ SUM 函数和 AVG 函数求和和平均值。

⑥ HAVING 子句对聚合以后的结果进行过滤，HAVING 类似于 WHERE，区别在于 WHERE 过滤行，HAVING 过滤组。

【实训内容】

① 计算 employee 表中所有的员工的总数。

② 计算员工表中软件开发工程师的数量。

③ 查询公司所有的岗位名称，并统计岗位的数量。

④ 统计员工表中每个岗位上各有多少名员工。

⑤ 查询员工表中最高工资是多少。

⑥ 查询各个岗位的最高工资和最低工资是多少。

⑦ 统计每个岗位的工资总和和平均工资是多少。

⑧ 统计软件开发工程师岗位上的员工数量。

【实训步骤】

使用实训 3 的步骤进行 employee 表的建表和添加数据操作。查询表中数据，进行数据准备。

```
mysql> select * from employee;
+-------+---------+--------+----------------+--------+------------+
| empId | empName | gender | jobTitle       | salary | hire_date  |
+-------+---------+--------+----------------+--------+------------+
|     1 | 邱子石  | 男     | 软件开发工程师 |   9560 | 2019-10-12 |
|     2 | 张信瑞  | 女     | 测试工程师     |   5500 | 2018-12-02 |
|     3 | 萧正业  | 男     | 软件开发工程师 |  20000 | 2019-05-08 |
|     4 | 侯鹏运  | 男     | 软件开发工程师 |   9200 | 2020-08-31 |
|     5 | 周媛雪  | 女     | 文案           |   5000 | 2021-07-18 |
|     6 | 沈备    | 男     | 产品经理       |   7000 | 2020-09-13 |
|     7 | 任凯康  | 男     | 设计           |   6000 | 2018-10-19 |
|     8 | 孙安康  | 男     | 销售员         |  10000 | 2019-10-12 |
|     9 | 王云溪  | 女     | 销售员         |  10000 | 2019-11-23 |
|    10 | 张小雨  | 女     | 销售员         |  10000 | 2017-10-04 |
|    11 | 赵棠离  | 女     | 测试工程师     |   6500 | 2018-10-05 |
|    12 | 李甜甜  | 女     | 软件开发工程师 |   8800 | 2019-05-31 |
+-------+---------+--------+----------------+--------+------------+
12 rows in set (0.00 sec)
```

① 使用 COUNT 函数计算 employee 表中所有的员工的总数。

```
mysql> select count(*) as "员工总数" from employee;
+----------+
| 员工总数 |
+----------+
|       12 |
+----------+
1 row in set (0.00 sec)
```

② 查看软件开发工程师的员工信息。

```
mysql> select * from employee where jobTitle = "软件开发工程师";
+-------+---------+--------+----------------+--------+------------+
| empId | empName | gender | jobTitle       | salary | hire_date  |
+-------+---------+--------+----------------+--------+------------+
|     1 | 邱子石  | 男     | 软件开发工程师 |   9560 | 2019-10-12 |
|     3 | 萧正业  | 男     | 软件开发工程师 |  20000 | 2019-05-08 |
|     4 | 侯鹏运  | 男     | 软件开发工程师 |   9200 | 2020-08-31 |
|    12 | 李甜甜  | 女     | 软件开发工程师 |   8800 | 2019-05-31 |
+-------+---------+--------+----------------+--------+------------+
4 rows in set (0.00 sec)
```

使用 COUNT 函数计算员工表中软件开发工程师的数量。

```
mysql> select count(*) from employee where jobTitle = "软件开发工程师";
+----------+
| count(*) |
+----------+
|        4 |
+----------+
1 row in set (0.00 sec)
```

可以看到，共有 4 位软件开发工程师。

③ 查询公司所有的岗位名称，并用 DISTINCT 去除重复的数据。

```
mysql> select distinct jobTitle from employee;
+----------------+
| jobTitle       |
+----------------+
| 软件开发工程师 |
| 测试工程师     |
| 文案           |
| 产品经理       |
| 设计           |
| 销售员         |
+----------------+
6 rows in set (0.00 sec)
```

统计一共有多少个岗位。

```
mysql> select count(distinct jobTitle) from employee;
+--------------------------+
| count(distinct jobTitle) |
+--------------------------+
|                        6 |
+--------------------------+
1 row in set (0.00 sec)
```

④ 使用 GROUP BY 统计员工表中每个岗位上各有多少名员工。

```
mysql> select jobTitle,count(empName) from employee group by jobTitle;
+----------------------+----------------+
| jobTitle             | count(empName) |
+----------------------+----------------+
| 产品经理             |              1 |
| 文案                 |              1 |
| 测试工程师           |              2 |
| 设计                 |              1 |
| 软件开发工程师       |              4 |
| 销售员               |              3 |
+----------------------+----------------+
6 rows in set (0.00 sec)
```

⑤ 使用 MAX 函数查询员工表中最高工资。

```
mysql> select empName,max(salary) from employee;
+---------+-------------+
| empName | max(salary) |
+---------+-------------+
| 邱子石  |        9800 |
+---------+-------------+
1 row in set (0.00 sec)
```

⑥ 查询各个岗位的最高工资和最低工资。

```
mysql> select jobTitle,max(salary),min(salary) from employee group by jobTitle;
+----------------------+-------------+-------------+
| jobTitle             | max(salary) | min(salary) |
+----------------------+-------------+-------------+
| 产品经理             |        7000 |        7000 |
| 文案                 |        5000 |        5000 |
| 测试工程师           |        6500 |        5500 |
| 设计                 |        6000 |        6000 |
| 软件开发工程师       |        9800 |        8800 |
| 销售员               |        8500 |        5000 |
+----------------------+-------------+-------------+
6 rows in set (0.00 sec)
```

⑦ 统计每个岗位的工资总和和平均工资。

```
mysql> select jobTitle,sum(salary),avg(salary) from employee group by jobTitle;
+----------------------+-------------+-------------+
| jobTitle             | sum(salary) | avg(salary) |
+----------------------+-------------+-------------+
| 产品经理             |        7000 |        7000 |
| 文案                 |        5000 |        5000 |
| 测试工程师           |       12000 |        6000 |
| 设计                 |        6000 |        6000 |
| 软件开发工程师       |       37360 |        9340 |
| 销售员               |       19500 |        6500 |
+----------------------+-------------+-------------+
6 rows in set (0.00 sec)
```

⑧ 统计软件开发工程师岗位上的员工数量。

```
mysql> select jobTitle,count(empName) from employee group by jobTitle having jobTitle = "软
件开发工程师";
+------------------+----------------+
| jobTitle         | count(empName) |
+------------------+----------------+
| 软件开发工程师    |              4 |
+------------------+----------------+
1 row in set (0.00 sec)
```

实训 7

多 表 查 询

【实训目的】

掌握使用表的连接进行多表查询。

① INNER JOIN(内连接):获取两个表中字段匹配关系的记录。

```
SELECT <字段名> FROM <表 1> INNER JOIN <表 2>
<ON 子句>
```

内连接中可以省略 INNER 关键字,只用关键字 JOIN。<ON 子句>用来设置内连接的连接条件。

② LEFT JOIN(左连接):获取左表所有记录,即使右表没有对应匹配的记录。

```
SELECT <字段名> FROM <表 1> LEFT JOIN <表 2>
<ON 子句>
```

左连接查询时,可以查询出<表 1>中的所有记录和<表 2>中匹配连接条件的记录。如果<表 1>的某行在<表 2>中没有匹配行,那么在返回结果中,<表 2>的字段值均为 NULL。

③ RIGHT JOIN(右连接):与 LEFT JOIN 相反,用于获取右表所有记录,即使左表没有对应匹配的记录。

```
SELECT <字段名> FROM <表 1> RIGHT JOIN <表 2>
<ON 子句>
```

与左连接相反,右连接查询时,可以查询出<表 2>中的所有记录和<表 1>中匹配连接条件的记录。如果<表 2>的某行在<表 1>中没有匹配行,那么在返回结果中,<表 1>的字段值均为 NULL。

【实训内容】

① 在数据库 demo 中创建部门表 department,表中加入三个字段 deptId、deptName、

city,分别表示部门编号、部门名称、所在城市。向表中插入如下数据。

```
+----+----------+--------+
| id | deptName | city   |
+----+----------+--------+
|  1 | 研发部    | 上海    |
|  2 | 研发部    | 深圳    |
|  3 | 研发部    | 北京    |
|  4 | 研发部    | 杭州    |
|  5 | 市场部    | 西安    |
|  6 | 销售部    | 北京    |
|  7 | 销售部    | 广州    |
|  8 | 客服部    | 上海    |
+----+----------+--------+
```

② 创建员工表 employee,表中加入字段 empId、empName、gender、jobTitle、salary、hire_date、deptId 分别表示员工编号、员工姓名、性别、职位、薪水、入职日期以及所在部门。向表中插入如下数据。

```
+-------+---------+--------+--------------+--------+------------+--------+
| empId | empName | gender | jobTitle     | salary | hire_date  | deptId |
+-------+---------+--------+--------------+--------+------------+--------+
|     1 | 邱子石   | 男     | 软件开发工程师 |   9560 | 2019-10-12 |      1 |
|     2 | 张信瑞   | 女     | 测试工程师     |   5500 | 2018-12-02 |      2 |
|     3 | 萧正业   | 男     | 软件开发工程师 |   9800 | 2019-05-08 |      2 |
|     4 | 侯鹏运   | 男     | 软件开发工程师 |   9200 | 2020-08-31 |      3 |
|     5 | 周媛雪   | 女     | 文案           |   5000 | 2021-07-18 |      5 |
|     6 | 沈备     | 男     | 产品经理       |   7000 | 2020-09-13 |      3 |
|     7 | 任凯康   | 男     | 设计           |   6000 | 2018-10-19 |      5 |
|     8 | 孙安康   | 男     | 销售员         |   5000 | 2019-10-12 |      6 |
|     9 | 王云溪   | 女     | 销售员         |   6000 | 2019-11-23 |      6 |
|    10 | 张小雨   | 女     | 销售员         |   8500 | 2017-10-04 |     10 |
|    11 | 赵棠离   | 女     | 测试工程师     |   6500 | 2018-10-05 |      4 |
|    12 | 李甜甜   | 女     | 软件开发工程师 |   8800 | 2019-05-31 |      4 |
+-------+---------+--------+--------------+--------+------------+--------+
```

③ 将 department 表和 employee 表按照所在部门编号进行内连接,并显示连接后的表。

④ 将 department 表和 employee 表按照所在部门编号进行左连接,并显示连接后的表。

⑤ 将 department 表和 employee 表按照所在部门编号进行右连接,并显示连接后的表。

⑥ 创建兴趣小组表 interestclub,表中加入字段 id、title、manager、wechat 分别表示小组编号、小组名称、组长、微信号。向表中插入如下数据。

```
+----+----------+---------+-------------+
| id | title    | manager | wechat      |
+----+----------+---------+-------------+
|  1 | 篮球组    | 潘智仁   | 13485983467 |
|  2 | 羽毛球组  | 殷小红   | 13567342561 |
|  3 | 乒乓球组  | 王咏惠   | 13937261523 |
|  4 | 户外徒步组| 吴杰     | 13653726633 |
|  5 | 文学影视组| 许凯旋   | 18834532345 |
+----+----------+---------+-------------+
```

⑦ 创建 joinclub 表来描述新入职员工加入兴趣小组的关系,表中加入字段 id、empId、clubId、join_date 分别表示名单编号、员工 id、兴趣小组 id、加入日期。一个员工可以加入多个兴趣小组,一个兴趣小组也可以接收多名员工。向表中插入如下数据。

```
+----+-------+--------+------------+
| id | empId | clubId | join_date  |
+----+-------+--------+------------+
|  1 |     1 |      1 | 2019-11-12 |
|  2 |     1 |      2 | 2020-01-10 |
|  3 |     1 |      3 | 2019-12-03 |
|  4 |     2 |      1 | 2018-12-24 |
|  5 |     2 |      3 | 2019-12-03 |
|  6 |     2 |      5 | 2019-05-30 |
|  7 |     3 |      4 | 2019-05-30 |
|  8 |     3 |      5 | 2019-06-13 |
|  9 |     4 |      1 | 2020-09-13 |
| 10 |     4 |      2 | 2020-09-13 |
| 11 |     4 |      4 | 2020-09-13 |
| 12 |     5 |      4 | 2021-08-30 |
| 13 |     5 |      5 | 2020-11-11 |
| 14 |     6 |      3 | 2020-11-22 |
| 15 |     7 |      5 | 2019-03-20 |
| 16 |     8 |      3 | 2020-10-10 |
| 17 |     8 |      4 | 2020-01-04 |
| 18 |     8 |      5 | 2021-02-14 |
| 19 |     9 |      5 | 2019-12-14 |
| 20 |    10 |      2 | 2018-03-24 |
| 21 |    10 |      4 | 2018-03-24 |
| 22 |    11 |      1 | 2019-06-26 |
| 23 |    12 |      2 | 2020-04-20 |
| 24 |    12 |      3 | 2020-04-20 |
+----+-------+--------+------------+
```

⑧ 查询员工周媛雪所在的部门名称以及城市。

⑨ 查询所有在北京的员工信息。

⑩ 统计每个城市的员工数量。

⑪ 统计每个员工参加的兴趣小组数量。

⑫ 查询员工周媛雪参加的所有兴趣小组的名称。

⑬ 统计加入每个兴趣小组的员工人数。

⑭ 查询参加文学影视组的所有员工姓名。

⑮ 查询同时参加三个兴趣小组的员工编号和姓名。

【实训步骤】

数据往往是由多张表组成的，有时我们需要从多张表中查找数据，这时候就会用到表的连接，首先将 demo 中的数据表删除，重新准备两张数据表。

① 在数据库 demo 中创建部门表 department。

```
mysql> CREATE TABLE 'department' (
    -> 'id'  INT AUTO_INCREMENT PRIMARY KEY ,
    -> 'deptName' varchar(50) NOT NULL,
    -> 'city' varchar(50) NOT NULL
    -> );
Query OK, 0 rows affected (0.36 sec)
```

向 department 表中插入数据。

```
mysql> insert   into 'department'
    -> ('deptName','city')
    -> values
    -> ("研发部","上海"),
```

```
    -> ("研发部","深圳"),
    -> ("研发部","北京"),
    -> ("研发部","杭州"),
    -> ("市场部","西安"),
    -> ("销售部","北京"),
    -> ("销售部","广州"),
    -> ("客服部","上海");
Query OK, 8 rows affected (0.07 sec)
Records: 8  Duplicates: 0  Warnings: 0
```

查询 department 表中的所有信息。

```
mysql> select * from department;
+----+----------+--------+
| id | deptName | city   |
+----+----------+--------+
|  1 | 研发部   | 上海   |
|  2 | 研发部   | 深圳   |
|  3 | 研发部   | 北京   |
|  4 | 研发部   | 杭州   |
|  5 | 市场部   | 西安   |
|  6 | 销售部   | 北京   |
|  7 | 销售部   | 广州   |
|  8 | 客服部   | 上海   |
+----+----------+--------+
8 rows in set (0.00 sec)
```

② 创建员工表 employee。

```
mysql> CREATE TABLE employee (
    -> empId INT AUTO_INCREMENT PRIMARY KEY,
    -> empName varchar(50)  NOT NULL,
    -> gender ENUM("男","女") NOT NULL,
    -> jobTitle varchar(50) NOT NULL,
    -> salary DOUBLE NOT NULL,
    -> hire_date DATE NOT NULL DEFAULT "2018-01-01",
    -> deptId INT NOT NULL
    -> );
Query OK, 0 rows affected (0.34 sec)
```

向 employee 表中插入数据。

```
mysql> insert   into 'employee'
    -> ('empName','gender','jobTitle', 'salary','hire_date','deptId')
```

```
    -> values
    -> ('邱子石','男','软件开发工程师',9560,'2019-10-12',1),
    -> ('张信瑞','女','测试工程师',5500,'2018-12-2',2),
    -> ('萧正业','男','软件开发工程师',9800,'2019-5-8',2),
    -> ('侯鹏运','男','软件开发工程师',9200,'2020-8-31',3),
    -> ('周媛雪','女','文案',5000,'2021-7-18',5),
    -> ('沈备','男','产品经理',7000,'2020-9-13',3),
    -> ('任凯康','男','设计',6000,'2018-10-19',5),
    -> ('孙安康','男','销售员',5000,'2019-10-12',6),
    -> ('王云溪','女','销售员',6000,'2019-11-23',6),
    -> ('张小雨','女','销售员',8500,'2017-10-4',10),
    -> ('赵棠离','女','测试工程师',6500,'2018-10-5',4),
    -> ('李甜甜','女','软件开发工程师',8800,'2019-5-31',4);
Query OK, 12 rows affected (0.07 sec)
Records: 12  Duplicates: 0  Warnings: 0
```

查询 employee 表中的所有信息。

```
mysql> select * from employee;
+-------+---------+--------+--------------------+--------+------------+--------+
| empId | empName | gender | jobTitle           | salary | hire_date  | deptId |
+-------+---------+--------+--------------------+--------+------------+--------+
|     1 | 邱子石  | 男     | 软件开发工程师     |   9560 | 2019-10-12 |      1 |
|     2 | 张信瑞  | 女     | 测试工程师         |   5500 | 2018-12-02 |      2 |
|     3 | 萧正业  | 男     | 软件开发工程师     |   9800 | 2019-05-08 |      2 |
|     4 | 侯鹏运  | 男     | 软件开发工程师     |   9200 | 2020-08-31 |      3 |
|     5 | 周媛雪  | 女     | 文案               |   5000 | 2021-07-18 |      5 |
|     6 | 沈备    | 男     | 产品经理           |   7000 | 2020-09-13 |      3 |
|     7 | 任凯康  | 男     | 设计               |   6000 | 2018-10-19 |      5 |
|     8 | 孙安康  | 男     | 销售员             |   5000 | 2019-10-12 |      6 |
|     9 | 王云溪  | 女     | 销售员             |   6000 | 2019-11-23 |      6 |
|    10 | 张小雨  | 女     | 销售员             |   8500 | 2017-10-04 |     10 |
|    11 | 赵棠离  | 女     | 测试工程师         |   6500 | 2018-10-05 |      4 |
|    12 | 李甜甜  | 女     | 软件开发工程师     |   8800 | 2019-05-31 |      4 |
+-------+---------+--------+--------------------+--------+------------+--------+
12 rows in set (0.00 sec)
```

③ 将 department 表和 employee 表按照所在部门编号进行内连接。

```
mysql> select * from department inner join employee
    -> on department.id = employee.deptId;
+----+----------+------+-------+---------+--------+--------------------+--------+------------+--------+
| id | deptName | city | empId | empName | gender | jobTitle           | salary | hire_date  | deptId |
+----+----------+------+-------+---------+--------+--------------------+--------+------------+--------+
|  1 | 研发部   | 上海 |     1 | 邱子石  | 男     | 软件开发工程师     |   9560 | 2019-10-12 |      1 |
|  2 | 研发部   | 深圳 |     2 | 张信瑞  | 女     | 测试工程师         |   5500 | 2018-12-02 |      2 |
|  2 | 研发部   | 深圳 |     3 | 萧正业  | 男     | 软件开发工程师     |   9800 | 2019-05-08 |      2 |
|  3 | 研发部   | 北京 |     4 | 侯鹏运  | 男     | 软件开发工程师     |   9200 | 2020-08-31 |      3 |
|  5 | 市场部   | 西安 |     5 | 周媛雪  | 女     | 文案               |   5000 | 2021-07-18 |      5 |
|  3 | 研发部   | 北京 |     6 | 沈备    | 男     | 产品经理           |   7000 | 2020-09-13 |      3 |
|  5 | 市场部   | 西安 |     7 | 任凯康  | 男     | 设计               |   6000 | 2018-10-19 |      5 |
|  6 | 销售部   | 北京 |     8 | 孙安康  | 男     | 销售员             |   5000 | 2019-10-12 |      6 |
|  6 | 销售部   | 北京 |     9 | 王云溪  | 女     | 销售员             |   6000 | 2019-11-23 |      6 |
|  4 | 研发部   | 杭州 |    11 | 赵棠离  | 女     | 测试工程师         |   6500 | 2018-10-05 |      4 |
|  4 | 研发部   | 杭州 |    12 | 李甜甜  | 女     | 软件开发工程师     |   8800 | 2019-05-31 |      4 |
+----+----------+------+-------+---------+--------+--------------------+--------+------------+--------+
11 rows in set (0.11 sec)
```

④ 将 department 表和 employee 表按照所在部门编号进行左连接。

```
mysql> select * from department left join employee
    -> on department.id = employee.deptId;
+----+----------+------+-------+---------+--------+----------------+--------+------------+--------+
| id | deptName | city | empId | empName | gender | jobTitle       | salary | hire_date  | deptId |
+----+----------+------+-------+---------+--------+----------------+--------+------------+--------+
|  1 | 研发部   | 上海 |     1 | 邱予石  | 男     | 软件开发工程师 |   9560 | 2019-10-12 |      1 |
|  2 | 研发部   | 深圳 |     2 | 张信瑞  | 女     | 测试工程师     |   5500 | 2018-12-02 |      2 |
|  2 | 研发部   | 深圳 |     3 | 萧正业  | 男     | 软件开发工程师 |   9800 | 2019-05-08 |      2 |
|  3 | 研发部   | 北京 |     4 | 侯鹏运  | 男     | 软件开发工程师 |   9200 | 2020-08-31 |      3 |
|  5 | 市场部   | 西安 |     5 | 周媛雪  | 女     | 文案           |   5000 | 2021-07-18 |      5 |
|  3 | 研发部   | 北京 |     6 | 沈备    | 男     | 产品经理       |   7000 | 2020-09-13 |      3 |
|  5 | 市场部   | 西安 |     7 | 任凯康  | 男     | 设计           |   6000 | 2018-10-19 |      5 |
|  6 | 销售部   | 北京 |     8 | 孙安康  | 男     | 销售员         |   5000 | 2019-10-12 |      6 |
|  6 | 销售部   | 北京 |     9 | 王云溪  | 女     | 销售员         |   6000 | 2019-11-23 |      6 |
|  4 | 研发部   | 杭州 |    11 | 赵棠离  | 女     | 测试工程师     |   6500 | 2018-10-05 |      4 |
|  4 | 研发部   | 杭州 |    12 | 李甜甜  | 女     | 软件开发工程师 |   8800 | 2019-05-31 |      4 |
|  7 | 销售部   | 广州 |  NULL | NULL    | NULL   | NULL           |   NULL | NULL       |   NULL |
|  8 | 客服部   | 上海 |  NULL | NULL    | NULL   | NULL           |   NULL | NULL       |   NULL |
+----+----------+------+-------+---------+--------+----------------+--------+------------+--------+
13 rows in set (0.11 sec)
```

可以看到，左连接后多了两条记录，因为 department 中编号为 7 和 8 的部门并没有员工，所以左连接后员工的字段信息均为 NULL。

⑤ 将 department 表和 employee 表按照所在部门编号进行右连接。

```
mysql> select * from department right join employee
    -> on department.id = employee.deptId;
+------+----------+------+-------+---------+--------+----------------+--------+------------+--------+
| id   | deptName | city | empId | empName | gender | jobTitle       | salary | hire_date  | deptId |
+------+----------+------+-------+---------+--------+----------------+--------+------------+--------+
|    1 | 研发部   | 上海 |     1 | 邱予石  | 男     | 软件开发工程师 |   9560 | 2019-10-12 |      1 |
|    2 | 研发部   | 深圳 |     2 | 张信瑞  | 女     | 测试工程师     |   5500 | 2018-12-02 |      2 |
|    2 | 研发部   | 深圳 |     3 | 萧正业  | 男     | 软件开发工程师 |   9800 | 2019-05-08 |      2 |
|    3 | 研发部   | 北京 |     4 | 侯鹏运  | 男     | 软件开发工程师 |   9200 | 2020-08-31 |      3 |
|    5 | 市场部   | 西安 |     5 | 周媛雪  | 女     | 文案           |   5000 | 2021-07-18 |      5 |
|    3 | 研发部   | 北京 |     6 | 沈备    | 男     | 产品经理       |   7000 | 2020-09-13 |      3 |
|    5 | 市场部   | 西安 |     7 | 任凯康  | 男     | 设计           |   6000 | 2018-10-19 |      5 |
|    6 | 销售部   | 北京 |     8 | 孙安康  | 男     | 销售员         |   5000 | 2019-10-12 |      6 |
|    6 | 销售部   | 北京 |     9 | 王云溪  | 女     | 销售员         |   6000 | 2019-11-23 |      6 |
| NULL | NULL     | NULL |    10 | 张小雨  | 女     | 销售员         |   8500 | 2017-10-04 |     10 |
|    4 | 研发部   | 杭州 |    11 | 赵棠离  | 女     | 测试工程师     |   6500 | 2018-10-05 |      4 |
|    4 | 研发部   | 杭州 |    12 | 李甜甜  | 女     | 软件开发工程师 |   8800 | 2019-05-31 |      4 |
+------+----------+------+-------+---------+--------+----------------+--------+------------+--------+
12 rows in set (0.00 sec)
```

可以看到，右连接后多了一条记录，因为 employee 中张小雨所在部门编号为 10，而 department 中没有此部门信息，所以右连接后部门的字段信息均为 NULL。

⑥ 在数据库中创建兴趣小组表 interestclub。

```
mysql> CREATE TABLE interestclub (
    -> id INT AUTO_INCREMENT PRIMARY KEY,
    -> title varchar(50) NOT NULL,
    -> manager varchar(50) NOT NULL,
    -> wechat varchar(50) NOT NULL
    -> );
Query OK, 0 rows affected (0.41 sec)
```

向 interestclub 表中插入数据。

```
mysql> insert into interestclub
    -> (title,manager,wechat)
    -> values
    -> ("篮球组","潘智仁","13485983467"),
    -> ("羽毛球组","殷小红","13567342561"),
    -> ("乒乓球组","王咏惠","13937261523"),
    -> ("户外徒步组","吴杰","13653726633"),
    -> ("文学影视组","许凯旋","18834532345");
Query OK, 5 rows affected (0.07 sec)
Records: 5  Duplicates: 0  Warnings: 0
```

查询 interestclub 表中的所有信息。

```
mysql> select * from interestclub;
+----+-----------+---------+-------------+
| id | title     | manager | wechat      |
+----+-----------+---------+-------------+
|  1 | 篮球组    | 潘智仁  | 13485983467 |
|  2 | 羽毛球组  | 殷小红  | 13567342561 |
|  3 | 乒乓球组  | 王咏惠  | 13937261523 |
|  4 | 户外徒步组| 吴杰    | 13653726633 |
|  5 | 文学影视组| 许凯旋  | 18834532345 |
+----+-----------+---------+-------------+
5 rows in set (0.00 sec)
```

⑦ 创建 joinclub 表。

```
mysql> CREATE TABLE joinclub (
    -> id INT AUTO_INCREMENT PRIMARY KEY,
    -> empid INT NOT NULL,
    -> clubid INT NOT NULL,
    -> join_date DATE NOT NULL
    -> );
Query OK, 0 rows affected (0.45 sec)
```

向 joinclub 表中插入数据。

```
mysql> insert into joinclub
    -> (empid,clubid,join_date)
    -> values
    -> (1,1,"2019-11-12"),
    -> (1,2,"2020-1-10"),
    -> (1,3,"2019-12-3"),
    -> (2,1,"2018-12-24"),
    -> (2,3,"2019-12-3"),
```

```
    -> (2,5,"2019-5-30"),
    -> (3,4,"2019-5-30"),
    -> (3,5,"2019-6-13"),
    -> (4,1,"2020-9-13"),
    -> (4,2,"2020-9-13"),
    -> (4,4,"2020-9-13"),
    -> (5,4,"2021-8-30"),
    -> (5,5,"2020-11-11"),
    -> (6,3,"2020-11-22"),
    -> (7,5,"2019-3-20"),
    -> (8,3,"2020-10-10"),
    -> (8,4,"2020-1-4"),
    -> (8,5,"2021-2-14"),
    -> (9,5,"2019-12-14"),
    -> (10,2,"2018-3-24"),
    -> (10,4,"2018-3-24"),
    -> (11,1,"2019-6-26"),
    -> (12,2,"2020-4-20"),
    -> (12,3,"2020-4-20");
Query OK, 24 rows affected (0.15 sec)
Records: 24  Duplicates: 0  Warnings: 0
```

查询 joinclub 表的所有信息。

```
mysql> select * from joinclub;
+----+-------+--------+------------+
| id | empId | clubId | join_date  |
+----+-------+--------+------------+
|  1 |   1   |   1    | 2019-11-12 |
|  2 |   1   |   2    | 2020-01-10 |
|  3 |   1   |   3    | 2019-12-03 |
|  4 |   2   |   1    | 2018-12-24 |
|  5 |   2   |   3    | 2019-12-03 |
|  6 |   2   |   5    | 2019-05-30 |
|  7 |   3   |   4    | 2019-05-30 |
|  8 |   3   |   5    | 2019-06-13 |
|  9 |   4   |   1    | 2020-09-13 |
| 10 |   4   |   2    | 2020-09-13 |
| 11 |   4   |   4    | 2020-09-13 |
| 12 |   5   |   4    | 2021-08-30 |
| 13 |   5   |   5    | 2020-11-11 |
| 14 |   6   |   3    | 2020-11-22 |
| 15 |   7   |   5    | 2019-03-20 |
```

```
| 16 |    8 |    3 | 2020-10-10 |
| 17 |    8 |    4 | 2020-01-04 |
| 18 |    8 |    5 | 2021-02-14 |
| 19 |    9 |    5 | 2019-12-14 |
| 20 |   10 |    2 | 2018-03-24 |
| 21 |   10 |    4 | 2018-03-24 |
| 22 |   11 |    1 | 2019-06-26 |
| 23 |   12 |    2 | 2020-04-20 |
| 24 |   12 |    3 | 2020-04-20 |
+----+------+------+------------+
24 rows in set (0.00 sec)
```

⑧ 查询员工周媛雪所在的部门名称以及城市。需要涉及 employee 表和 department 两个表，将两个表按照部门编号连接之后做查询。

```
mysql> select e.empId,e.empName,e.jobTitle,d.deptName,d.city
    -> from department as d inner join employee as e
    -> on d.id = e.deptId
    -> where e.empName = "周媛雪";
+-------+---------+----------+----------+------+
| empId | empName | jobTitle | deptName | city |
+-------+---------+----------+----------+------+
|     5 | 周媛雪  | 文案     | 市场部   | 西安 |
+-------+---------+----------+----------+------+
1 row in set (0.00 sec)
```

⑨ 查询所有在北京的员工信息。将 employee 表和 department 表按照部门编号连接之后做条件查询。

```
mysql> select e.empId,e.empName,e.jobTitle,e.salary,d.deptName,d.city
    -> from department as d inner join employee as e
    -> on d.id = e.deptId
    -> where d.city = "北京";
+-------+---------+-------------------+--------+----------+------+
| empId | empName | jobTitle          | salary | deptName | city |
+-------+---------+-------------------+--------+----------+------+
|     4 | 侯鹏运  | 软件开发工程师    |   9200 | 研发部   | 北京 |
|     6 | 沈备    | 产品经理          |   7000 | 研发部   | 北京 |
|     8 | 孙安康  | 销售员            |   5000 | 销售部   | 北京 |
|     9 | 王云溪  | 销售员            |   6000 | 销售部   | 北京 |
+-------+---------+-------------------+--------+----------+------+
4 rows in set (0.00 sec)
```

⑩ 统计每个城市的员工数量。将 employee 表和 department 表连接之后按照城市名

称进行统计。

```
mysql> select d.city, count(e.empName) as "人数"
    -> from department as d inner join employee as e
    -> on d.id = e.deptId
    -> group by d.city;
```

city	人数
上海	1
北京	4
杭州	2
深圳	2
西安	2

5 rows in set (0.00 sec)

⑪ 统计每个员工参加的兴趣小组数量。需要将 employee 表和 joinclub 表连接，然后按照员工编号进行分组查询。

```
mysql> select j.empId, e.empName, count(j.clubId) as "参加的兴趣小组数量"
    -> from joinclub as j
    -> inner join employee as e
    -> on j.empId = e.empId
    -> group by j.empId;
```

empId	empName	参加的兴趣小组数量
1	邱予石	3
2	张信瑞	3
3	萧正业	2
4	侯鹏运	3
5	周媛雪	2
6	沈备	1
7	任凯康	1
8	孙安康	3
9	王云溪	1
10	张小雨	2
11	赵棠离	1
12	李甜甜	2

12 rows in set (0.00 sec)

可以看到每个员工参加的兴趣小组数量。

⑫ 查询员工周媛雪参加的所有兴趣小组的名称。涉及员工姓名以及兴趣小组名称，这里需要将 joinclub 表、employee 表和 interestclub 表连接起来。

```
mysql> select c.title as 兴趣小组
    -> from joinclub as j
    -> inner join employee as e
    -> inner join interestclub as c
    -> on j.empId = e.empId and j.clubId = c.id
    -> where e.empName = "周媛雪";
+----------+
| 兴趣小组 |
+----------+
| 户外徒步组 |
| 文学影视组 |
+----------+
2 rows in set (0.00 sec)
```

⑬ 统计加入每个兴趣小组的员工人数。需要将 interestclub 表和 joinclub 表连接，然后按照兴趣小组编号进行分组查询。

```
mysql> select j.clubId,c.title, count(j.empId) as "员工数量"
    -> from joinclub as j
    -> inner join interestclub as c
    -> on j.clubId = c.id
    -> group by j.clubId;
+--------+----------+----------+
| clubId | title    | 员工数量 |
+--------+----------+----------+
|      1 | 篮球组   |        4 |
|      2 | 羽毛球组 |        4 |
|      3 | 乒乓球组 |        5 |
|      4 | 户外徒步组 |      5 |
|      5 | 文学影视组 |      6 |
+--------+----------+----------+
5 rows in set (0.00 sec)
```

⑭ 查询参加文学影视组的所有员工姓名。涉及员工姓名以及兴趣小组名称，这里也需要将 joinclub 表、employee 表和 interestclub 表连接起来。

```
mysql> select e.empName
    -> from joinclub as j
    -> inner join employee as e
    -> inner join interestclub as c
    -> on j.empId = e.empId and j.clubId = c.id
    -> where c.title = "文学影视组";
```

```
+-----------+
| empName   |
+-----------+
| 张信瑞    |
| 萧正业    |
| 周媛雪    |
| 任凯康    |
| 孙安康    |
| 王云溪    |
+-----------+
6 rows in set (0.00 sec)
```

⑮ 查询同时参加三个兴趣小组的员工编号和姓名。需要将 employee 表和 joinclub 表连接,然后在分组查询的基础上设置条件。

```
mysql> select j.empId,e.empName
    -> from joinclub as j
    -> inner join employee as e
    -> on j.empId = e.empId
    -> group by j.empId
    -> having count(j.clubId) = 3;
+-------+---------+
| empId | empName |
+-------+---------+
|   1   | 邱子石  |
|   2   | 张信瑞  |
|   4   | 侯鹏运  |
|   8   | 孙安康  |
+-------+---------+
4 rows in set (0.00 sec)
```

实训 8

外 键 约 束

【实训目的】

掌握添加外键约束和删除外键约束的方法,了解使用了外键约束的关联表操作。外键约束常用操作如下。

① 添加外键约束,一般有两种常用方法。

a. 在创建数据表(CREATE TABLE 语句)时添加外键约束。

```
CREATE TABLE <表名>(
...
[CONSTRAINT <外键约束名>] FOREIGN KEY [<索引名>](<列名>)
REFERENCES <主表名>(<列名>)
);
```

b. 修改表结构(ALTER TABLE 语句)添加外键约束。

```
ALTER TABLE <表名>
ADD [CONSTRAINT <外键约束名>] FOREIGN KEY [<索引名>] (<列名>)
REFERENCES <主表名>(<列名>);
```

② 通过 SHOW CREATE TABLE 语句查看外键约束。

```
SHOW CREATE TABLE <表名>;
```

③ 删除外键约束。

```
ALTER TABLE <表名>
DROP FOREIGN KEY <外键约束名>;
```

④ 关联表操作。

a. 在创建表时添加外键约束并设置级联操作。

```
CREATE TABLE <表名>(
...
[CONSTRAINT <外键约束名>] FOREIGN KEY [<索引名>](<列名>)
REFERENCES <主表名>(<列名>)
[ON DELETE {RESTRICT | CASCADE | SET NULL | NO ACTION}]
[ON UPDATE {RESTRICT | CASCADE | SET NULL | NO ACTION}]
);
```

b. 在修改表添加外键约束并设置级联操作。

```
ALTER TABLE <表名>
ADD [CONSTRAINT <外键约束名>] FOREIGN KEY [<索引名>] (<列名>)
REFERENCES <主表名>(<列名>)
[ON DELETE {RESTRICT | CASCADE | SET NULL | NO ACTION}]
[ON UPDATE {RESTRICT | CASCADE | SET NULL | NO ACTION}];
```

【实训内容】

数据库 demo 中已存在 employee 表和 department 表,查询表中数据如下。

```
mysql> select * from employee;
+-------+----------+--------+-----------------+--------+------------+--------+
| empId | empName  | gender | jobTitle        | salary | hire_date  | deptId |
+-------+----------+--------+-----------------+--------+------------+--------+
|     1 | 邱予石   | 男     | 软件开发工程师  |   9560 | 2019-10-12 |      1 |
|     2 | 张信瑞   | 女     | 测试工程师      |   5500 | 2018-12-02 |      2 |
|     3 | 萧正业   | 男     | 软件开发工程师  |   9800 | 2019-05-08 |      2 |
|     4 | 侯鹏运   | 男     | 软件开发工程师  |   9200 | 2020-08-31 |      3 |
|     5 | 周媛雪   | 女     | 文案            |   5000 | 2021-07-18 |      5 |
|     6 | 沈备     | 男     | 产品经理        |   7000 | 2020-09-13 |      3 |
|     7 | 任凯康   | 男     | 设计            |   6000 | 2018-10-19 |      5 |
|     8 | 孙安康   | 男     | 销售员          |   5000 | 2019-10-12 |      6 |
|     9 | 王云溟   | 女     | 销售员          |   6000 | 2019-11-23 |      6 |
|    10 | 张小雨   | 女     | 销售员          |   8500 | 2017-10-04 |     10 |
|    11 | 赵棠离   | 女     | 测试工程师      |   6500 | 2018-10-05 |      4 |
|    12 | 李甜甜   | 女     | 软件开发工程师  |   8800 | 2019-05-31 |      4 |
+-------+----------+--------+-----------------+--------+------------+--------+

mysql> select * from department;
+----+----------+-------+
| id | deptName | city  |
+----+----------+-------+
|  1 | 研发部   | 上海  |
|  2 | 研发部   | 深圳  |
|  3 | 研发部   | 北京  |
|  4 | 研发部   | 杭州  |
|  5 | 市场部   | 西安  |
|  6 | 销售部   | 北京  |
|  7 | 销售部   | 广州  |
|  8 | 客服部   | 上海  |
+----+----------+-------+
```

可以发现,employee 表中员工姓名为张小雨所在部门编号为 10,但 department 表中并没有此部门编号,这时就出现数据信息不对等的情况。要让 employee 表只能插入 department 表中已存在的部门编号,保证数据的一致性,可以在表中添加外键约束。

① 修改 employee 表,添加外键约束 emp_dept_fk_1,将字段 deptId 关联到 department 表的 id 字段。

② 删除 employee 表中的外键约束 emp_dept_fk_1。

③ 重新创建 joinclub 表,表中加入字段 id、empId、clubId、join_date 分别表示名单编号、员工 id、兴趣小组 id、加入日期。在创建表的时候添加外键约束 emp_club_fk_1,将字段 empId 关联到 employee 表的 empId 字段,以及外键约束 emp_club_fk_2,将字段 clubId 关联到 interestclub 表的 id 字段。创建表完成后重新插入数据。

④ 向 joinclub 表中插入一条数据(1,10,"2019-12-15"),验证是否插入成功。

⑤ 向 interestclub 表中删除 id 为 1 的数据,验证是否删除成功。

⑥ 修改 joinclub 表的外键约束 emp_club_fk_2,将外键约束设置为级联删除。向 interestclub 表中删除 id 为 1 的数据,重新验证删除操作是否成功。

【实训步骤】

① 修改 employee 表,添加外键约束 emp_dept_fk_1,将字段 deptId 关联到 department 表的 id 字段。这里 employee 表是从表,department 表是主表。直接使用 ALTER TABLE 语句添加外键约束。

```
mysql> ALTER TABLE 'employee'
    -> ADD
    -> CONSTRAINT 'emp_dept_fk_1'
    -> FOREIGN KEY ('deptId')
    -> REFERENCES 'department' ('id');
ERROR 1452 (23000): Cannot add or update a child row: a foreign key constraint fails ('demo'.
#sql-a48_6', CONSTRAINT 'emp_dept_fk_1' FOREIGN KEY ('deptId') REFERENCES
'department' ('id'))
```

这里提示创建外键约束失败,这是因为,在为已经创建好的数据表添加外键约束时,要确保添加外键约束的列的值全部来自主键列。因从表 employee 中员工姓名为张小雨所在部门编号为 10,但主表 department 中并没有此部门编号,所以外键约束创建不成功。修改 employee 表中张小雨的部门编号为 8。

```
mysql> update employee set deptId = 8
    -> where empName = "张小雨";
Query OK, 1 row affected (0.16 sec)
Rows matched: 1  Changed: 1  Warnings: 0
```

修改完成之后,重新使用 ALTER TABLE 语句添加外键约束。

```
mysql> ALTER TABLE 'employee'
    -> ADD
    -> CONSTRAINT 'emp_dept_fk_1'
    -> FOREIGN KEY ('deptId')
    -> REFERENCES 'department' ('id');
Query OK, 12 rows affected (1.13 sec)
Records: 12  Duplicates: 0  Warnings: 0
```

查看表 employee 中的外键约束。

```
mysql> show create table employee \G
*************************** 1. row ***************************
      Table: employee
```

```
Create Table:
CREATE TABLE 'employee' (
  'empId' int(11) NOT NULL AUTO_INCREMENT,
  'empName' varchar(50) NOT NULL,
  'gender' enum('男','女') NOT NULL,
  'jobTitle' varchar(50) NOT NULL,
  'salary' double NOT NULL,
  'hire_date' date NOT NULL DEFAULT '2018-01-01',
  'deptId' int(11) NOT NULL,
  PRIMARY KEY ('empId'),
  KEY 'emp_dept_fk_1' ('deptId'),
  CONSTRAINT 'emp_dept_fk_1' FOREIGN KEY ('deptId') REFERENCES 'department' ('id')
) ENGINE = InnoDB AUTO_INCREMENT = 13 DEFAULT CHARSET = utf8
1 row in set (0.00 sec)
```

可以看到，外键约束创建成功。

② 删除 employee 表中的外键约束 emp_dept_fk_1。

```
mysql> ALTER TABLE employee
    -> DROP FOREIGN KEY emp_dept_fk_1;
Query OK, 0 rows affected (0.25 sec)
Records: 0  Duplicates: 0  Warnings: 0
```

③ 删除 joinclub 表。

```
mysql> drop table joinclub;
Query OK, 0 rows affected (0.30 sec)
```

重新创建 joinclub 表，表中加入字段 id、empId、clubId、join_date 分别表示名单编号、员工 id、兴趣小组 id、加入日期，并在创建表的时候添加外键约束，将字段 empId 关联到 employee 表的 empId 字段，字段 clubId 关联到 interestclub 表的 id 字段。在 CREATE TABLE 语句时添加外键约束。

```
mysql> CREATE TABLE joinclub (
    -> id INT AUTO_INCREMENT PRIMARY KEY,
    -> empId INT NOT NULL,
    -> clubId INT NOT NULL,
    -> join_date DATE NOT NULL,
    -> CONSTRAINT 'emp_club_fk_1' FOREIGN KEY ('empId') REFERENCES 'employee' ('empId'),
    -> CONSTRAINT 'emp_club_fk_2' FOREIGN KEY ('clubId') REFERENCES 'interestclub' ('id')
    -> );
Query OK, 0 rows affected (0.52 sec)
```

查看表 joinclub 中的外键约束。

```
mysql> show create table joinclub \G
*************************** 1. row ***************************
       Table: joinclub
Create Table: CREATE TABLE `joinclub` (
  `id` int(11) NOT NULL AUTO_INCREMENT,
  `empId` int(11) NOT NULL,
  `clubId` int(11) NOT NULL,
  `join_date` date NOT NULL,
  PRIMARY KEY (`id`),
  KEY `emp_club_fk_1` (`empId`),
  KEY `emp_club_fk_2` (`clubId`),
  CONSTRAINT `emp_club_fk_1` FOREIGN KEY (`empId`) REFERENCES `employee` (`empId`),
  CONSTRAINT `emp_club_fk_2` FOREIGN KEY (`clubId`) REFERENCES `interestclub` (`id`)
) ENGINE=InnoDB DEFAULT CHARSET=utf8
1 row in set (0.01 sec)
```

可以看到，joinclub 表中创建了两个外键约束。重新插入数据。

```
mysql> insert into joinclub
    -> (empId,clubId,join_date)
    -> values
    -> (1,1,"2019-11-12"),
    -> (1,2,"2020-1-10"),
    -> (1,3,"2019-12-3"),
    -> (2,1,"2018-12-24"),
    -> (2,3,"2019-12-3"),
    -> (2,5,"2019-5-30"),
    -> (3,4,"2019-5-30"),
    -> (3,5,"2019-6-13"),
    -> (4,1,"2020-9-13"),
    -> (4,2,"2020-9-13"),
    -> (4,4,"2020-9-13"),
    -> (5,4,"2021-8-30"),
    -> (5,5,"2020-11-11"),
    -> (6,3,"2020-11-22"),
    -> (7,5,"2019-3-20"),
    -> (8,3,"2020-10-10"),
    -> (8,4,"2020-1-4"),
    -> (8,5,"2021-2-14"),
    -> (9,5,"2019-12-14"),
    -> (10,2,"2018-3-24"),
    -> (10,4,"2018-3-24"),
    -> (11,1,"2019-6-26"),
```

```
    -> (12,2,"2020-4-20"),
    -> (12,3,"2020-4-20");
Query OK, 24 rows affected (0.39 sec)
Records: 24  Duplicates: 0  Warnings: 0
```

可以看到,数据插入成功。

④ 向 joinclub 表中插入一条数据(1,10,"2019-12-15")。

```
mysql> insert into joinclub
    -> (empId,clubId,join_date)
    -> values
    -> (1,10,"2019-12-15");
ERROR 1452 (23000): Cannot add or update a child row: a foreign key constraint fails ('demo'.
'joinclub', CONSTRAINT 'emp_club_fk_2' FOREIGN KEY ('clubId') REFERENCES 'interestclub'
('id'))
```

可以看到插入操作未成功。这是因为设置了外键约束,在给从表添加数据时,从表外键字段不能插入主表中不存在的数据。由于 interestclub 表中不存在 clubId 为 10 的记录,故插入失败。

⑤ 向 interestclub 表中删除 id 为 1 的数据。

```
mysql> delete from interestclub
    -> where id = 1;
ERROR 1451 (23000): Cannot delete or update a parent row: a foreign key constraint fails
('demo'.'joinclub', CONSTRAINT 'emp_club_fk_2' FOREIGN KEY ('clubId') REFERENCES
'interestclub' ('id'))
```

可以看到删除操作未成功。这是因为设置了外键约束,当主表进行删除操作时,若从表中的外键字段有关联记录,就会阻止主表的删除操作。由于 joinclub 表中引用了 interestclub 表中 id 为 1 的记录,故删除操作失败。

⑥ 修改 joinclub 表的外键约束 emp_club_fk_2,将外键约束设置为级联删除。

首先删除外键 emp_club_fk_2。

```
mysql> ALTER TABLE joinclub
    -> DROP FOREIGN KEY emp_club_fk_2;
Query OK, 0 rows affected (0.20 sec)
Records: 0  Duplicates: 0  Warnings: 0
```

重新添加外键 emp_club_fk_2,并设置级联删除。

```
mysql> ALTER TABLE 'joinclub'
    -> ADD
    -> CONSTRAINT 'emp_club_fk_2'
    -> FOREIGN KEY ('clubId')
```

```
    -> REFERENCES 'interestclub' ('id')
    -> ON DELETE CASCADE;
Query OK, 24 rows affected (1.15 sec)
    Records: 24  Duplicates: 0  Warnings: 0
```

重新查看 joinclub 表的外键信息。

```
mysql> show create table joinclub \G
*************************** 1. row ***************************
       Table: joinclub
Create Table: CREATE TABLE 'joinclub' (
  'id' int(11) NOT NULL AUTO_INCREMENT,
  'empId' int(11) NOT NULL,
  'clubId' int(11) NOT NULL,
  'join_date' date NOT NULL,
  PRIMARY KEY ('id'),
  KEY 'emp_club_fk_1' ('empId'),
  KEY 'emp_club_fk_2' ('clubId'),
  CONSTRAINT 'emp_club_fk_1' FOREIGN KEY ('empId') REFERENCES 'employee' ('empId'),
  CONSTRAINT 'emp_club_fk_2' FOREIGN KEY ('clubId') REFERENCES 'interestclub' ('id') ON
DELETE CASCADE
) ENGINE=InnoDB AUTO_INCREMENT=26 DEFAULT CHARSET=utf8
1 row in set (0.00 sec)
```

向 interestclub 表中删除 id 为 1 的数据。

```
mysql> delete from interestclub
    -> where id = 1;
Query OK, 1 row affected (0.16 sec)
```

可以看到，interestclub 表中的记录删除成功。并且，重新查看 joinclub 中 clubId 为 1 的记录。

```
mysql> select * from joinclub
    -> where clubId = 1;
Empty set (0.00 sec)
```

查看结果可知，添加外键约束时设置了级联删除 ON DELETE CASCADE，当主表进行删除操作时，若从表中的外键字段有关联记录，则从表中与之对应的记录也会被同步删除。

实训 9

索引操作

【实训目的】

掌握创建索引、查看索引、删除索引的方法。

① 创建索引。常见的创建索引有三种方式。

a. 使用 CREATE INDEX 命令直接创建索引。

```
CREATE [UNIQUE | FULLTEXT | SPATIAL] INDEX〈索引名称〉
ON〈表名〉(〈列名〉[(〈长度〉)] [ ASC | DESC]);
```

b. 使用 CREATE TABLE 时创建索引。

```
CREATE TABLE〈表名〉(
...
[UNIQUE | FULLTEXT | SPATIAL] [ INDEX | KEY ] [〈索引名称〉](〈列名〉,…)
);
```

c. 使用 ALTER TABLE 命令创建索引。

```
ALTER TABLE〈表名〉
ADD [UNIQUE | FULLTEXT | SPATIAL ] INDEX [〈索引名称〉](〈列名〉,…);
```

② 查看索引。

```
SHOW INDEX FROM〈表名〉;
```

③ 删除索引,有两种方法。

a. 使用 DROP INDEX 命令直接创建索引。

```
DROP INDEX〈索引名称〉ON〈表名〉;
```

b. 使用 ALTER TABLE 命令删除索引。

```
ALTER TABLE〈表名〉DROP INDEX〈索引名称〉;
```

【实训内容】

① 对 employee 表的职位列创建普通索引 index_employee_jobtitle。

② 对 employee 表的员工姓名列创建唯一索引 index_employee_name。

③ 删除 employee 表上的索引 index_employee_jobtitle 和索引 index_employee_name。

④ 查看 joinclub 表上的所有索引的详细信息,删除默认的外键索引。

⑤ 对 joinclub 表中的员工编号和兴趣小组编号创建复合索引 index_joinclub_emp_club。

⑥ 删除 joinclub 表中的索引 index_joinclub_emp_club。

【实训步骤】

① 通过 CREATE INDEX 关键字为字段 jobTitle 添加普通索引。

```
mysql> CREATE INDEX index_employee_jobtitle on employee (jobTitle);
Query OK, 0 rows affected (0.59 sec)
Records: 0  Duplicates: 0  Warnings: 0
```

查看 employee 表上的索引。

```
mysql> show index from employee;
+----------+------------+------------------------+--------------+-------------+-----------+-------------+----------+--------+------+------------+---------+---------------+
| Table    | Non_unique | Key_name               | Seq_in_index | Column_name | Collation | Cardinality | Sub_part | Packed | Null | Index_type | Comment | Index_comment |
+----------+------------+------------------------+--------------+-------------+-----------+-------------+----------+--------+------+------------+---------+---------------+
| employee |          0 | PRIMARY                |            1 | empId       | A         |          12 |     NULL |   NULL |      | BTREE      |         |               |
| employee |          1 | emp_dept_fk_1          |            1 | deptId      | A         |           7 |     NULL |   NULL |      | BTREE      |         |               |
| employee |          1 | index_employee_jobtitle|            1 | jobTitle    | A         |           6 |     NULL |   NULL |      | BTREE      |         |               |
+----------+------------+------------------------+--------------+-------------+-----------+-------------+----------+--------+------+------------+---------+---------------+
3 rows in set (0.00 sec)
```

可以看到，employee 表中除了自动创建的主键索引和外键索引，还创建了普通索引 index_employee_jobtitle。

② 对 employee 表的员工姓名列创建唯一索引 index_employee_name。

```
mysql> create unique index index_employee_name on employee (empName);
Query OK, 0 rows affected (0.48 sec)
Records: 0  Duplicates: 0  Warnings: 0
```

查看 employee 表上的索引。

```
mysql> show index from employee;
+----------+------------+------------------------+--------------+-------------+-----------+-------------+----------+--------+------+------------+---------+---------------+
| Table    | Non_unique | Key_name               | Seq_in_index | Column_name | Collation | Cardinality | Sub_part | Packed | Null | Index_type | Comment | Index_comment |
+----------+------------+------------------------+--------------+-------------+-----------+-------------+----------+--------+------+------------+---------+---------------+
| employee |          0 | PRIMARY                |            1 | empId       | A         |          12 |     NULL |   NULL |      | BTREE      |         |               |
| employee |          0 | index_employee_name    |            1 | empName     | A         |          12 |     NULL |   NULL |      | BTREE      |         |               |
| employee |          1 | emp_dept_fk_1          |            1 | deptId      | A         |           7 |     NULL |   NULL |      | BTREE      |         |               |
| employee |          1 | index_employee_jobtitle|            1 | jobTitle    | A         |           6 |     NULL |   NULL |      | BTREE      |         |               |
+----------+------------+------------------------+--------------+-------------+-----------+-------------+----------+--------+------+------------+---------+---------------+
4 rows in set (0.00 sec)
```

可以看到，employee 表中为字段 empName 添加的唯一索引 index_employee_name 创建成功。

③ 删除 employee 表上的索引 index_employee_jobtitle。

```
mysql> DROP INDEX index_employee_jobtitle ON employee;
Query OK, 0 rows affected (0.31 sec)
Records: 0  Duplicates: 0  Warnings: 0
```

删除索引 index_employee_name。

```
mysql> DROP INDEX index_employee_name ON employee;
Query OK, 0 rows affected (0.34 sec)
Records: 0  Duplicates: 0  Warnings: 0
```

重新查看 employee 表上的索引。

```
mysql> show index from employee;
```

Table	Non_unique	Key_name	Seq_in_index	Column_name	Collation	Cardinality	Sub_part	Packed	Null	Index_type	Comment	Index_comment
employee	0	PRIMARY	1	empId	A	12	NULL	NULL		BTREE		
employee	1	emp_dept_fk_1	1	deptId	A	7	NULL	NULL		BTREE		

```
2 rows in set (0.00 sec)
```

可以看到,索引 index_employee_jobtitle 和 index_employee_name 删除成功。

④ 查看 joinclub 表上的所有索引的详细信息。

```
mysql> show index from joinclub;
```

Table	Non_unique	Key_name	Seq_in_index	Column_name	Collation	Cardinality	Sub_part	Packed	Null	Index_type	Comment	Index_comment
joinclub	0	PRIMARY	1	Id	A	0	NULL	NULL		BTREE		
joinclub	1	emp_club_fk_1	1	empId	A	0	NULL	NULL		BTREE		
joinclub	1	emp_club_fk_2	1	clubId	A	0	NULL	NULL		BTREE		

```
3 rows in set (0.02 sec)
```

可以看到,joinclub 表中有自动创建的主键索引和外键索引。删除外键索引首先需要删除外键。

```
mysql> ALTER TABLE joinclub
    -> DROP FOREIGN KEY emp_club_fk_1;
Query OK, 0 rows affected (0.13 sec)
Records: 0  Duplicates: 0  Warnings: 0
mysql> ALTER TABLE joinclub
    -> DROP FOREIGN KEY emp_club_fk_2;
Query OK, 0 rows affected (0.16 sec)
Records: 0  Duplicates: 0  Warnings: 0
```

接下来删除索引。

```
mysql> DROP INDEX emp_club_fk_1 ON joinclub;
Query OK, 0 rows affected (0.33 sec)
Records: 0  Duplicates: 0  Warnings: 0
mysql> DROP INDEX emp_club_fk_2 ON joinclub;
Query OK, 0 rows affected (0.18 sec)
Records: 0  Duplicates: 0  Warnings: 0
```

查看joinclub表中的索引。

```
mysql> show index from joinclub;
+---------+------------+----------+--------------+-------------+-----------+-------------+----------+--------+------+------------+---------+---------------+
| Table   | Non_unique | Key_name | Seq_in_index | Column_name | Collation | Cardinality | Sub_part | Packed | Null | Index_type | Comment | Index_comment |
+---------+------------+----------+--------------+-------------+-----------+-------------+----------+--------+------+------------+---------+---------------+
| joinclub|      0     | PRIMARY  |       1      |     id      |     A     |     20      |   NULL   |  NULL  |      |   BTREE    |         |               |
+---------+------------+----------+--------------+-------------+-----------+-------------+----------+--------+------+------------+---------+---------------+
1 row in set (0.00 sec)
```

可以看到，删除索引成功。

⑤ 对joinclub表中的员工编号和兴趣小组编号创建复合索引index_joinclub_emp_club。

```
mysql> ALTER TABLE joinclub
    -> ADD INDEX index_joinclub_emp_club(empId,clubId);
Query OK, 0 rows affected (0.41 sec)
Records: 0  Duplicates: 0  Warnings: 0
```

查看joinclub表中的索引。

```
mysql> show index from joinclub;
+---------+------------+-------------------------+--------------+-------------+-----------+-------------+----------+--------+------+------------+---------+---------------+
| Table   | Non_unique | Key_name                | Seq_in_index | Column_name | Collation | Cardinality | Sub_part | Packed | Null | Index_type | Comment | Index_comment |
+---------+------------+-------------------------+--------------+-------------+-----------+-------------+----------+--------+------+------------+---------+---------------+
| joinclub|      0     | PRIMARY                 |       1      |     id      |     A     |     20      |   NULL   |  NULL  |      |   BTREE    |         |               |
| joinclub|      1     | index_joinclub_emp_club |       1      |   empId     |     A     |     20      |   NULL   |  NULL  |      |   BTREE    |         |               |
| joinclub|      1     | index_joinclub_emp_club |       2      |   clubId    |     A     |     11      |   NULL   |  NULL  |      |   BTREE    |         |               |
+---------+------------+-------------------------+--------------+-------------+-----------+-------------+----------+--------+------+------------+---------+---------------+
3 rows in set (0.00 sec)
```

可以看到，复合索引index_joinclub_emp_club创建成功。

⑥ 使用ALTER TABLE语句删除joinclub表中的索引index_joinclub_emp_club。

```
mysql> ALTER TABLE joinclub
    -> DROP INDEX index_joinclub_emp_club;
Query OK, 0 rows affected (0.28 sec)
Records: 0  Duplicates: 0  Warnings: 0
```

重新查看joinclub表中的索引。

```
mysql> show index from joinclub;
+---------+------------+----------+--------------+-------------+-----------+-------------+----------+--------+------+------------+---------+---------------+
| Table   | Non_unique | Key_name | Seq_in_index | Column_name | Collation | Cardinality | Sub_part | Packed | Null | Index_type | Comment | Index_comment |
+---------+------------+----------+--------------+-------------+-----------+-------------+----------+--------+------+------------+---------+---------------+
| joinclub|      0     | PRIMARY  |       1      |     id      |     A     |     20      |   NULL   |  NULL  |      |   BTREE    |         |               |
+---------+------------+----------+--------------+-------------+-----------+-------------+----------+--------+------+------------+---------+---------------+
1 row in set (0.00 sec)
```

可以看到，复合索引index_joinclub_emp_club删除成功。

实训 10

视 图 应 用

【实训目的】

掌握视图的创建、查看、修改和删除操作。

① 创建视图。

```
CREATE VIEW 〈视图名称〉
AS
SELECT 语句;
```

② 查看视图。

a. 直接使用 DESCRIBE 语句查看视图的字段信息。

```
DESC〈视图名称〉;
```

b. 使用 SHOW TABLE STATUS 语句查看视图的基本信息。

```
SHOW TABLE STATUS LIKE '〈视图名称〉';
```

c. 使用 SHOW CREATE VIEW 语句查看创建视图时的定义语句。

```
SHOW CREATE VIEW〈视图名称〉;
```

③ 修改视图。

```
ALTER VIEW〈视图名称〉
AS
SELECT 语句;
```

④ 删除视图。

```
DROP VIEW〈视图名称〉;
```

【实训内容】

① 在 demo 数据库中创建视图 emp_dept_view,将 employee 表和 department 表两个表进行连接,显示员工编号、员工姓名、职位、薪水、所在部门编号、部门名称、所在城市等字段。基于此视图中统计每个城市的平均工资。

② 创建视图 join_club_view,将 joinclub 表、employee 表和 interestclub 表三个表进行连接,显示员工编号、员工姓名、性别、职位、薪水、所在部门编号、参加兴趣小组编号、兴趣小组名称、加入日期等字段。基于此视图中查询参加文学影视组的男员工和女员工各多少人。

③ 修改视图 join_club_view,将 joinclub 表、employee 表和 interestclub 表三个表进行

连接,显示员工编号、员工姓名、性别、职位、薪水、所在部门编号、入职日期、参加兴趣小组编号、兴趣小组名称等字段。基于此视图查询同时参加篮球组和羽毛球组的员工编号和姓名。

④ 查看视图 emp_dept_view 和 join_club_view 的字段信息和详细定义。

⑤ 删除视图 emp_dept_view 和 join_club_view。

【实训步骤】

① 统计每个城市的平均工资,需要将 employee 表和 department 表两个表进行连接,直接做查询。

```
mysql> select d.city,avg(e.salary)
    -> from employee as e
    -> inner join department as d
    -> on e.deptId = d.id
    -> group by d.city ;
+--------+---------------+
| city   | avg(e.salary) |
+--------+---------------+
| 上海   |          9030 |
| 北京   |          6800 |
| 杭州   |          7650 |
| 深圳   |          7650 |
| 西安   |          5500 |
+--------+---------------+
5 rows in set (0.11 sec)
```

创建视图 emp_dept_view,将 employee 表和 department 表两个表的连接查询保存到视图中。

```
mysql> create view emp_dept_view
    -> as
    -> select
e.empId,e.empName,e.jobTitle,e.salary,e.deptId,d.deptName,d.city from employee  as e
    -> inner join department as d
    -> on e.deptId = d.id;
Query OK, 0 rows affected (0.09 sec)
```

视图创建成功后,查询视图中的数据。

```
mysql> select * from emp_dept_view;
+-------+---------+-------------------+--------+--------+----------+------+
| empId | empName | jobTitle          | salary | deptId | deptName | city |
+-------+---------+-------------------+--------+--------+----------+------+
|     1 | 邱子石  | 软件开发工程师    |   9560 |      1 | 研发部   | 上海 |
|     2 | 张信瑞  | 测试工程师        |   5500 |      2 | 研发部   | 深圳 |
|     3 | 萧正业  | 软件开发工程师    |   9800 |      2 | 研发部   | 深圳 |
|     4 | 侯鹏运  | 软件开发工程师    |   9200 |      3 | 研发部   | 北京 |
|     5 | 周媛雪  | 文案              |   5000 |      5 | 市场部   | 西安 |
|     6 | 沈备    | 产品经理          |   7000 |      3 | 研发部   | 北京 |
|     7 | 任凯康  | 设计              |   6000 |      5 | 市场部   | 西安 |
|     8 | 孙安康  | 销售员            |   5000 |      6 | 销售部   | 北京 |
|     9 | 王云溪  | 销售员            |   6000 |      6 | 销售部   | 北京 |
|    10 | 张小雨  | 销售员            |   8500 |      8 | 客服部   | 上海 |
|    11 | 赵棠离  | 测试工程师        |   6500 |      4 | 研发部   | 杭州 |
|    12 | 李甜甜  | 软件开发工程师    |   8800 |      4 | 研发部   | 杭州 |
+-------+---------+-------------------+--------+--------+----------+------+
12 rows in set (0.02 sec)
```

通过视图统计每个城市的平均工资。

```
mysql> select city,avg(salary)
    -> from emp_dept_view
    -> group by city;
+------+-------------+
| city | avg(salary) |
+------+-------------+
| 上海 |        9030 |
| 北京 |        6800 |
| 杭州 |        7650 |
| 深圳 |        7650 |
| 西安 |        5500 |
+------+-------------+
5 rows in set (0.00 sec)
```

可以看到，统计结果与直接做查询结果一样。

② 创建视图 join_club_view，将 joinclub 表、employee 表和 interestclub 表进行连接并保存在视图中。

```
mysql> create view join_club_view
    -> as
    -> select e.empId,e.empName,e.gender,e.jobTitle,e.salary,e.deptId,j.clubId,c.title,j.join_date
    -> from joinclub as j
    -> inner join employee as e
    -> inner join interestclub as c
    -> on j.empId = e.empId and j.clubId = c.id;
Query OK, 0 rows affected (0.14 sec)
```

查看此视图中的数据。

```
mysql> select * from join_club_view;
+-------+---------+--------+----------------+--------+--------+--------+--------------+------------+
| empId | empName | gender | jobTitle       | salary | deptId | clubId | title        | join_date  |
+-------+---------+--------+----------------+--------+--------+--------+--------------+------------+
|     1 | 邱子石  | 男     | 软件开发工程师 |   9560 |      1 |      1 | 篮球组       | 2019-11-12 |
|     2 | 张信瑞  | 女     | 测试工程师     |   5500 |      2 |      1 | 篮球组       | 2018-12-24 |
|     4 | 侯鹏运  | 男     | 软件开发工程师 |   9200 |      3 |      1 | 篮球组       | 2020-09-13 |
|    11 | 赵棠离  | 女     | 测试工程师     |   6500 |      4 |      1 | 篮球组       | 2019-06-26 |
|     1 | 邱子石  | 男     | 软件开发工程师 |   9560 |      1 |      1 | 羽毛球组     | 2020-01-10 |
|     4 | 侯鹏运  | 男     | 软件开发工程师 |   9200 |      3 |      2 | 羽毛球组     | 2020-09-13 |
|    10 | 张小雨  | 女     | 销售员         |   8500 |      8 |      2 | 羽毛球组     | 2018-03-24 |
|    12 | 李甜甜  | 女     | 软件开发工程师 |   8800 |      4 |      2 | 羽毛球组     | 2020-04-20 |
|     1 | 邱子石  | 男     | 软件开发工程师 |   9560 |      1 |      2 | 乒乓球组     | 2019-12-03 |
|     2 | 张信瑞  | 女     | 测试工程师     |   5500 |      2 |      3 | 乒乓球组     | 2019-12-03 |
|     6 | 沈备    | 男     | 产品经理       |   7000 |      3 |      3 | 乒乓球组     | 2020-11-22 |
|     8 | 孙安康  | 男     | 销售员         |   5000 |      6 |      3 | 乒乓球组     | 2020-10-10 |
|    12 | 李甜甜  | 女     | 软件开发工程师 |   8800 |      4 |      3 | 乒乓球组     | 2020-04-20 |
|     3 | 萧正业  | 男     | 软件开发工程师 |   9800 |      2 |      3 | 户外徒步组   | 2019-05-30 |
|     4 | 侯鹏运  | 男     | 软件开发工程师 |   9200 |      3 |      4 | 户外徒步组   | 2020-09-13 |
|     5 | 周媛雪  | 女     | 文案           |   5000 |      5 |      4 | 户外徒步组   | 2021-08-30 |
|     8 | 孙安康  | 男     | 销售员         |   5000 |      6 |      4 | 户外徒步组   | 2020-01-04 |
|    10 | 张小雨  | 女     | 销售员         |   8500 |      8 |      4 | 户外徒步组   | 2018-03-24 |
|     2 | 张信瑞  | 女     | 测试工程师     |   5500 |      2 |      4 | 文学影视组   | 2019-05-30 |
|     3 | 萧正业  | 男     | 软件开发工程师 |   9800 |      2 |      5 | 文学影视组   | 2019-06-13 |
|     5 | 周媛雪  | 女     | 文案           |   5000 |      5 |      5 | 文学影视组   | 2020-11-11 |
|     7 | 任凯康  | 男     | 设计           |   6000 |      5 |      5 | 文学影视组   | 2019-03-20 |
|     8 | 孙安康  | 男     | 销售员         |   5000 |      6 |      5 | 文学影视组   | 2021-02-14 |
|     9 | 王云溪  | 女     | 销售员         |   6000 |      6 |      5 | 文学影视组   | 2019-12-14 |
+-------+---------+--------+----------------+--------+--------+--------+--------------+------------+
24 rows in set (0.00 sec)
```

基于此视图做分组查询。

```
mysql> select gender,count(empName)
    -> from join_club_view
    -> where title = "文学影视组"
    -> group by gender;
+--------+----------------+
| gender | count(empName) |
+--------+----------------+
| 男     |              3 |
| 女     |              3 |
+--------+----------------+
2 rows in set (0.00 sec)
```

可以查询到参加文学影视组的男员工和女员工各3人。

③ 修改视图 join_club_view。

```
mysql> alter view join_club_view
    -> as
    -> select e.empId,e.empName,e.gender,e.jobTitle,e.salary,
    -> e.deptId,e.hire_date,j.clubId,c.title
    -> from joinclub as j
    -> inner join employee as e
    -> inner join interestclub as c
    -> on j.empId = e.empId and j.clubId = c.id;
Query OK, 0 rows affected (0.14 sec)
```

查看此视图中的数据。

```
mysql> select * from join_club_view;
+-------+---------+--------+----------------+--------+--------+------------+--------+-----------+
| empId | empName | gender | jobTitle       | salary | deptId | hire_date  | clubId | title     |
+-------+---------+--------+----------------+--------+--------+------------+--------+-----------+
|     1 | 邱予石  | 男     | 软件开发工程师 |   9560 |      1 | 2019-10-12 |      1 | 篮球组    |
|     2 | 张信瑞  | 女     | 测试工程师     |   5500 |      2 | 2018-12-02 |      1 | 篮球组    |
|     4 | 侯鹏运  | 男     | 软件开发工程师 |   9200 |      3 | 2020-08-31 |      1 | 篮球组    |
|    11 | 赵棠离  | 女     | 测试工程师     |   6500 |      4 | 2018-10-05 |      1 | 篮球组    |
|     1 | 邱予石  | 男     | 软件开发工程师 |   9560 |      1 | 2019-10-12 |      2 | 羽毛球组  |
|     4 | 侯鹏运  | 男     | 软件开发工程师 |   9200 |      3 | 2020-08-31 |      2 | 羽毛球组  |
|    10 | 张小雨  | 女     | 销售员         |   8500 |      8 | 2017-10-04 |      2 | 羽毛球组  |
|    12 | 李甜甜  | 女     | 软件开发工程师 |   8800 |      4 | 2019-05-31 |      2 | 羽毛球组  |
|     1 | 邱予石  | 男     | 软件开发工程师 |   9560 |      1 | 2019-10-12 |      3 | 乒乓球组  |
|     2 | 张信瑞  | 女     | 测试工程师     |   5500 |      2 | 2018-12-02 |      3 | 乒乓球组  |
|     6 | 沈备    | 男     | 产品经理       |   7000 |      3 | 2020-09-13 |      3 | 乒乓球组  |
|     8 | 孙安康  | 男     | 销售员         |   5000 |      6 | 2019-10-12 |      3 | 乒乓球组  |
|    12 | 李甜甜  | 女     | 软件开发工程师 |   8800 |      4 | 2019-05-31 |      3 | 乒乓球组  |
|     3 | 萧正业  | 男     | 软件开发工程师 |   9800 |      2 | 2019-05-08 |      4 | 户外徒步组 |
|     4 | 侯鹏运  | 男     | 软件开发工程师 |   9200 |      3 | 2020-08-31 |      4 | 户外徒步组 |
|     5 | 周媛雪  | 女     | 文案           |   5000 |      5 | 2021-07-18 |      4 | 户外徒步组 |
|     8 | 孙安康  | 男     | 销售员         |   5000 |      6 | 2019-10-12 |      4 | 户外徒步组 |
|    10 | 张小雨  | 女     | 销售员         |   8500 |      8 | 2017-10-04 |      4 | 户外徒步组 |
|     2 | 张信瑞  | 女     | 测试工程师     |   5500 |      2 | 2018-12-02 |      5 | 文学影视组 |
|     3 | 萧正业  | 男     | 软件开发工程师 |   9800 |      2 | 2019-05-08 |      5 | 文学影视组 |
|     5 | 周媛雪  | 女     | 文案           |   5000 |      5 | 2021-07-18 |      5 | 文学影视组 |
|     7 | 任凯康  | 男     | 设计           |   6000 |      5 | 2018-10-19 |      5 | 文学影视组 |
|     8 | 孙安康  | 男     | 销售员         |   5000 |      6 | 2019-10-12 |      5 | 文学影视组 |
|     9 | 王云溪  | 女     | 销售员         |   6000 |      6 | 2019-11-23 |      5 | 文学影视组 |
+-------+---------+--------+----------------+--------+--------+------------+--------+-----------+
24 rows in set (0.00 sec)
```

基于此视图查询同时参加篮球组和羽毛球组的员工编号和姓名。

```
mysql> select empId,empName,gender,jobTitle,salary,deptId,hire_date
    -> from join_club_view
    -> where title in ('篮球组','羽毛球组')
    -> group by empId
    -> having count(empId) > 1;
+-------+---------+--------+----------------+--------+--------+------------+
| empId | empName | gender | jobTitle       | salary | deptId | hire_date  |
+-------+---------+--------+----------------+--------+--------+------------+
|     1 | 邱予石  | 男     | 软件开发工程师 |   9560 |      1 | 2019-10-12 |
|     4 | 侯鹏运  | 男     | 软件开发工程师 |   9200 |      3 | 2020-08-31 |
+-------+---------+--------+----------------+--------+--------+------------+
2 rows in set (0.00 sec)
```

④ 查看视图 join_club_view 的字段信息。

```
mysql> desc join_club_view;
+-----------+---------------+------+-----+------------+-------+
| Field     | Type          | Null | Key | Default    | Extra |
+-----------+---------------+------+-----+------------+-------+
| empId     | int(11)       | NO   |     | 0          |       |
| empName   | varchar(50)   | NO   |     | NULL       |       |
| gender    | enum('男','女')| NO   |     | NULL       |       |
| jobTitle  | varchar(50)   | NO   |     | NULL       |       |
| salary    | double        | NO   |     | NULL       |       |
| deptId    | int(11)       | NO   |     | NULL       |       |
| hire_date | date          | NO   |     | 2018-01-01 |       |
| clubId    | int(11)       | NO   |     | NULL       |       |
| title     | varchar(50)   | NO   |     | NULL       |       |
+-----------+---------------+------+-----+------------+-------+
9 rows in set (0.18 sec)
```

使用 SHOW TABLE STATUS 查看视图的基本信息。

```
mysql> SHOW TABLE STATUS WHERE comment='view' \G
*************************** 1. row ***************************
           Name: emp_dept_view
         Engine: NULL
        Version: NULL
     Row_format: NULL
           Rows: NULL
 Avg_row_length: NULL
    Data_length: NULL
Max_data_length: NULL
   Index_length: NULL
      Data_free: NULL
 Auto_increment: NULL
    Create_time: NULL
    Update_time: NULL
     Check_time: NULL
      Collation: NULL
       Checksum: NULL
 Create_options: NULL
        Comment: VIEW
*************************** 2. row ***************************
           Name: join_club_view
         Engine: NULL
        Version: NULL
     Row_format: NULL
           Rows: NULL
 Avg_row_length: NULL
    Data_length: NULL
Max_data_length: NULL
   Index_length: NULL
      Data_free: NULL
 Auto_increment: NULL
    Create_time: NULL
    Update_time: NULL
     Check_time: NULL
      Collation: NULL
       Checksum: NULL
 Create_options: NULL
        Comment: VIEW
2 rows in set (0.00 sec)
```

可以看到数据库 demo 中存在两个视图，分别是 emp_dept_view 和 join_club_view。使用 SHOW CREATE VIEW 查看视图 emp_dept_view 的详细信息。

```
mysql> show create view emp_dept_view \G
*************************** 1. row ***************************
                View: emp_dept_view
         Create View: CREATE ALGORITHM=UNDEFINED
DEFINER=`root`@`localhost` SQL SECURITY DEFINER VIEW `emp_dept_view` AS
select `e`.`empId` AS `empId`,`e`.`empName` AS `empName`,`e`.`jobTitle` AS
`jobTitle`,`e`.`salary` AS `salary`,`e`.`deptId` AS
`deptId`,`d`.`deptName` AS `deptName`,`d`.`city` AS `city` from (`employee` `e` join `department`
`d` on((`e`.`deptId` = `d`.`id`)))
character_set_client: utf8
collation_connection: utf8_general_ci
1 row in set (0.00 sec)
```

可以查看到视图 emp_dept_view 的详细定义。

⑤ 删除视图 emp_dept_view 和 join_club_view。

```
mysql> drop view emp_dept_view;
Query OK, 0 rows affected (0.00 sec)
mysql> drop view join_club_view;
Query OK, 0 rows affected (0.00 sec)
```

重新查看数据库中的视图。

```
mysql> SHOW TABLE status WHERE comment='view';
Empty set (0.00 sec)
```

可以看到，视图 emp_dept_view 和 join_club_view 删除成功。

实训 11

存 储 过 程

【实训目的】

掌握存储过程的创建和调用，以及存储过程的查看、修改和删除。

① 创建存储过程。

```
DELIMITER //
CREATE PROCEDURE 〈存储过程名〉(参数列表)
BEGIN
sql 语句;
END //
DELIMITER ;
```

② 调用存储过程。

```
CALL <存储过程名>(参数列表);
```

③ 查看存储过程。

a. 使用 SHOW PROCEDURE STATUS 查看数据库中的存储过程。

```
SHOW PROCEDURE STATUS WHERE db='<数据库名>';
```

b. 使用 SHOW CREATE PROCEDURE 查看存储过程的定义。

```
SHOW CREATE PROCEDURE  <存储过程名>;
```

④ 删除存储过程。

```
DROP PROCEDURE IF EXISTS <存储过程名>;
```

【实训内容】

① 在 demo 数据库中定义一个存储过程 GetClubEmployeeName，将兴趣小组名称的值作为 IN 输入参数传入，根据小组名称求出参加此兴趣小组的所有员工姓名，调用存储过程查看篮球组和羽毛球组的成员名单。

② 定义一个存储过程 GetCityEmplyeeCount，将城市名称作为 IN 输入参数传入，根据所在城市求出该城市的员工人数，调用存储过程查看北京和上海的员工人数。

③ 查看数据库 demo 中的所有存储过程，并查看存储过程 GetClubEmployeeName 的详细信息。

④ 删除存储过程 GetClubEmployeeName 和 GetCityEmplyeeCount。

【实训步骤】

① 定义一个存储过程 GetClubEmployeeName，将兴趣小组名称的值作为 IN 输入参数传入，根据小组名称求出参加此兴趣小组的所有员工姓名。

```
mysql> DELIMITER //
mysql> CREATE PROCEDURE GetClubEmployeeName(
    ->   IN club_name_in VARCHAR(50)
    -> )
    -> BEGIN
    ->     select e.empName
    ->     from joinclub as j
    ->     inner join employee as e
    ->     inner join interestclub as c
    ->     on j.empId = e.empId and j.clubId = c.id
    ->     where c.title = club_name_in;
    -> END //
Query OK, 0 rows affected (0.00 sec)
mysql> DELIMITER ;
```

调用存储过程查看篮球组的成员名单。

```
mysql> CALL GetClubEmployeeName("篮球组");
+---------+
| empName |
+---------+
| 邱子石  |
| 张信瑞  |
| 侯鹏运  |
| 赵棠离  |
+---------+
4 rows in set (0.00 sec)
```

调用存储过程查看羽毛球组的成员名单。

```
mysql> CALL GetClubEmployeeName("羽毛球组");
+---------+
| empName |
+---------+
| 邱子石  |
| 侯鹏运  |
| 张小雨  |
| 李甜甜  |
+---------+
4 rows in set (0.00 sec)
```

② 定义一个存储过程 GetCityEmplyeeCount,将城市名称作为 IN 输入参数传入,根据所在城市求出该城市的员工人数。

```
mysql> DELIMITER //
mysql> CREATE PROCEDURE GetCityEmplyeeCount(
    ->   IN city_name_in VARCHAR(50),
    ->   OUT counts INT(11)
    -> )
    -> BEGIN
    ->     select count(e.empName) into counts
    ->     from department as d inner join employee as e
    ->     on d.id = e.deptId
    ->     where d.city = city_name_in;
    -> END //
Query OK, 0 rows affected (0.00 sec)
mysql> DELIMITER ;
```

调用存储过程查看北京的员工人数。

```
mysql> CALL GetCityEmplyeeCount("北京",@count);
Query OK, 1 row affected (0.00 sec)
mysql> select @count;
+--------+
| @count |
+--------+
|      4 |
+--------+
1 row in set (0.00 sec)
```

调用存储过程查看上海的员工人数。

```
mysql> CALL GetCityEmplyeeCount("上海",@count);
Query OK, 1 row affected (0.00 sec)

mysql> select @count;
+--------+
| @count |
+--------+
|      2 |
+--------+
1 row in set (0.00 sec)
```

③ 查看数据库 demo 中的所有存储过程。

```
mysql> SHOW PROCEDURE STATUS WHERE db='demo'\G
*************************** 1. row ***************************
                  Db: demo
                Name: GetCityEmplyeeCount
                Type: PROCEDURE
             Definer: root@localhost
            Modified: 2021-08-01 12:58:38
             Created: 2021-08-01 12:58:38
       Security_type: DEFINER
             Comment:
character_set_client: utf8
collation_connection: utf8_general_ci
  Database Collation: utf8_general_ci
*************************** 2. row ***************************
                  Db: demo
                Name: GetClubEmployeeName
                Type: PROCEDURE
             Definer: root@localhost
            Modified: 2021-08-01 12:55:50
             Created: 2021-08-01 12:55:50
       Security_type: DEFINER
             Comment:
character_set_client: utf8
collation_connection: utf8_general_ci
  Database Collation: utf8_general_ci
2 rows in set (0.00 sec)
```

使用 SHOW CREATE PROCEDURE 查看存储过程 GetClubEmployeeName 的详细信息。

```
mysql> SHOW CREATE PROCEDURE  GetClubEmployeeName  \G
*************************** 1. row ***************************
           Procedure: GetClubEmployeeName
            sql_mode:
STRICT_TRANS_TABLES,NO_AUTO_CREATE_USER,NO_ENGINE_SUBSTITUTION
     Create  Procedure: CREATE   DEFINER = 'root' @ 'localhost' PROCEDURE
'GetClubEmployeeName'(
  IN club_name_in VARCHAR(50)
)
BEGIN
    select e.empName
    from joinclub as j
    inner join employee as e
    inner join interestclub as c
    on j.empId = e.empId and j.clubId = c.id
    where c.title = club_name_in;
END
character_set_client: utf8
collation_connection: utf8_general_ci
  Database Collation: utf8_general_ci
1 row in set (0.00 sec)
```

④ 删除存储过程 GetClubEmployeeName。

```
mysql> DROP PROCEDURE IF EXISTS GetClubEmployeeName;
Query OK, 0 rows affected (0.00 sec)
```

删除存储过程 GetCityEmplyeeCount。

```
mysql> DROP PROCEDURE IF EXISTS GetCityEmplyeeCount;
Query OK, 0 rows affected (0.00 sec)
```

重新查询数据库 demo 中的所有存储过程。

```
mysql> SHOW PROCEDURE STATUS WHERE db='demo';
Empty set (0.00 sec)
```

可以看到,存储过程删除成功。

实训 12

触 发 器

【实训目的】

掌握触发器的创建与执行,以及查看触发器、删除触发器的方法。

① 创建触发器。

```
CREATE TRIGGER〈触发器名称〉〈触发时间〉〈触发事件〉
ON〈表名〉
FOR EACH ROW
BEGIN
    语句列表
END
```

② 执行触发器。

触发器在创建完成后,若要执行触发器,则需要让触发器指定的数据表执行对应的操作。

③ 查看触发器。

```
SHOW TRIGGERS [ IN〈数据库名称〉]
[LIKE '匹配模式' | | WHERE 条件表达式];
```

④ 删除触发器。

```
DROP TRIGGER [IF EXISTS] [〈数据库名〉.]〈触发器名称〉
```

【实训内容】

① 在 demo 数据库中给 employee 表创建一个 AFTER DELETE 型触发器,用来实现当向 employee 表中删除一条员工记录时,将删除的学生信息自动备份到 employee_del 表中。

② 给 department 表创建一个 AFTER UPDATE 型触发器。用来实现当修改 department 中的部门编号时,同步修改 employee 表中引用了此部门编号的记录。

③ 查看 demo 数据库中的触发器。

④ 删除触发器 after_employee_delete 和 after_department_update。

【实训步骤】

① 首先创建 employee_del 表,用来存放删除员工的信息,包含 id、员工编号、员工姓名、性别、职位、删除日期等信息。

```
mysql> CREATE TABLE employee_del (
    -> id INT AUTO_INCREMENT PRIMARY KEY,
    -> empId INT,
    -> empName varchar(50) NOT NULL,
    -> gender ENUM("男","女") NOT NULL ,
    -> jobTitle varchar(50) NOT NULL,
    -> del_date DATE NOT NULL
    -> );
Query OK, 0 rows affected (0.78 sec)
```

给 employee 表创建 AFTER DELETE 型触发器,实现当向 employee 表中删除一条员工记录时,将删除的学生信息插入 employee_del 表中。

```
mysql> DELIMITER //
mysql>
mysql> CREATE TRIGGER after_employee_delete after delete
    -> ON employee
    -> FOR EACH ROW
    -> BEGIN
    ->   insert into employee_del(empId,empName,gender,jobTitle,del_date)
    ->   values
    ->   (OLD.empId,OLD.empName,OLD.gender,OLD.jobTitle,NOW());
    -> END //
Query OK, 0 rows affected (0.31 sec)
mysql>
mysql> DELIMITER ;
```

接下来在 employee 表中删除员工编号为 10 的记录。

```
mysql> delete from employee
    -> where empId = 10;
Query OK, 1 row affected (0.31 sec)
```

删除成功,再查看 employee_del 表。

```
mysql> select * from employee_del;
+----+-------+---------+--------+----------+------------+
| id | empId | empName | gender | jobTitle | del_date   |
+----+-------+---------+--------+----------+------------+
|  1 |    10 | 张小雨  | 女     | 销售员   | 2021-08-01 |
+----+-------+---------+--------+----------+------------+
1 row in set (0.00 sec)
```

可以看到,触发器被执行,数据备份成功。

② 给 department 表创建 AFTER UPDATE 型触发器,用来实现当修改 department 中

的部门编号时,同步修改 employee 表中引用了此部门编号的记录。

```
mysql> DELIMITER //
mysql>
mysql> CREATE TRIGGER after_department_update after update
    -> ON department
    -> FOR EACH ROW
    -> BEGIN
    ->   UPDATE employee SET deptId = NEW.id
    ->   where deptId = OLD.id;
    -> END //
Query OK, 0 rows affected (0.27 sec)
mysql>
mysql> DELIMITER ;
```

接下来在 department 表中将部门编号 1 修改为 11。

```
mysql> UPDATE department SET id = 11 WHERE id = 1;
Query OK, 1 row affected (0.07 sec)
Rows matched: 1  Changed: 1  Warnings: 0
```

修改成功,再查看 employee 表中的记录。

```
mysql> select * from employee;
+-------+---------+--------+-----------------+--------+------------+--------+
| empId | empName | gender | jobTitle        | salary | hire_date  | deptId |
+-------+---------+--------+-----------------+--------+------------+--------+
|     1 | 邱予石  | 男     | 软件开发工程师  |   9560 | 2019-10-12 |     11 |
|     2 | 张信瑞  | 女     | 测试工程师      |   5500 | 2018-12-02 |      2 |
|     3 | 萧正业  | 男     | 软件开发工程师  |   9800 | 2019-05-08 |      2 |
|     4 | 侯鹏运  | 男     | 软件开发工程师  |   9200 | 2020-08-31 |      3 |
|     5 | 周媛雪  | 女     | 文案            |   5000 | 2021-07-18 |      5 |
|     6 | 沈备    | 男     | 产品经理        |   7000 | 2020-09-13 |      3 |
|     7 | 任凯康  | 男     | 设计            |   6000 | 2018-10-19 |      5 |
|     8 | 孙安康  | 男     | 销售员          |   5000 | 2019-10-12 |      6 |
|     9 | 王云溪  | 女     | 销售员          |   6000 | 2019-11-23 |      6 |
|    10 | 张小雨  | 女     | 销售员          |   8500 | 2017-10-04 |      8 |
|    11 | 赵棠离  | 女     | 测试工程师      |   6500 | 2018-10-05 |      4 |
|    12 | 李甜甜  | 女     | 软件开发工程师  |   8800 | 2019-05-31 |      4 |
+-------+---------+--------+-----------------+--------+------------+--------+
11 rows in set (0.00 sec)
```

可以看到,employee 表中同步将部门编号 1 修改为 11,触发器被执行。

③ 查看 demo 数据库中的触发器。

```
mysql> SHOW TRIGGERS  IN demo \G
*************************** 1. row ***************************
         Trigger: after_department_update
           Event: UPDATE
            Table: department
        Statement: BEGIN
  UPDATE employee SET deptId = NEW.id
```

```
          where deptId = OLD.id;
END
              Timing: AFTER
            Created: 2021-08-01 15:11:10.31
           sql_mode:
STRICT_TRANS_TABLES,NO_AUTO_CREATE_USER,NO_ENGINE_SUBSTITUTION
            Definer: root@localhost
character_set_client: utf8
collation_connection: utf8_general_ci
  Database Collation: utf8_general_ci
*************************** 2. row ***************************
             Trigger: after_employee_delete
               Event: DELETE
               Table: employee
           Statement: BEGIN
  insert into employee_del(empId,empName,gender,jobTitle,del_date)
  values
  (OLD.empId,OLD.empName,OLD.gender,OLD.jobTitle,NOW());
END
              Timing: AFTER
            Created: 2021-08-01 15:06:58.32
           sql_mode:
STRICT_TRANS_TABLES,NO_AUTO_CREATE_USER,NO_ENGINE_SUBSTITUTION
            Definer: root@localhost
character_set_client: utf8
collation_connection: utf8_general_ci
  Database Collation: utf8_general_ci
2 rows in set (0.00 sec)
```

④ 删除触发器 after_employee_delete 和 after_department_update。

```
mysql> DROP TRIGGER IF EXISTS after_employee_delete;
Query OK, 0 rows affected (0.00 sec)
```

删除 after_department_update。

```
mysql> DROP TRIGGER IF EXISTS after_department_update;
Query OK, 0 rows affected (0.00 sec)
```

重新查询数据库中存在的触发器。

```
mysql> SHOW TRIGGERS  IN demo;
Empty set (0.00 sec)
```

可以看到,触发器删除成功。

实训 13

事 务

【实训目的】

掌握事务的开启、提交和回滚操作以及保存点设置。

① 开启事务。

```
START TRANSACTION;
```

② 提交事务。

```
COMMIT;
```

③ 回滚事务。

```
ROLLBACK;
```

④ 保存点设置。

a. 创建保存点。

```
SAVEPOINT <保存点名>;
```

b. 删除保存点。

```
RELEASE SAVEPOINT <保存点名>;
```

c. 回滚到保存点。

```
ROLLBACK TO <保存点名>;
```

【实训内容】

① 在数据库 demo 中,如果 department 表中修改了部门编号信息,那么 employee 表中的记录需要同步修改,否则就会造成数据的不一致,将这两个操作作为一个事务进行管理,使用 COMMIT 语句提交事务。

② 利用事务实现删除 employee 表中的记录,发现误删之后使用 ROLLBACK 回滚取消删除。

③ 利用事务实现在 employee 表中添加一条新记录,更新员工编号为 3 的员工岗位为项目经理,然后删除员工编号为 4 的记录,为每一步操作设置保存点,让事务可以回滚到保存点。

【实训步骤】

① 在数据库 demo 中,如果 department 表中修改了部门编号信息,那么 employee 表引用了此部门编号的记录需要同步修改,否则就会造成数据的不一致,将这两个操作作为一

个事务进行管理。

开启事务。

```
mysql> START TRANSACTION;
Query OK, 0 rows affected (0.11 sec)
```

将 department 中部门编号 2 修改为 22。

```
mysql> UPDATE department SET id = 22 WHERE id = 2;
Query OK, 1 row affected (0.11 sec)
Rows matched: 1  Changed: 1  Warnings: 0
```

在 employee 表中同步修改部门编号。

```
mysql> UPDATE employee SET deptId = 22 WHERE deptId = 2;
Query OK, 2 rows affected (0.00 sec)
Rows matched: 2  Changed: 2  Warnings: 0
```

使用 COMMIT 语句提交事务。

```
mysql> COMMIT;
Query OK, 0 rows affected (0.13 sec)
```

② 使用 START TRANSACTION 语句开启事务。

```
mysql> START TRANSACTION;
Query OK, 0 rows affected (0.00 sec)
```

在事务中执行操作，删除 employee 表中的周媛雪的记录。

```
mysql> delete from employee
    -> where empName = "周媛雪";
Query OK, 1 row affected (0.11 sec)
```

这时查询该员工记录。

```
mysql> select * from employee
    -> where empName = "周媛雪";
Empty set (0.00 sec)
```

可以看到记录被删除。如果这时我们发现进行了误删操作，那么就可以使用 ROLLBACK 语句回滚事务。

```
mysql> ROLLBACK;
Query OK, 0 rows affected (0.05 sec)
```

重新查询该员工记录。

```
mysql> select * from employee
    -> where empName = "周媛雪";
```

```
+-------+---------+--------+----------+--------+------------+--------+
| empId | empName | gender | jobTitle | salary | hire_date  | deptId |
+-------+---------+--------+----------+--------+------------+--------+
|   5   | 周媛雪  |   女   |   文案   |  5000  | 2021-07-18 |   5    |
+-------+---------+--------+----------+--------+------------+--------+
1 row in set (0.00 sec)
```

可以看到，删除操作被撤销。

③ 利用事务实现在 employee 表中添加一条新记录，更新员工编号为 3 的员工岗位为项目经理，然后删除员工编号为 4 的记录，为每一步操作设置保存点，让事务可以回滚到保存点。

查看员工初始记录。

```
mysql> select * from employee;
+-------+---------+--------+----------------+--------+------------+--------+
|empId  |empName  |gender  |jobTitle        |salary  |hire_date   |deptId  |
+-------+---------+--------+----------------+--------+------------+--------+
|   1   |邱子石   |  男    |软件开发工程师  | 9560   |2019-10-12  |  11    |
|   2   |张信瑞   |  女    |测试工程师      | 5500   |2018-12-02  |  22    |
|   3   |萧正业   |  男    |软件开发工程师  | 9800   |2019-05-08  |  22    |
|   4   |侯鹏运   |  男    |软件开发工程师  | 9200   |2020-08-31  |   3    |
|   5   |周媛雪   |  女    |文案            | 5000   |2021-07-18  |   5    |
|   6   |沈备     |  男    |产品经理        | 7000   |2020-09-13  |   3    |
|   7   |任凯康   |  男    |设计            | 6000   |2018-10-19  |   5    |
|   8   |孙安康   |  男    |销售员          | 5000   |2019-10-12  |   6    |
|   9   |王云溪   |  女    |销售员          | 6000   |2019-11-23  |   6    |
|  11   |赵棠离   |  女    |测试工程师      | 6500   |2018-10-05  |   4    |
|  12   |李甜甜   |  女    |软件开发工程师  | 8800   |2019-05-31  |   4    |
+-------+---------+--------+----------------+--------+------------+--------+
11 rows in set (0.00 sec)
```

使用 START TRANSACTION 语句开启事务，并使用保存点记下初始状态并设置保存点名为 initial。

```
mysql> START TRANSACTION;
Query OK, 0 rows affected (0.00 sec)
mysql> SAVEPOINT initial;
Query OK, 0 rows affected (0.00 sec)
```

在事务中执行操作，将添加新的员工记录、更新员工职位、删除一条记录作为一个事务进行处理，并为每一步设置保存点。

```
mysql> insert into 'employee'
    -> ('empName','gender','jobTitle','salary','hire_date','deptId')
    -> values
    -> ('王习习','女','美工',7020,'2021-2-12',2);
Query OK, 1 row affected (0.02 sec)
mysql> SAVEPOINT ins;
Query OK, 0 rows affected (0.00 sec)

mysql> UPDATE employee SET jobTitle = "项目经理" WHERE empId = 3;
Query OK, 1 row affected (0.01 sec)
```

```
Rows matched: 1  Changed: 1  Warnings: 0
mysql> SAVEPOINT upd;
Query OK, 0 rows affected (0.00 sec)

mysql> delete from employee where empId = 4;
Query OK, 1 row affected (0.00 sec)
mysql> SAVEPOINT del;
Query OK, 0 rows affected (0.00 sec)
```

可以看到,我们设置了三个保存点,分别是 ins、upd、del。这时查看表的临时状态。

```
mysql> select * from employee;
+-------+---------+--------+--------------+--------+------------+--------+
| empId | empName | gender | jobTitle     | salary | hire_date  | deptId |
+-------+---------+--------+--------------+--------+------------+--------+
|     1 | 邱予石  | 男     | 软件开发工程师 |   9560 | 2019-10-12 |     11 |
|     2 | 张信瑞  | 女     | 测试工程师    |   5500 | 2018-12-02 |     22 |
|     3 | 萧正业  | 男     | 项目经理      |   9800 | 2019-05-08 |     22 |
|     5 | 周媛雪  | 女     | 文案          |   5000 | 2021-07-18 |      5 |
|     6 | 沈备    | 男     | 产品经理      |   7000 | 2020-09-13 |      3 |
|     7 | 任凯康  | 男     | 设计          |   6000 | 2018-10-19 |      5 |
|     8 | 孙安康  | 男     | 销售员        |   5000 | 2019-10-12 |      6 |
|     9 | 王云溪  | 女     | 销售员        |   6000 | 2019-11-23 |      6 |
|    11 | 赵棠离  | 女     | 测试工程师    |   6500 | 2018-10-05 |      4 |
|    12 | 李甜甜  | 女     | 软件开发工程师 |   8800 | 2019-05-31 |      4 |
|    13 | 王习习  | 女     | 美工          |   7020 | 2021-02-12 |      2 |
+-------+---------+--------+--------------+--------+------------+--------+
11 rows in set (0.00 sec)
```

如果我们想让事务回到保存点 ins 的状态,也就是执行完插入操作,但还未执行更新和删除操作的状态。使用 ROLLBACK TO 返回到此保存点。

```
mysql> ROLLBACK TO ins;
Query OK, 0 rows affected (0.00 sec)
```

重新查看 employee。

```
mysql> select * from employee;
+-------+---------+--------+--------------+--------+------------+--------+
| empId | empName | gender | jobTitle     | salary | hire_date  | deptId |
+-------+---------+--------+--------------+--------+------------+--------+
|     1 | 邱予石  | 男     | 软件开发工程师 |   9560 | 2019-10-12 |     11 |
|     2 | 张信瑞  | 女     | 测试工程师    |   5500 | 2018-12-02 |     22 |
|     3 | 萧正业  | 男     | 软件开发工程师 |   9800 | 2019-05-08 |     22 |
|     4 | 侯鹏运  | 男     | 软件开发工程师 |   9200 | 2020-08-31 |      3 |
|     5 | 周媛雪  | 女     | 文案          |   5000 | 2021-07-18 |      5 |
|     6 | 沈备    | 男     | 产品经理      |   7000 | 2020-09-13 |      3 |
|     7 | 任凯康  | 男     | 设计          |   6000 | 2018-10-19 |      5 |
|     8 | 孙安康  | 男     | 销售员        |   5000 | 2019-10-12 |      6 |
|     9 | 王云溪  | 女     | 销售员        |   6000 | 2019-11-23 |      6 |
|    11 | 赵棠离  | 女     | 测试工程师    |   6500 | 2018-10-05 |      4 |
|    12 | 李甜甜  | 女     | 软件开发工程师 |   8800 | 2019-05-31 |      4 |
|    13 | 王习习  | 女     | 美工          |   7020 | 2021-02-12 |      2 |
+-------+---------+--------+--------------+--------+------------+--------+
12 rows in set (0.00 sec)
```

可以看到,事务撤销了更新和删除操作,回退到了插入成功的状态。

参 考 文 献

[1] 黑马程序员. MySQL 数据库原理设计与应用[M]. 北京:清华大学出版社,2019.
[2] 任丽娜. MySQL 数据库实用教程[M]. 北京:清华大学出版社,2022.
[3] 郭水泉. MySQL 数据库项目式教程[M]. 陕西:西安电子科技大学出版社,2022.
[4] 李锡辉,王敏. MySQL 数据库技术与项目应用教程(微课版)[M]. 2 版. 北京:人民邮电出版社,2022.
[5] 肖睿. MySQL 数据库应用技术及实战[M]. 2 版. 北京:人民邮电出版社,2022.
[6] 李士勇. MySQL 数据库应用技术[M]. 北京:北京邮电大学出版社,2019.